HARVEY FRIEDMAN'S RESEARCH ON THE FOUNDATIONS OF MATHEMATICS

STUDIES IN LOGIC

AND

THE FOUNDATIONS OF MATHEMATICS

VOLUME 117

NORTH-HOLLAND
AMSTERDAM • NEW YORK • OXFORD

HARVEY FRIEDMAN'S RESEARCH ON THE FOUNDATIONS OF MATHEMATICS

Edited by

L. A. HARRINGTON
University of California
Berkeley, CA, U.S.A.

M. D. MORLEY
Cornell University
Ithaca, NY, U.S.A.

A. SČĚDROV
University of Pennsylvania
Philadelphia, PA, U.S.A.

S. G. SIMPSON
Pennsylvania State University
University Park, PA, U.S.A.

1985

NORTH-HOLLAND
AMSTERDAM • NEW YORK • OXFORD

ISBN: 0 444 87834 3

Published by:

Elsevier Science Publishers B.V.
P.O. Box 1991
1000 BZ Amsterdam
The Netherlands

Sole distributors for the U.S.A. and Canada:

Elsevier Science Publishing Company, Inc.
52 Vanderbilt Avenue
New York, N.Y. 10017
U.S.A.

Library of Congress Cataloging-in-Publication Data
Main entry under title:

Harvey Friedman's research on the foundations of
mathematics.

(Studies in logic and the foundations of
mathematics ; v. 117)
"Harvey Friedman's publications"--P.
1. Mathematics--Philosophy--Addresses, essays,
lectures. 2. Friedman, Harvey, 1948- . I. Friedman,
Harvey, 1948- . II. Harrington, L.A. III. Series.
QA8.6.H37 1985 511.3 85-16169
ISBN 0-444-87834-3 (U.S.)

PRINTED IN THE NETHERLANDS

PREFACE

This volume discusses various aspects of Friedman's research in the foundations of mathematics over the past fifteen years. We felt that it was especially worthwhile to present this volume to a wide audience of mathematicians, computer scientists, and mathematically oriented philosophers. Indeed, much of Friedman's work enjoys a quality of broad accessibility and communicability to anyone who is even slightly interested in the foundations of mathematics.

We were particularly interested in this project because we were disturbed by the recent trend in mathematical logic toward increasingly technical problems which are of interest only to small groups of specialists. A result of this tendency is that the subject is drifting away from its historical roots in mathematics and philosophy. For a number of years now, Harvey Friedman has been the most outspoken critic of this trend. Moreover, Friedman's insistence of applying standards of general scientific significance to his own research has led him to produce many of the most interesting recent results. Not the least of Friedman's achievements has been to remind mathematical logicians that truly valid work of a foundational character is still possible.

The Editors

INTRODUCTION

Mathematical logic began in the late XIX century as a mathematical study of mathematical proof, and came into prominence in the 1930's largely through the major work of Kurt Gödel. Since then, our understanding of this subject has expanded and deepened.

Like many other areas of mathematical science, the subject has gradually drifted away from the original concerns addressed in the early part of the century. Much of this later work can be viewed as technical background information, which has served to facilitate major advances. In some cases, this work actually appears to have been essential.

Harvey Friedman is one of only a handful of logicians who have consistently stated that this primarily technical development has outlived its usefulness, and have called for a return to the kind of basic intellectual issues that gave mathematical logic its impetus earlier in the century. His striking success in basing his own scientific research on this view is both rare and remarkable. His work is distinguished by its focus on the central questions concerning the fundamental concepts of the subject. It is a decisive example of how an intellectual viewpoint at odds with current research practice can, in the hands of a gifted young researcher, lead to work of unusual scope, originality, and growing influence. In many cases, topics within mathematical logic which used to be considered exotic or moribund have attained new importance in light of Friedman's work. In fact, the work is sufficiently original and systematic as to be identifiable as the major force behind an ongoing reorientation of the field. This kind of contribution by a young scientist is virtually without precedent in the postwar period.

Logic is customarily divided into set theory, model theory, recursion theory, proof theory, and intuitionism. In this Introduction, we will not try to give a scholarly history of these areas. We will give some very brief historical remarks about their origins, and discuss only Friedman's role in them. Bear in mind that we are discussing much ongoing research, so we cannot attempt to be either comprehensive or entirely up to date, and we limit ourselves to just some of the most remarkable aspects of Friedman's work.

Set theory originated with Cantor and became part of mathematical logic when it became clear that it was possible to cast arbitrary mathematical proofs in set-theoretic form. This aim was fully realized through the axiomatization of set theory by Zermelo, later modified by Skolem and Fraenkel. This axiomatization is referred to as Zermelo-Fraenkel set theory with the Axiom of Choice and is customarily abbreviated as ZFC. It forms a convincing upper bound for mathematical reasoning, in the sense that, for any commonly accepted proof of a mathematical theorem, its routine casting in set-theoretic terms results in a proof of that theorem within ZFC (or in a conservative extension such as the von Neumann-Gödel-Bernays theory of classes). For this reason, ZFC is the standard vehicle for casting results on the full undecidability of mathematical statements. It is not appropriate, however, for a more delicate study of the nature of mathematical proof. Standard mathematics is not inherently or peculiarly set-theoretic. The axioms of ZFC address issues that are remote from the preponderance of mathematical activity. This leads to the critical question of how much strength of ZFC is used or needed in order to prove theorems that can be cast in the more usual concrete terms of standard mathematics. One of the main contributions of Harvey Friedman has been to provide diverse examples of the necessary use of abstract set theory in relatively concrete mathematical situations. This work is a direct outgrowth of the work of Gödel, who showed that abstract set theory is necessary to prove the consistency of formal systems of abstract set theory.

In order to obtain such provably necessary uses of abstract set theory, Friedman must reach beyond the basic metamathematical techniques that are most often used, i.e., self-reference, inner models (both due to Gödel in the 1930's), and forcing (due to Paul J. Cohen in the early 1960's). This is primarily because of the role that Gödel's constructible sets, L, play in set theory. Specifically, it is known that self-reference, inner models, and forcing cannot be the main techniques for establishing the independence of a mathematical statement of sufficiently concrete character from ZFC.

The original setting for Friedman's necessary uses of abstract set theory has been in descriptive set theory. This area is mainly the study of the so-called projective hierarchy. In Friedman's view, the most important and mathematically relevant part of the projective hierarchy is the Borel sets (and some fragments thereof, e.g. F_σ and G_δ sets), and to some extent, analytic sets. These classes are related to the standard limit operations considered in analysis, and are the lowest levels of the hierarchy. Friedman has conjectured that all of the existing forcing style independence results would disappear if the independent statements would be recast in terms of Borel sets and functions. In particular, the Continuum Hypothesis so reformulated asserts that every Borel set of reals is either in Borel one-to-one correspondence with the integers or in Borel one-to-one correspondence with the reals. This is a well-known classical theorem in descriptive set theory. The Suslin's Hypothesis is another example. It is independent of ZFC, but Friedman reformulated it in Borel terms and gave a proof of it, jointly with S. Shelah. (This is discussed at the end of section 2 of C. Steinhorn's article). Much work in descriptive set theory, particularly at the lower levels of the projective hierarchy, has provided invaluable background information for all sorts of central investigations in logic, and Friedman has used this extensively.

Friedman has discovered a wide variety of examples of necessary uses of abstract set theory in the context of Borel measurable functions on various standard separable metric spaces. These results represent a twofold breakthrough over the uses of various extensions of ZFC by large cardinals or by assertions concerning infinite games in obtaining information about the higher projective hierarchy. Firstly, Friedman's necessary uses deal with substantially more concrete, regular objects of a definite mathematical character, and secondly, they are impervious to adjustments in the basic set-theoretic apparatus, such as reformulation in terms of L or even short initial segments thereof. (This work is discussed in L. Stanley's contribution.)

Friedman's most recent contribution in the direction of necessary uses of abstract set theory is a virtual translation of the most abstract infinite combinatorics (large cardinals) into the realm of the finite. This work of Friedman will appear soon in Advances of Mathematics.

Proof theory began as the central branch of mathematical logic. Gödel's first three major theorems started the modern evolution of this subject. The first of these is the Completeness Theorem for predicate calculus, discussed here in the section on model theory. He followed this with his First and Second Incompleteness Theorems. The first one asserts that every sufficiently extensive, consistent formal system (and almost all formal systems are sufficiently extensive) is incomplete in the sense that there exist sentences expressed within the system that cannot be decided within it. The second one provides additional information that consistency of such a system is a sentence of this kind. The next major development in proof theory was Gentzen's work relating formal elementary number theory (so-called Peano arithmetic, PA in short) and the ordinal ε_0 (the first ordinal α such that $\omega^\alpha = \alpha$).

It is generally recognized that Gentzen's work provides a significant insight into the nature of PA because of the basic combinatorial nature of ε_0. It also provides a kind of foundational reduction in the sense that Gentzen shows that any proof of a suitably simple formula (Π_2^0) within PA can be analyzed by quantifier-free transfinite induction and transfinite recursion on ε_0 (recall that PA is based on arbitrary finite alternations of quantifiers). In Friedman's view, these insights are very difficult to claim for the current extensions of Gentzen's work to stronger formal systems because of the ordinals and ordinal representations involved rapidly become exceedingly complicated and hard to visualize. At one time it was the case that such work was a major tool in the establishment of relations between formal systems (i.e. conservative extension results and consistency strength). Primarily through the work of Friedman, such arguments have been systematically replaced by the arguments of recursion-theoretic or model-theoretic nature, often involving nonstandard models (even with nonstandard integers). His methods in this connection established many new results of this kind. (They are discussed in C. Smoryński's longer article in this volume).

Through insights of Friedman, the value of extensions of Gentzen's work to stronger formal systems becomes problematic. A dramatic reversal has taken place, however, which suggests a permanent place for this kind of proof theory for independence results (cf. below).

Despite the seminal importance of Gödel's incompleteness theorems, they do not address the question of whether there are independence results from PA and related systems which are of a definite mathematical (as opposed to metamathematical) character. Friedman's work in 1968 on the independence of Borel determinacy from Zermelo set theory with Choice, and his 1976 work on the independence of Borel diagonalization from second-order arithmetic with Choice set the stage for the search of new independence phenomena. In 1977, Paris discovered the first palatable mathematical (finite combinatorial) independence result from PA, a result soon modified into an elegant, mathematical independence result known as the Paris-Harrington variant of Ramsey's theorem (PH). (This discovery grew out of Paris' earlier work over a period of years on nonstandard models of arithmetic, which was to some extent inspired by Friedman's still earlier work on the same subject). A few years later Friedman found another very important standard combinatorial theorem that is a source of new finite combinatorial independence results. PH is a modified finite version of Ramsey's theorem, the usual finite version being quite dependent. Friedman's discovery is based on Kruskal's theorem in graph theory, which is logically simpler than Ramsey's theorem (Π_1^1 versus Π_2^1). Friedman's finite combinatorial theorem is a straightforward finite form of Kruskal's theorem, requiring no modification. Aside from these advantages over PH, Friedman's finite form (FFF) is independent of vastly stronger systems, such as predicative analysis. Furthermore, FFF has a systematic series of simple parametric restrictions that illustrate a whole range of independence phenomena, including PH. Also, when one of the parameters is fixed at 12, the resulting statement asserts the existence of an integer which is so large that no proof of its existence can be given without either making some use of uncountable sets, or using 2^{100} pieces of paper. Moreover, the property defining this integer is algorithmically testable. (S. G. Simpson's first article in this volume deals with this topic. See also R. L. Smith's contribution as well as the shorter articles at the end of the volume.) This is the dramatic reversal alluded to in the discussion of the extensions of Gentzen's work (cf. above), because Friedman makes use of many achievements in this direction.

Clearly, any penetrating analysis of mathematical proof should be able to classify mathematical theorems according to the mathematical axioms used or needed to prove them. What is lacking is the methodology for establishing that a given axiom is needed to prove a given theorem. Friedman came to the realization that, in many cases, one can actually prove that the theorem is equivalent to the axiom (hence "reverse mathematics"!). In order to initiate such a study, Friedman had to isolate appropriate "base theories" and systems of mathematical axioms, together with basic illustrative examples. The general framework used for this study are the so-called subsystems of second-order arithmetic. (S. G. Simpson's second article in this volume discusses these issues.)

Model theory came about as a result of Frege's isolation of first-order predicate calculus as the common basis of all working formal systems. In his famous Completeness Theorem, Gödel established that a sentence can be derived from a formal theory (i.e., a set of sentences) if and only if that sentence is true in all models of that theory. (A thorough treatment of the concepts involved was given later by Tarski.)

Pure model theory is concerned with the detailed study of sentences and theories in predicate calculus and their models. This study is based on a general concept of model, i.e., an abstract set (domain) together with the system of relations and functions on this domain, and distinguished elements of the domain. It is generally regarded that much of uncountable model theory displays substantial amounts of set-theoretic pathology significantly beyond what is usual in mathematics. The countable case is very interesting, with a manageable level of abstraction, but it does not cover many important mathematical situations, particularly in natural mathematical contexts of the power of the continuum. Specifically, the context Friedman has in mind is that of Borel sets and Borel functions on complete separable metric spaces. These are so-called Borel structures of Friedman. His Borel model theory has a technical character similar to the study of countable structures but does capture the situations mentioned above withot being forced to deal with set-theoretic pathologies. (This topic is discussed in C. Steinhorn's contribution to this volume.)

One of the concepts in model theory emphasized by Tarski is translatability. After such basic notions as sentence and model, this is the singlemost versatile and fundamental concept in the whole of model theory. One of the concepts in proof theory that was basic to Gödel's work and naturally grew out of the failure of Hilbert's consistency program is that of relative consistency between formal systems. The only apparent relation between these two concepts is that if a formal system S is translatable into a formal system T, then S is consistent relative to T. Remarkably, Friedman has proved that under very general conditions on S and T, these two basic notions are equivalent! (This is discussed in section 5 of C. Smoryński's longer article in this volume.)

In Friedman's view, the second principal line of investigation in mathematical logic is that of direct applications to areas of mathematics outside logic. Through the work of Tarski and A. Robinson, applied model theory has been mainly concerned with the study of models arising in algebra (e.g., a group is a model of the group axioms), or finding alternative proofs and occasionally answering questions in various parts of mathematics outside logic. To date these have represented the bulk of direct applications of logic to mathematics. Sometimes this study of logic as a tool involves general results about proofs, e.g., both in Kreisel's and in Friedman's work on extracting upper bounds in Hilbert's XVII problem. In Friedman's view, however, it would be narrow-minded to assign principal importance to the applications. He has consistently stressed that they are secondary to the foundational studies, which best reflect the nature of logic as a fundamental intellectual discipline.

Recursion theory had its origins in the work of Turing and others on the mathematical treatment of the intuitive notion of algorithm. Vital auxiliary notions (such as partial recursive functions, recursively enumerable sets, etc.) were delineated and their basic properties proved in the 1930's and 1940's by Kleene and others. Several modifications of Turing's basic definitions were given and shown to be equivalent (so-called verification of Church's Thesis).

Much of the current work in recursion theory deals with the detailed structure of some of the most basic partial orderings that arise from these auxiliary notions. In Friedman's view, there is a lack of regularity that such orderings exhibit, the apparently intrinsic lack of critical examples, and the lack of connections with any of the motivating purposes of recursion theory, so he has consistently commented that this area is overdeveloped. There has also been considerable work on generalizations of these notions to many different contexts, and, in Friedman's view, much the same can be said here, with the additional objection that because of the infinitary character of the objects involved, the concept of algorithm loses its informal meaning. Nevertheless, this work has provided significant technical background information, perhaps most usefully through the Turing degrees and hyperdegrees, and the so-called admissible sets and admissible ordinals. This has organized and simplified arguments used to establish significant results, and Friedman has used this work extensively (cf. R. E. Byerly's article in this volume).

Clearly the most interesting work in recursion theory is the recent fundamental work on algorithms in which computational resources (such as the amount of computer time and space used or required) play a central role. This area is called computational complexity theory. Friedman has done some influential work in foundations of complexity theory. He extended the concept of algorithm to arbitrary relational structures in such a way that the finite computational content of algorithms is the central focus. This work anticipates the computer science literature not only on program schemata (cf. J. C. Shepherdson's first article in this volume), but also on pebbling. (A. J. Kfoury's article in this volume deals with the latter topic.) Friedman has also considered computational complexity issues in the context of classical analysis, related to the discrete case. This work has just begun to be extended to the realm of ordinary differential equations. (These issues are discussed in J. C. Shepherdson's second contribution to this volume.)

Intuitionism originated in the 1900's when Brouwer rejected the usual principles of mathematical reasoning as extended to the realm of the infinite in favor of the principles of constructive (intuitionistic) mathematical reasoning. This led Brouwer to his intuitionistic reconstruction of many areas of mathematics, especially analysis. He continued his investigations throughout the first part of the century, but his philosophical orthodoxy prevented his particular approach to his subject from having any considerable mathematical influence.

The connection with mathematical logic was first established by Heyting in 1930. He described a formal logic of constructive reasoning, which he obtained from the classical propositional calculus by deleting the Law of Excluded Middle. Soon afterwards, Gödel gave the so-called negative interpretation of PA into first-order formal number theory based on Heyting's predicate calculus (Heyting's arithmetic, HA in short), showing that PA is an equiconsistent subsystem of HA. The renewal of interest in such intuitionistic systems in the 1940's and 1950's was due largely to the work of Kleene and Kreisel on HA and elementary intuitionistic formal analysis. These and other much stronger formal systems based on Heyting's predicate calculus (e.g. intuitionistic ZF) are now known to have rather elegant metamathematical properties, as well as a multitude of interpretations of independent mathematical interest. The roots of many of these interpretations lie in Grothendieck's work in algebraic geometry in the late 1950's.

Two of the most basic properties of intuitionistic formal systems are the disjunction property (if $A \vee B$ is provable, then A is provable or B is provable), and the numerical existence property (if $\exists n.A(n)$ is provable, then for some numeral $\bar{n}$, $A(\bar{n})$ is provable). By a remarkable application of Gödel's self-reference method used to establish Gödel's second incompleteness theorem, Friedman obtained the rather startling result that these two properties are equivalent for any intuitionistic system satisfying roughly the same weak hypotheses that Gödel used for his theorem. (These topics are discussed in D. Leivant's contribution to this volume).

Metamathematical study of strong intuitionistic formal systems was initiated by Friedman (cf. A. Ščedrov's contribution to this volume). Friedman extended the methods and results of Gödel and Kleene on HA to higher-order arithmetic and to set theory, and gave a simple, uniform proof that they have the same provably recursive functions and provable ordinals as their classical counterparts (obtained by restoring the Law of Excluded Middle to their logic). In ZF, the schema of Replacement is equivalent to the schema of Collection, and this is a basic and important fact. In the context of intuitionistic set theory, the problem of equivalence is also fundamental, and has been considered for a decade. Recently, Friedman gave an elegant negative answer using constructions from the theory of ordinary ZFC, and the proof is likely to open up several new lines of investigations. Friedman also recently gave a concise and uniform proof (jointly with Ščedrov) that, for virtually all of the usual intuitionistic systems, every arithmetic theorem is provable by some instance of arithmetic transfinite induction on some primitive recursively presented well ordering.

We thank all of the authors for their contributions. In these articles, the reader will find many more probing historical discussions of these areas of mathematical logic, Friedman's role in their development, and the relations of his work to the work of others.

The Editors

BIOGRAPHY OF HARVEY FRIEDMAN

Harvey Friedman was born in Chicago, Illinois, on September 23, 1948. He attended the Massachusetts Institute of Technology, where he received a Ph.D. degree in Mathematics in August, 1967.

He was appointed to an Assistant Professorship in the Philosophy Department at Stanford, effective September 1967, and then promoted to an Associate Professorship there from 1969 until 1973. During that time he also held two additional positions, Associate Professor of Mathematics at the University of Wisconsin during 1970-1971, and Visiting Professor of Mathematics at State University of New York at Buffalo during 1972-1973. From 1973 to 1977 he was Professor of Mathematics at State University of New York at Buffalo. Since 1977 he has held his present position of Professor of Mathematics at The Ohio State University. He has served as a consultant to AT&T Bell Laboratories and as a Visiting Scientist at IBM.

In 1984, Dr. Friedman was presented with the ninth annual Alan T. Waterman Award. This award has been given annually since 1976 by the National Science Foundation "to the outstanding young scientist in the United States". The award was granted to Dr. Friedman "for his revitalization of the foundations of mathematics, his penetrating investigations into the Gödel incompleteness phenomena, and his fundamental contributions to virtually all areas of mathematical logic".

Dr. Friedman is an accomplished amateur pianist. He married Betty McClanahan in 1973, with whom he resides in Columbus, Ohio.

TABLE OF CONTENTS

HARVEY FRIEDMAN'S RESEARCH ON THE FOUNDATIONS
OF MATHEMATICS, L.A. Harrington et al. (editors)
Elsevier Science Publishers B.V. (North-Holland), 1985

The Work of Harvey Friedman[1)]

Anil Nerode and Leo A. Harrington

Mathematical logician Harvey Friedman was recently awarded the National Science Foundation's annual Waterman Prize, honoring the most outstanding American scientist under thirty-five years of age in all fields of science and engineering. When a mathematician wins such an award, the mathematical community naturally wishes to understand the underlying achievements, and their implications. Friedman continues the great tradition of Frege, Russel, and Gödel. This can be characterized as the exercise of acute philosophical perspective to distill exact mathematical definitions and questions from important foundational issues. These questions in turn give rise to mathematical subjects and theorems of depth and beauty. Friedman's contributions span all branches of mathematical logic (recursion theory, proof theory, model theory, set theory, theory of computation). He is a generalist in an age of specialization, yet his theorems often require extraordinary technical virtuosity. We discuss only a few selected highlights.

Friedman's ideas have yielded radically new kinds of independence results. The kinds of statements proved independent before Friedman were mostly disguised properties of formal systems (such as Gödel's theorem on unprovability of consistency) or assertions about abstract sets (such as the continuum hypothesis or Souslin's hypothesis). In constrast, Friedman's independence results are about questions of a more concrete nature involving, for example, Borel functions or the Hilbert cube.

1) Reprinted from the Notices of the American Mathematical Society, (1984) "The Work of Harvey Friedman", by Anil Nerode and Leo Harrington, Volume 31, pp. 563-566, by permission of the American Mathematical Society.

In [1] Friedman showed that Borel determinacy cannot be proved in Zermelo set theory with the axiom of choice (ZC), or any of the usual formal systems associated with at most countably many iterations of the power set operation, after Martin [17] proved Borel determinacy from certain reasonable extensions of Zermelo Fraenkel set theory with the axiom of choice (ZFC). Subsequently, Martin [18] gave a proof of Borel determinacy using uncountably many iterations of the power set operation.

These results reached their mature form in [9] as follows:

THEOREM 1 (FRIEDMAN [9]). It is necessary and sufficient to use uncountably many iterations of the power set operation to prove the following. Every symmetric Borel subset of the unit square contains or is disjoint from the graph of a Borel function. In particular, this assertion is provable in ZFC but not in ZC.

In Friedman [4,9], the classical theorem of Cantor that the unit interval I is uncountable is analyzed. Cantor's argument produces an $x \in I$ which is not a term in any given infinite sequence $y_1, y_2, \ldots \in I$. Specifically, there is a Borel diagonalization function $F : \overline{Q} \to I$ such that no $F(y)$ is a coordinate of y (here $\overline{Q} = I^{\omega}$ is the Hilbert cube). Friedman observes that the value of $F(y)$ depends on the order in which the coordinates of y are given, at least for the F coming from Cantor's argument. Let $x \approx y$ mean that $x, y \in \overline{Q}$ have the same coordinates, and $x \sim y$ mean that y is obtained by permuting finitely many coordinates of x.

THEOREM 2 (FRIEDMAN [4,9]). If $F : \overline{Q} \to I$ is a Borel function satisfying the invariance condition $x \approx y \to F(x) = F(y)$, then some $F(x)$ is a coordinate of x. Furthermore, this is provable in ZC but not in ZFC with the power set axiom deleted (even for the weaker theorem with $\approx$ replaced by $\sim$).

Subsequently, Friedman [12] gave closely related one-dimensional theorems with the same meta-mathematical properties. We present two of these.

Let $K = \{0,1\}^{\omega}$ be the Cantor set. Define the shift $s : K \to K$ by $s(x_n) = x_{n+1}$. Define the "square" $x^{(2)}$ for $x \in K$ by $x^{(2)}_n = x_{n^2}$. Let T be the circle group (i.e., $[0,1)$ with addition modulo 1). We say that $F : K \to K$

is shift invariant if $F(sx) = F(x)$. We say that $F : T \to T$ is doubling invariant if $F(2x) = F(x)$.

THEOREM 3 (FRIEDMAN [12]). Every shift invariant Borel function on K is somewhere its square. There is a Borel (in fact, continuous) function on T which agrees somewhere with every doubling invariant Borel function on T . Furthermore, these (three) theorems are provable in ZC but not in ZFC with the power set axiom deleted.

Friedman [9] develops extensions of Theorem 2, which proved using uncountably many iterations of the power set operation. Recently, Friedman has extended this idea as follows: Let Q be the rational numbers and $P(Q)$ be the Cantor space of all subsets of Q . Let G be the Baire space of all products defined on ω .

THEOREM 4 (FRIEDMAN [11]). If $F : G \to G$ is a Borel function such that isomorphic elements go to isomorphic values, then some $F(x)$ is isomorphically imbeddable in x . Furthermore, this theorem is provable in ZC but not in ZFC with the power set axiom deleted.

THEOREM 5 (FRIEDMAN [11]). If $F : P(Q) \to P(Q)$ is a Borel function such that order isomorphic arguments go to order isomorphic values, then some $F(A)$ is isomorphic to an interval in A (even of the form (a,b) where a,b are points in A) . Furthermore, this theorem requires uncountably many iterations of the power set operation to prove. In particular, it is provable in ZFC but not in ZC.

Friedman [9] also develops far-reaching extensions of Theorem 2 by combining these ideas with Ramsey's theorem. This leads to independence results from full ZFC, and in fact to theorems about Borel functions on the Hilbert cube for which it is necessary and sufficient to use "large cardinals" to prove.

The large cardinals involved are as follows: A cardinal is inaccessible if $\lambda < \kappa$ implies $2^{\lambda} < \kappa$, and κ is not the sup of fewer than κ cardinals below κ . We say that κ is a Mahlo cardinal if κ is inaccessible and every closed unbounded subset of κ contains an inaccessible cardinal.

The 1-Mahlo cardinals are the Mahlo cardinals. The (n+1)-Mahlo cardinals are the n-Mahlo cardinals in which every closed unbounded subset contains an n-Mahlo cardinal.

The group of all permutations of ω which fix all but finitely many numbers acts on $\overline{Q}$ by permuting coordinates. This group also acts diagonally on any $\overline{Q}^n$ by $g \cdot (x_1,\ldots,x_n) = (gx_1,\ldots,gx_n)$. For $x,y \in \overline{Q}^n$ let $x \sim y$ indicate that x,y are in the same orbit under this action.

THEOREM 6 (FRIEDMAN [9]). Let $F : \overline{Q} \times \overline{Q}^n \to I$ be a Borel function such that if $x \in \overline{Q}$, $y,z \in \overline{Q}^n$, and $y \sim z$, then $F(x,y) = F(x,z)$. Then there is an infinite sequence $\{x_k\}$ from $\overline{Q}$ such that for all indices $s < t_1 < \ldots < t_n$, $F(x_s, x_{t_1}, \ldots, x_{t_n})$ is the first coordinate of x_{s+1}. In order to prove this for all n, it is necessary and sufficient to use the existence of n-Mahlo cardinals for all n.

These ideas have been extended in Friedman [11] to much larger cardinals. A measurable cardinal is a cardinal κ which carries a κ-additive $\{0,1\}$-valued measure on all subsets of κ. A Ramsey cardinal is a cardinal κ such that for every set E of finite subsets of κ, there is an unbounded $B \subseteq \kappa$ such that for finite subsets x of B, membership in E depends only on the cardinality of x.

Let G_0 be the Baire space of finitely generated products on ω.

THEOREM 7 (FRIEDMAN [11]). Let $F : G_0^\omega \to G_0$ be a Borel function mapping pointwise isomorphic arguments to isomorphic values. Then for some $x \in G_0^\omega$, $F(x)$ is isomorphically imbeddable into a coordinate of x. Furthermore, this theorem is provable in ZFC but not in ZC.

THEOREM 8 (FRIEDMAN [11]). Let $F : G_0^\omega \to G_0$ be a Borel function mapping pointwise isomorphic arguments to isomorphic values. Then for some $x \in G_0^\omega$, for all subsequences y of x, $F(y)$ is isomorphically imbeddable into a coordinate of y. This is provable in ZFC + "there is a measurable cardinal," but not in ZFC + "there is a Ramsey cardinal."

In Paris-Harrington [21] an interesting example of a theorem stated in finite set theory but not provable in finite set theory, is given. Also see [16], [23], [24].

THEOREM 9 (PARIS-HARRINGTON [21]). For each k,r,s, there is an n so large that if all subsets of $[1,n]$ of size k are colored with r colors, then there is a set $E \subseteq [1,n]$ such that all subsets of E of size k have the same color and the size of E is at least s and the minimum of E. Furthermore, this is provable in finite set theory augmented with definition by recursion on ω, but not in finite set theory.

Subsequently, Friedman [7] found an interesting example of a finite theorem which is conceptually even clearer and independent of much stronger systems. Just as Theorem 9 is based on Ramsey's theorem, Friedman's work is based on the following theorem of J. B. Kruskal [17].

A tree is a nonempty partial ordering with a least element, such that the set of predecessors of any element is linearly order. If T_1, T_2 are finite trees then $h : T_1 \to T_2$ is said to be a homeomorphic imbedding if $a \leq_{T_1} b$ iff $h(a) \leq_{T_2} h(b)$, and $h(\inf(a,b)) = \inf(h(a),h(b))$. We write $T_1 \leq T_2$ if such an h exists.

THEOREM 10 (KRUSKAL [19]), NASH-WILLIAMS [20]). If $T_1, T_2, \ldots$ are finite trees then for some $i < j$, $T_i \leq T_j$.

The formal system ATR is obtained from finite set theory by introducing countably infinite sets with the principle of definition by transfinite recursion on countable well orderings. This system goes just beyond what is referred to as predicative analysis.

THEOREM 11 (FRIEDMAN [7,13]; SIMPSON [22]). For all k there is an n so large that for all finite trees $T_1, \ldots, T_n$ and $\operatorname{card}(T_i) \leq i$, there are $i_1 < \ldots < i_k$ such that $T_{i_1} \leq \ldots \leq T_{i_k}$. This theorem, as well as Kruskal's Theorem 10 above, are provable in ZFC without the power set axiom, but not in ATR. Also see [16,23,24].

THEOREM 12 (FRIEDMAN [8,13]). Theorem 11 for $k = 12$ is provable in ZFC without the power set axiom in a few pages, but any proof in ATR must use at least $2^{[1000]}$ pages.

Friedman [10] has extended Kruskal's theorem in an interesting way so that the theorem has yet stronger metamathematical properties. He considers the class $Tr_\infty(n)$ of finite trees with n distinct labels, and defines $T_1 \leq_r T_2$ if and only if there is an $h : T_1 \to T_2$ which is a label-preserving homeomorphic imbedding, preserving left to rightness, with the additional crucial condition that if b is an immediate successor of a in T_1 and $h(a) < c < h(b)$ in T_2, then $l(c) \geq l(h(b))$ ($l(c)$ is the label of c). The subscript "r" means "restricted".

THEOREM 13 (FRIEDMAN [10], SIMPSON [22]). For every $T_1, T_2, \ldots \in Tr_\infty(n)$, there are $i < j$ such that $T_i \leq_r T_j$. For all k,n there is an m so large that for all $T_1, \ldots, T_m \in Tr_\infty(n)$ with each $\mathrm{card}(T_i) \leq i$, there are $i_1 < \ldots < i_k$ such that $T_{i_1} \leq_r T_{i_2} \leq_r \ldots \leq_r T_{i_k}$. These theorems are provable in the single set quantifier comprehension axiom system with the full scheme of induction (Π^1_1-CA), but not in Π^1_1-CA with set induction instead of the full scheme of induction (Π^1_1-CA_0).

Two of the most fundamental concepts in mathematical logic are the model theoretic concept of translatability and the proof theoretic concept of relative consistency. Friedman [6] proves that under surprisingly general conditions, these concepts coincide. Specifically, let S,T be finitely axiomatized first-order theories. We say that S is translatable into T if there are first-order definitions of the symbols in S in terms of those in T, such that every axiom of S, when so translated, becomes a theorem of T.

There are a few somewhat different ways of defining "S is consistent relative to T," considered in [6]. We focus attention on the following one: Let EFA (exponential function arithmetic) be the standard weak system of arithmetic based on $0,1$, $<$, $=$, $+$, $\cdot$, and exponentiation, where induction is applied to bounded formulas only. The quantifier complexity of a formula is a standard measure of the number of alternations of quantifiers that are present.

We say that S is consistent relative to T if for some fixed n, the following is provable in EFA. If there is an inconsistency proof in S then there is an inconsistency proof in T which uses only formulas whose quantifier complexity is at most n more than the greatest quantifier complexity of formulas used in the given inconsistency proof in S.

It is straightforward to see that translatability implies relative consistency.

THEOREM 14 (FRIEDMAN [6]). Let S,T be finitely axiomatized theories containing EFA (and a weak theory of finite sequences of objects other than natural numbers, if applicable). Then S is translatable into T if and only if S is consistent relative to T.

Friedman [5] has initiated an interesting new branch of model theory called Borel model theory. A totally Borel model is a structure whose domain is $\mathbb{R}$ and every relation that is definable (in the language considered) over the structure is Borel. Friedman [5] considers the three fundamental quantifiers Q_m (almost all in the sense of Lebesgue measure), Q_c (almost all in the sense of Baire category), and Q_{ω_1} (uncountably many), in addition, of course, to the usual quantifiers $\forall, \exists$.

Friedman proves the following completeness and duality theorem.

THEOREM 15 (FRIEDMAN [5], STEINHORN [25]). Let ϕ be a sentence based on Q_m (or Q_m and $\forall,\exists$). Then ϕ is true in all totally Borel models if and only if ϕ can be proved using, roughly speaking, the following principles: singletons are of measure 0, subsets of measure 0 are of measure 0, the union of two sets of measure 0 is of measure 0, $\mathbb{R}$ is not of measure 0, and almost all vertical cross sections of a two-dimensional set are of measure 0 if and only if almost all horizontal cross sections are of measure 0 (Fubini's theorem). If Q_c is used instead of Q_m, then this is true if "measure 0" is replaced by "meager." As a consequence we have that ϕ is true in all totally Borel structures for Q_m (or $(Q_m,\forall,\exists)$) if and only if ϕ^* is true in all totally Borel structures for Q_c (or $(Q_{c_1} \forall, \exists)$) where ϕ^* is obtained from ϕ by replacing Q_m by Q_c (duality).

Friedman has obtained a number of fundamental results in intuitionistic set theory. The usual axioms for ZF are extensionality, pairing, union, infinity, foundation, power set, comprehension, and finally replacement. The axiom scheme of collection is an alternative to the axiom scheme of replacement, but it is a fundamental theorem of set theory that these are equivalent.

However, if we use intuitionistic logic instead of ordinary logic, then the proof that replacement implies collection breaks down. Thus we let ZFIR be ZF with intuitionistic logic formulated with the scheme of replacement, and ZFIC be ZF with intuitionistic logic formulated with the scheme of collection.

THEOREM 16 (FRIEDMAN [2]); FRIEDMAN-SCEDROV [15]). ZFIR does not imply ZFIC. It is provable in a weak system of arithmetic that ordinary ZFC is consistent if and only if ZFIC is consistent.

Two basic desirable properties of intuitionistic formal systems (which almost never hold for ordinary formal systems) are the disjunction property, which asserts that if a disjunction $A \vee B$ is provable then one of the disjuncts is provable; and the numerical existence property, which asserts that if $(\exists n)(\forall n)$ is provable then for some $n, A\bar{n}$ is provable. DP trivially follows from NEP.

Friedman proves the following highly surprising theorem via a mysterious application of Gödel self-reference.

THEOREM 17 (FRIEDMAN [3]). Let T be a recursively axiomatized intuitionistic formal system subject to, roughly, the same weak hypotheses commonly used in Gödel's incompleteness theorems. Then T has the numerical existence property if and only if T has the disjunction property.

SELECTED REFERENCES

[1] H. Friedman, Higher set theory and mathematical practice, Ann. Math. Logic 2 (1971), pages 326-357.

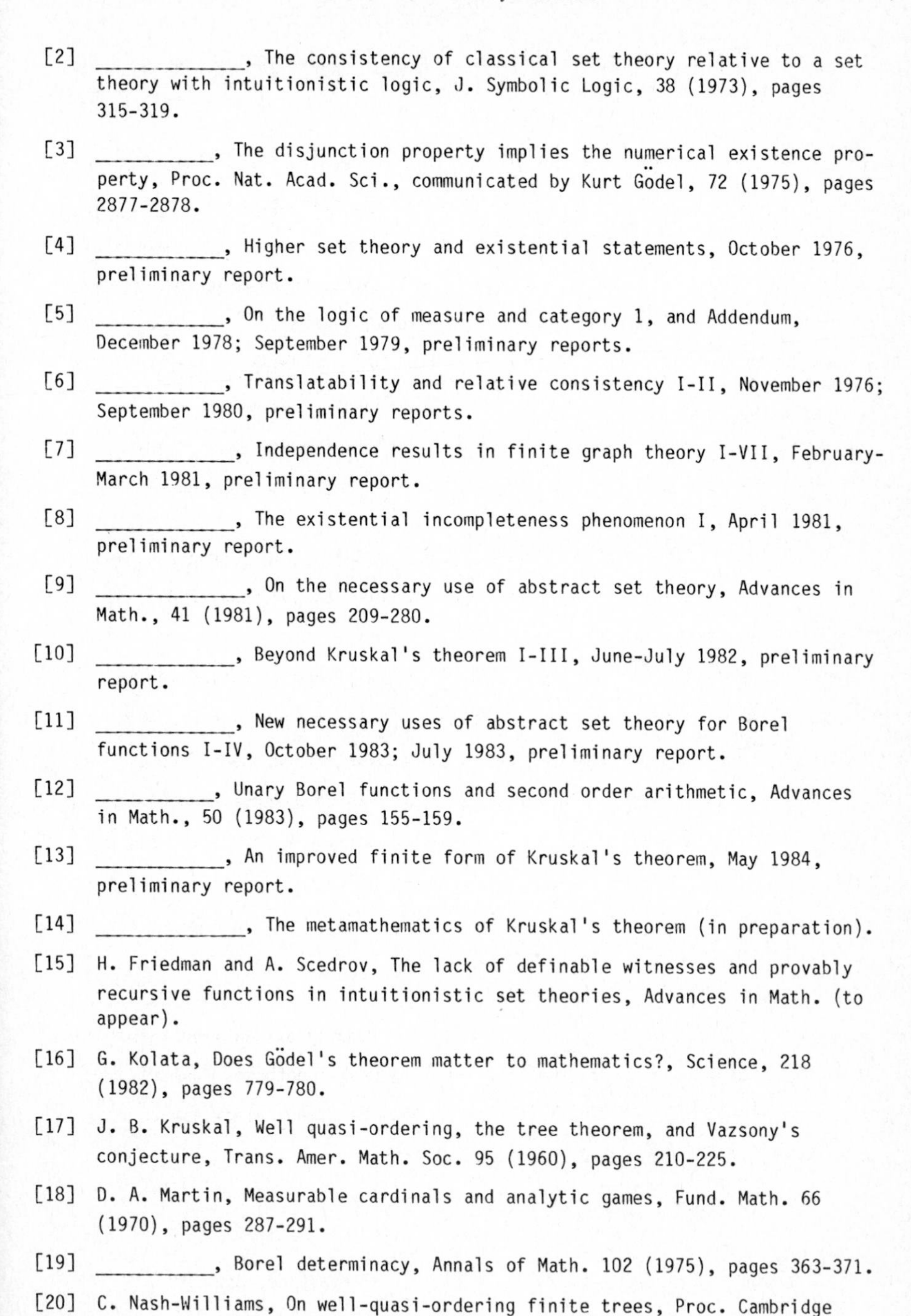

[2] ______________, The consistency of classical set theory relative to a set theory with intuitionistic logic, J. Symbolic Logic, 38 (1973), pages 315-319.

[3] __________, The disjunction property implies the numerical existence property, Proc. Nat. Acad. Sci., communicated by Kurt Gödel, 72 (1975), pages 2877-2878.

[4] ___________, Higher set theory and existential statements, October 1976, preliminary report.

[5] ___________, On the logic of measure and category 1, and Addendum, December 1978; September 1979, preliminary reports.

[6] ___________, Translatability and relative consistency I-II, November 1976; September 1980, preliminary reports.

[7] ____________, Independence results in finite graph theory I-VII, February-March 1981, preliminary report.

[8] ____________, The existential incompleteness phenomenon I, April 1981, preliminary report.

[9] _____________, On the necessary use of abstract set theory, Advances in Math., 41 (1981), pages 209-280.

[10] ____________, Beyond Kruskal's theorem I-III, June-July 1982, preliminary report.

[11] ____________, New necessary uses of abstract set theory for Borel functions I-IV, October 1983; July 1983, preliminary report.

[12] __________, Unary Borel functions and second order arithmetic, Advances in Math., 50 (1983), pages 155-159.

[13] ___________, An improved finite form of Kruskal's theorem, May 1984, preliminary report.

[14] _____________, The metamathematics of Kruskal's theorem (in preparation).

[15] H. Friedman and A. Scedrov, The lack of definable witnesses and provably recursive functions in intuitionistic set theories, Advances in Math. (to appear).

[16] G. Kolata, Does Gödel's theorem matter to mathematics?, Science, 218 (1982), pages 779-780.

[17] J. B. Kruskal, Well quasi-ordering, the tree theorem, and Vazsony's conjecture, Trans. Amer. Math. Soc. 95 (1960), pages 210-225.

[18] D. A. Martin, Measurable cardinals and analytic games, Fund. Math. 66 (1970), pages 287-291.

[19] __________, Borel determinacy, Annals of Math. 102 (1975), pages 363-371.

[20] C. Nash-Williams, On well-quasi-ordering finite trees, Proc. Cambridge Phil. Soc. 59 (1963), pages 833-835.

[21] J. Paris and L. Harrington, A mathematical incompleteness in Peano arithmetic, in Handbook of Mathematical Logic (Jon Barwise, Ed.), North-Holland, Amsterdam, 1977, pages 1133-1142.

[22] S. Simpson, Unprovability of certain combinatorial properties of finite trees (to appear).

[23] C. Smorynski, The varieties of arboreal experience, Math. Intelligencer, 4 (1982), pages 182-189.

[24] ___________, "Big" news from Archimedes to Friedman, Notices Amer. Math. Soc. 30 (1983), pages 251-256.

[25] C. Steinhorn, Borel structures and measure and category logic, Model-Theoretic Logics (J. Barwise and S. Feferman, Ed.), North-Holland, Amsterdam (to appear).

HARVEY FRIEDMAN'S RESEARCH ON THE FOUNDATIONS
OF MATHEMATICS, L.A. Harrington et al. (editors)

BOREL DIAGONALIZATION AND ABSTRACT SET THEORY: RECENT RESULTS OF HARVEY FRIEDMAN

Lee J. Stanley
Department of Mathematics
Lehigh University
Bethlehem, Pennsylvania
U.S.A.

Recently, Friedman has greatly improved the results of [04] by formulating a new family of propositions concerning Borel diagonalization properties of the spaces of groups, semigroups, linear orderings, products (all with underlying set ω) and Turing degrees. These propositions have certain metamathematical properties in common with those of [04], but exhibit a greater range of consistency strength and, arguably, they are more concrete, intelligible and natural. This paper presents an account of this work of Friedman, providing detailed proofs.

INTRODUCTION

While acknowledging the technical virtuosity and potential importance (for other branches of mathematics and science) of high-level work in contemporary set theory, Harvey Friedman has long stressed that it could and should be "grounded", connected to more concrete questions which are more closely related to non set-theoretic mathematics. In [04], he realized an important advance in this ambitious program (which can be traced back at least to [03]). As he remarked in the introduction to [04], the results there differ in several important respects from the Godel and Cohen style independence proofs: the propositions involved are not overtly metamathematical, nor do their statements involve unduly abstract notions such as uncountable ordinals or arbitrary sets of real numbers. Ideally such propositions will be equivalent, over weak base theories, to a standard subsystem, or extension by strong axioms of infinity, of set theory; in practice we shall often only be able to situate such propositions between two such systems. Further, while independent, their status cannot be settled by specializing the notion of set, i.e. by "cutting down" the set-theoretic universe, since one aspect of their concreteness is that they are highly absolute.

An important measure of the success of Friedman's program at any given time is the extent to which the existing stock of such propositions are (and are recognized as) concrete, intelligible and natural. By this standard, Friedman's work since [04], on propositions concerning Borel diagonalization properties of the spaces of groups, semigroups, linear orderings, products (all with underlying set ω) and Turing degrees, represents substantial progress. During the same period, Friedman also obtained remarkable results connecting set theory with finite mathematics, for which see [06]. Taking a still larger view, these two directions are similar in spirit to a vast body of work on what has come to be called "reverse mathematics", see [16]. An excellent survey of this area appears in [11], where many of Friedman's new results in this area were first reported. This paper is an account of Friedman's new results related to Borel diagonalization properties, providing detailed proofs.

In §1, we consider the basic Borel diagonalization property for the spaces of groups and products. We prove, in ZF minus the power set axiom + the axiom of choice for countable families of non-empty sets (henceforth denoted $ZF + AC_\omega - P$) + "the power set of ω exists", that Borel diagonalization <u>fails</u> (Theorems 1, 1'); we also prove that this failure <u>cannot</u> <u>be</u> <u>proved</u> in $ZF + AC_\omega - P$ alone (Corollaries 1, 1'). Here, and throughout the paper, the provability results are labelled "THEOREM n" (or some variant of this), in §n, the unprovability results are labelled "COROLLARY n" (or some variant), again in §n, except in §2. Corollary n is always (except in §2) obtained via a result asserting the existence of a model of an appropriate theory; this result is always labelled "REVERSAL n" (or some variant); in §2, Reversal 2 also plays the role of Corollary 2. Theorem 1' and Corollary 1' together correspond to Theorem 4 of [11]; since this is not completely obvious from their statements, we shall comment on this and related questions below. What is at issue here is the possibility of transferring results back and forth between theories formulated in first order languages appropriate for set theories and theories formulated in second order languages appropriate for second order arithmetic.

In §2, we consider the spaces $\mathbb{R}$, $\mathbb{R}^\omega$ (actually, as usual, we really work with Baire space, B, and B^ω), and we consider Borel measurable functions, F, from $\mathbb{R}^\omega$ to $\mathbb{R}$, which are invariant for a given analytic equivalence relation, E, on $\mathbb{R}$. Here, invariance means that two infinite sequences which are termwise equivalent for E give E-equivalent F-values. In this setting, the failure of Borel diagonalization means that for any analytic equivalence relation, E, and any Borel, E-invariant F mapping B^ω to B there's $\vec{x} \in B^\omega$ and $i \in \omega$ such that $F(\vec{x})$ is E-equivalent to x_i. We prove (Theorem 2), that this failure is provable in (ZF-P) + "ω_1 iterations of the power set operation". Using the notion of the dense mix of countably many countable linear orderings, we show (Lemma 2) that the failure of Borel diagonalization implies that whenever F maps the power set of the rationals to itself and sends order isomorphic arguments to order isomorphic values, there's a set, A, of rationals whose F-value is order isomorphic to an interval of A (this is Proposition E). We then show (Reversal 2) that "ω_1 iterations of the power set operation" are necessary to prove Proposition E. Theorem 2, Lemma 2 and Reversal 2 together correspond to Theorem 5 of [11].

In §3, we consider three spaces of structures on ω: groups, semigroups and products. Actually, we work with semigroups with identity element, but this is no loss of generality since we are considering embedding properties of semigroups and any semigroup embeds in a semigroup with identity element: just add a new element; make it the identity element of the new semigroup. We shall denote each of these spaces by G and we shall denote by G_f the subspace of finitely generated points of G. For each of our spaces G, we also consider two families of Propositions. Each family has a weak version (Propositions F(G), F'(G)) and a strong version (Propositions G(G), G'(G)). The strong version is obtained from the weak version by adding in a Ramsey-type property of the sort considered by Galvin and Prikry, [07] and Silver [13].

For Propositions F(G), G(G), we consider Borel functions F mapping G_f^ω to G_f, which map termwise isomorphic sequences to isomorphic values. Proposition F(G) expresses the failure of the basic Borel diagonalization property in this setting: for all such F there's $\vec{x} \in B^\omega$ and $n \in \omega$ such that $F(\vec{x})$ is isomorphically embeddable in x_n. Proposition G(G) requires that for every subsequence $\vec{y}$, of $\vec{x}$, $F(\vec{y})$ is isomorphically embeddable in a term of $\vec{y}$. For Propositions F'(G), G'(G), we consider Borel functions F mapping G_f^ω to G, which map termwise isomorphic sequences to isomorphic values. Proposition F'(G) states the existence of $\vec{x}$ such that $F(\vec{x})$ is isomorphically embeddable in a direct limit of $\vec{x}$, while Proposition G'(G) requires that for all subsequences $\vec{y}$, of $\vec{x}$, $F(\vec{y})$ is isomorphically embeddable in a direct limit of $\vec{y}$. We also consider variants of Propositions G(G), G'(G). It will be easy to show that Proposition F'(G) => Proposition F(G) and Proposition G'(G) => Proposition G(G).

We prove (Theorem 3) that Proposition F'(G) is provable in ZF + AC_ω - P + "$V_{\omega+\omega}$ exists" and that Proposition G'(G) is provable in ZFC + "there exists a measurable cardinal", and (Corollary 3) that Proposition F(G) is unprovable in Zermelo set theory with the axiom of choice plus the axiom of constructibility and that Proposition G(G) is unprovable in ZFC plus "for all sets x, $x^\#$ exists". Our method is to prove similar statements (Theorem 3', Corollary 3') with Baire space, B, in place of G, with Turing-reducibility in place of isomorphic embeddability and for Borel functions F mapping B^ω to B, which map termwise Turing-equivalent sequences to Turing-equivalent values. We then obtain the results for our spaces G by developing a theory permitting us to transfer such results back and forth between B, with the recursion-theoretic notions and the various G, with the algebraic notions. This is done in (3.4) and (3.5); the techniques involved seem to be of independent interest especially when G = groups. Theorem 3 and Corollary 3 correspond to Theorems 7, 8 of [11].

In §4, we prove some results connecting determinacy with a Proposition similar to the recursion-theoretic version of Proposition G(G), but with separating sequences instead of subsequences. A separating sequence for a sequence $\vec{x} \in B^\omega$ is a sequence $\vec{y} \in B^\omega$ such that for some subsequence $\vec{z}$ of $\vec{x}$ and all $n \in \omega$ z_{2n} is recursive in y_n is recursive in z_{2n+1}. Proposition H states that whenever F is a Borel function mapping B^ω to B there's $\vec{x} \in B^\omega$, which is increasing for the "recursive-in" partial preorder and such that for any separating subsequence $\vec{y}$ for $\vec{x}$, there's a term of $\vec{y}$ in which $F(\vec{y})$ is recursive. We prove (Theorem 4) that Proposition H is provable in ZFC + the axiom of determinacy for sets of reals constructible from a real (AD(L[$\mathbb{R}$]) and (Corollary 4) that it is not provable in ZFC + the axiom of determinacy for projective sets of reals (PD).

Theorem 2 is proved using Borel determinacy to embody the ω_1 iterations of the power set operation; the proof closely parallels that of Theorem 3.2 of [04]. Theorems 1, 3, 4 are proved by forcing over models of sufficiently rich fragments of the appropriate theory. We work with countable transitive models but this is purely for convenience: the forcing could be done over countable ill-founded ω-models. This is important since if such models exist at all, they exist even inside very small initial segments of the set-theoretic universe, satisfying rather weak or restrictive set existence axioms, e.g. closure under hyperjump or $V = L$. This provides strong absoluteness for our Propositions, despite their independence and, in §3, §4, their considerable consistency strength. Frequently, the extension is by a set generic for a countable fragment of an infinitary language.

In §1, the forcing adds a one-to-one collapse of the continuum to be countable. In §3, for the recursion-theoretic variant of the Propositions $F'(G)$, the forcing adds a generic sequence of collapsing maps, $(f_\alpha: \alpha < \aleph_\omega)$, each f_α mapping ω onto $L_\alpha[u]$, where u is a real in the ground model which is a Borel code of a Borel, Turing-invariant F. The proof of the recursion-theoretic variant of the Propositions $G(G)$ uses a partial ordering which adds a generic sequence of collapsing maps, $(f_\alpha: \alpha < \kappa)$, where κ is a measurable cardinal of the ground model, and each f_α maps ω onto the ground model's version of V_α. A technical novelty is that the generic sequence is chosen generic, not only over the ground model, but over the extension of the ground model by adding a Prikry sequence through κ. We show that this can be done by "adjusting", from the outside, a generic sequence which does not have this property. Both proofs use the idea of superimposing one forcing condition on another or on a generic sequence. Iterating the latter operation ω times in a suitable fashion produces a generic sequence which is <u>similar</u> to the given generic sequence in that any bounded segment of the new sequence differs from the corresponding bounded segment of the given one by a finite amount of information. This permits us to prove that the F-value of the b from B^ω produced from the generic sequence is determined by a bounded segment of the generic sequence, since otherwise we could form a perfect set of generic sequences, each similar to the one we chose, which yield distinct F-values for the b' which they produce. Since F is Turing-invariant, this, however, contradicts that similar sequences produce termwise Turing-equivalent b. Such methods were already used in [04].

In §4, the forcing is by a partial ordering which adds an

ω-sequence of reals cofinal in the "recursive-in" partial preordering of the ground model. There is a strong analogy between this forcing and Prikry forcing: the Martin measure on sets of reals replaces the normal measure on a measurable cardinal. At the end of §4, we present a geometric proposition, involving lines of rational slope in $\mathbb{R}^2$ (these play the role of cones of Turing degrees), which Friedman, in very recent work, has shown to be equivalent to determinacy; no proofs are given.

The proofs of the Reversals all follow a common pattern: we define, usually by cases, a Borel, appropriately invariant function. We associate to each x in the domain of the function a set of reals, in a Borel fashion. The function is defined in such a way that if the associated set of reals fails to model an appropriate theory formulated in the language of second order arithmetic then the x which gave rise to it cannot satisfy the conclusion of the Proposition which is being "reversed". Again, this approach goes back to [04].

We should elaborate, at this point, on the results which allow us to transfer back and forth between the second order number theories and the first order set theories, especially since the Reversals always produce models of the former and the Corollaries always state results in terms of the latter. This should also clarify the relationship between the theorems of this paper and their analogues in [11], as sketched above. Many of these equiconsistency results go back to the Mostowski school and some can only be referred to as part of the folklore. For others, [01] is a good reference and similar material can be extracted from [08]. More recent material appears in [12], [14], [15]. A common thread running through all this material is the coding of sets by well-founded trees.

The coding is made especially explicit in [14], where it is shown that the theory ATR_0 (for arithmetic transfinite recursion) is sufficient to carry out this coding. Here, this means the Reversals and Corollaries can be strengthened by replacing $ZF + AC_\omega - P$ by ATR_0. We elaborate on the intertranslatability between these two families of theories only in (3.3), where the details of the procedure really are germane. Also, most often, we won't verify that certain sets are Borel and that certain functions are Borel measurable. A good source for the descriptive set theory involved is [10].

The arguments of (3.3) are among the most subtle of the paper, involving developing the machinery of sharps in the context of ill-founded countable models. The apparatus developed in (3.4),

(3.5) to go back and forth between our spaces $\mathcal{G}$ and B involves a pair of maps, σ from B to $\mathcal{G}_f$ and τ from $\mathcal{G}_f$ to B. For $x \in \mathcal{G}$, $\tau(x)$ will be the equational theory of terms in a finite set of variables, corresponding to a canonical choice of a set of generators. For $\mathcal{G}$ = semigroups, $\sigma(b)$ will be the semigroup of partial b-recursive functions, with composition and the identity function as identity element, isomorphed in a canonical way onto ω. For $\mathcal{G}$ = products, $\sigma(b)$ will be defined using the "enumerative systems" and "basic recursive function theories" of [02]. The essential idea is that $n*m$ will be $\varphi_n^b(m)$, if this is defined and a conventional value, if not. Some technical twists and turns are necessary due to the asymmetry between the treatment of the first argument on the one hand, and the second argument and the value on the other hand; this is where the techniques of [02] come in. When $\mathcal{G}$ = groups, the work involved in defining σ is the most substantial and interesting. We prove a uniform and effective strengthening of Higman's theorem, [09]: for any $b \in B$, there is a finitely-generated group which isomorphically embeds any group on ω whose product operation is recursive in b. This is $\sigma(b)$. Similar results for semigroups and products fall out of the transfer mechanism for these spaces. This is stated in (3.4), (*) and proved in (3.5.8), for groups and in (3.5.9) for our other spaces.

This material is certainly very substantial on a purely technical level. The methods and intuitions come from an astonishingly wide range of subfields of logic and, in the case of the group theory, from other areas of mathematics, going so far as to establish methods and results which may well be of independent interest for these areas. Perhaps the central innovation is the systematic development of techniques of model construction for ill-founded models of various theories, be they first-order set theories or second order number theories. While this aspect of his work is certainly important, in Friedman's eyes, the conception of the program presented at the beginning of this Introduction and the successful formulation of the Propositions this program involves is of an even higher, in fact critical, level of importance and he regards this as his true breakthrough in this area. This happy marriage of high-level technical work and an overriding preoccupation with fundamental foundational issues, which orients and organizes the technique, has always been characteristic of Friedman's work.

§1. For $1 \leq n < \omega$, let B_n be n-place Baire space: ω^{ω^n} with the product of the discrete topology on ω. If $f,g \in B_n$, we say that $f \cong g$ iff $(\omega,f) \cong (\omega,g)$ as models. We say that $g \preccurlyeq f$ (g is isomorphically embeddable in f) iff (ω,g) is isomorphic to a substructure of (ω,f).

Let T_n state the failure of the Borel diagonalization property for B_n, i.e.:

> (T_n): Whenever $F: B_n \to B_n$ is Borel (i.e., Borel measurable) and F is isomorphically invariant (i.e. $f \cong g \Rightarrow F(f) \cong F(g)$), then, for some f, $F(f) \preccurlyeq f$.

<u>THEOREM 1</u>: $(\forall n<\omega)T_n$ is provable in $ZF + AC_\omega - P +$ "$\mathcal{P}(\omega)$ exists".

<u>REVERSAL 1</u>: The following is provable in $ZF + AC_\omega - P + T_2$: "there's an ω-model of Z_2 (second-order arithmetic)".

<u>COROLLARY 1</u>: T_2 is not provable in $ZF + AC_\omega - P$.

We now prove Theorem 1. Suppose φ is a partial function from $\mathcal{P}(\omega^n)$ to $\mathcal{P}(\omega^n)$. φ is <u>finitely generated</u> iff range $\varphi \subseteq$ dom φ and there's finite $a \subseteq$ dom φ such that dom φ is the closure of a under φ.

(1.1) <u>LEMMA</u>: The following is provable in $ZF + AC_\omega - P$:

For each $n \geq 1$, there's a Borel function H_n: $\mathcal{P}(\omega^n)$ $\mathcal{P}(\omega^n)$ with the following saturation property:

(*) suppose f,g are injective, finitely generated partial functions, $f \subseteq g \cap H$. Then g is embeddable in H by a function extendable to a Borel function on $\mathcal{P}(\omega^n)$ and which is the identity on dom f.

The proof requires the following:

<u>Sublemma</u>: Suppose $F: X \to \mathcal{P}(\omega^n)$ is Borel measurable, where $X \subseteq \mathcal{P}(\omega^n)$ and X, $\mathcal{P}(\omega^n) \setminus X$ are both uncountable (thus, both are Borel and therefore Borel isomorphic to $\mathcal{P}(\omega^n)$). Let Y be any uncountable Borel subset of $\mathcal{P}(\omega^n) \setminus X$. Then there's a Borel function $G: X \cup Y \to \mathcal{P}(\omega^n)$, $G \supseteq F$ such that for any countable subfunction $f \subseteq F$ and any countable extension $g \supseteq f$ with dom $g \subseteq X$, range $g \subseteq \mathcal{P}(\omega^n)$, g is embeddable in G by a function which is the identity on dom f.

The proof of the Sublemma is obvious by making disjoint copies of all the dom g $\setminus$ dom f, for f,g as in the statement of the Sublemma and coding the union into a function on Y.

The lemma is then proved by iterating the Sublemma ω times, starting from a partition $(X_n : n < \omega)$ of $P(\omega^n)$ into uncountable Borel subsets and an arbitrary Borel $F: X_0 \to P(\omega^n)$. At stage $n + 1$, we apply the sublemma with $F = F_n$, $X = X_0 \cup \ldots \cup X_n$, $Y = Y_{n+1}$ and we take $F_{n+1} = G$ as given by the Sublemma. The H required by the lemma is then just $\bigcup_n F_n$.

(1.2) Now let $F: B_n \to B_n$ be Borel and isomorphically invariant, and let H be as guaranteed by the Lemma. Fix Borel codes u for F and v for H, and let A be a countable admissible set with $u, v \in A$. Let X be countable and an $\in$-elementary substructure of $L_{\omega_1}[u,v]$ with respect to the infinitary language L_A, and with $u, v \in X$. Let $L_\alpha[u,v]$ be the transitive collapse of X; thus $L_\alpha[u,v]$ models L_A - ZFC - P (i.e. its a model of the replacement scheme with $\in$-formulas from L_A.

Let $\bar{H}$ be the Borel function of $L_\alpha[u,v]$ coded by v (so $\bar{H}$ is $L_\alpha[u,v]$'s version of H, and satisfies (*) of the Lemma inside $L_\alpha[u,v]$). Let $f: \omega^n \to P(\omega^n) \cap L_\alpha[u,v]$ be L_A-generic over $L_\alpha[u,v]$ for the partial ordering of finite injective conditions. $\bar{H}$ then is coded as $f^{-1} \circ \bar{H} \circ f: \omega^n \to \omega^n$, and then easily coded by $h^*: \omega^n \to \omega$, so h^* B_n. We'll show $F(h^*) \preccurlyeq h^*$.

Now h^* depends on which generic enumeration f of $P(\omega^n)$ is taken, but the h^*'s arising from different enumerations are naturally isomorphic. Thus, up to isomorphism, $F(h^*)$ doesn't depend on f, since F is isomorphically invariant. This also yields:

(1) $\forall a_1, \ldots, a_p \in \omega$ $E_{a_1 \ldots a_p} \in L_\alpha[u,v]$, where $E_{a_1 \ldots a_p}$ is the set of parameter-free equations in $v_1, \ldots v_p$ satisfied by $a_1, \ldots, a_p$ in $F(h^*)$.

(1) follows from the fact that $E_{a_1 \ldots a_p}$ is in <u>all</u> generic extensions. This in turn follows from the fact that if h^* arises from f, h' arises from f', then $F(h^*) \cong F(h')$, so, letting σ be an isomorphism, $E_{a_1 \ldots a_p}$ in $F(h^*) = E_{\sigma(a_1) \ldots \sigma(a_p)}$ in $F(h')$, so $E_{a_1 \ldots a_p}$ lies in the generic extension by h'.

Now, for each $n < \omega$, let W_n be the substructure of $F(h^*)$ generated by $\{0, \ldots, n\}$. We then show that $F(h^*) \preccurlyeq h^*$ by showing inductively that each $W_n \preccurlyeq h^*$ say by π_n, and further that the π_n's can be chosen with $\pi_n \subseteq \pi_{n+1}$. This uses the saturation property (*) of $\overline{H}$ with respect to $L_\alpha[u,v]$, and the fact that W_n is completely determined by $E_{0,\ldots,n}$ which, by (1) lies in $L_\alpha[u,v]$. The induction is straightforward.

<u>REMARK</u>: B_n corresponds to the choice of the similarity type ρ with exactly one n-place function. In fact, the proof of Theorem 1 goes over word for word if we replace ρ by any finite similarity type ρ' which either has no constant symbols or is trivial in that it consists <u>only</u> of constant symbols. The difficulty with having constant symbols in a non-trivial ρ' is that the proof of the Lemma breaks down.

(1.3) Before turning to the proof of REVERSAL 1, we note that we cannot replace T_2 by T_1, since we can prove in $ZF + AC_\omega - P$ that there is a universal $f \in B_1$; i.e., for all $g \in B_1$, $g \preccurlyeq f$. We briefly describe such an f. A $\mathbb{Z}$-tree is a (non-well-founded) tree in which each node has a countably infinite set of immediate successors and in which the order type of the set of predecessors of any node is ω^* - the reverse order on ω. If $k < \omega$, an (ω,k) tree consists of a simple cycle of order k, one of whose nodes is the root of a height ω tree in which every node has a countably infinite set of immediate successors. f then consists of a countably infinite set of disjoint $\mathbb{Z}$-trees and for each $k < \omega$, a countably infinite set of disjoint (ω,k)-trees, where, for $k \neq k'$ every (ω,k)-tree is disjoint from every (ω,k')-tree, with the interpretation $f(x)$ = the immediate predecessor of x for x not involved in a simple cycle, while for x involved in a simple cycle of order k, $\{f^i(x): i \leq k\}$ enumerates the cycle while $f^{k+1}(x) = x$.

In order to prove REVERSAL 1, if $1 \leq n < \omega$, $f \in B_2$ $x \in \omega$, we define x_f^n recursively: $x_f^1 = f(x,x)$, $x_f^{n+1} = f(x_f^n,x)$. When f is clear from context, we denote x_f^n by x^n. Then, if $f \in B_2$, we define $\tilde{f} \subseteq \omega$ by:

$i \in \tilde{f} \Leftrightarrow$ for some $x \in \omega$ $\{x^n: 1 \leq n < \omega\}$ has exactly $i+1$ elements. We also let $|f| = \{\tilde{g}: g$ is a finitely generated subfunction of $f\}$. Remark that if $f \preccurlyeq g$, then $|f| \subseteq |g|$.

Before defining a Borel, isomorphically invariant $F: B_2 \to B_2$ to which we apply T_2 to obtain an ω-model of Z_2, we need an auxiliary notion. If $f \in B_2$ and $(\omega, |f|, \in, \ldots)$ is not a model of Z_2, we let $|f|^*$ = the set of all subsets of ω which are first-order definable (as elements) in parameters from $|f|$ in $(\omega, P(\omega), \in, \ldots)$. Thus, $|f| \subsetneq |f|^*$. We then prove:

LEMMA: If $r \in |f|^*$, there's a finitely generated $g = g_r : \omega^2 \to \omega$ such that $\tilde{g} = r$. Further the map $r \to g$ can be taken to be Borel.

Proof of Lemma: For $i \in r$, we arrange for g to have disjoint simple cycles of order i, i.e. we introduce $x(i) \in \omega$ s.t. $\{x(i)^n_g : n < \omega\}$ has exactly $i+1$ elements. For $i \in \omega \setminus r$, we introduce $x(i) \in \omega$ s.t. $\{x(i)^n_g : n < \omega\}$ is infinite. The $g(x,y)$ which are still undefined are defined arbitrarily. Finally, to make g finitely generated, a new element, α, is introduced s.t. $\{\alpha^n_g : n < \omega\}$ is infinite and s.t. for $n < \omega$, $g(\alpha,n) = n+1$. Then g is coded as an element of B_2.

Finally, we define $F: B_2 \to B_2$ by: $F(f) = f$, if $(\omega, |f|, \in, \ldots)$ models Z_2, while, if not, $F(f)$ is the disjoint union of the g_r for $r \in |f|^*$, coded as a partial function from ω^2 to ω, and extended to a total function by defining $g(x,y)$ to be arbitrary, when x,y come from domains of different g_r. Note that, in this case, $|F(f)| \supseteq |f|^* \supsetneq |f|$, so $F(f)$ fails to be embeddable in f. But some $F(f) \preccurlyeq f$, so this must be for an f s.t. $(\omega, |f|, \in, \ldots)$ models Z_2, so Z_2 has an ω-model.

REMARK: In the Lemma, g_r can be taken to have the same Turing degree as r.

(1.4) We now obtain some versions of THEOREM 1, REVERSAL 1 and COROLLARY 1 for spaces other than the B_n. Let $B_{\phi;2,1;1}$ be the space of algebras with underlying set ω, with a binary and a unary operation and one constant, with the topology induced by the natural bijection with $B_2 \times B_1 \times \omega$. Let G be the closed set of groups. Let T_G state the Borel analogue of T_n for G.

THEOREM 1'. T_G is provable in $ZF + AC_\omega - P +$ "$P(\omega)$ exists".

REVERSAL 1': $ZF + AC_\omega - P + T_G$ proves: "there's an ω-model of Z_2".

COROLLARY 1': T_G is not provable in $ZF + AC_\omega - P$.

The proof of Theorem 1' is a routine rephrasing of the proof of Theorem 1, once we have an analogue of the Lemma of (1.1).

LEMMA: There is a Borel group structure, H^*, on $P(\omega)$ with the following saturation property:

(*) whenever E is a finitely generated group with underlying set a subset of $P(\omega)$, and G is a finitely generated subgroup of $E \cap H^*$, then E is embeddable in H^* by an embedding extendable to a Borel function on $P(\omega)$ and which is the identity on the underlying set of G.

The proof of the Lemma involves an analogue of the Sublemma of (1.1) whose proof is somewhat more substantial here, and involves the free amalgamation of groups.

Sublemma: Suppose H is a Borel group structure on $X \subseteq P(\omega)$, X an uncountable, co-uncountable Borel set. Let Y be an uncountable Borel subset of $P(\omega) \setminus X$. Then there's H', a Borel supergroup of H, with underlying set $X \cup Y$ such that whenever E is a countable group with underlying set a subset of X and whose identity is the identity of H and G is a subgroup of $E \cap H$, then E is embeddable in H', by a function extendable to a Borel function and which is the identity on the underlying set of G.

The proof of the Sublemma proceeds by making disjoint copies of the $E \setminus G$, each disjoint from X, taking the free amalgamation over H of these groups and then coding this into a group structure on $X \cup Y$ extending H. It's easily checked that this can be done in a Borel fashion, yielding H'.

The Lemma is then proved from the Sublemma, and the Theorem from the Lemma exactly as in (1.1).

(1.5) The proof of REVERSAL 1' requires some further new group-theoretic ingredients, among which, the following which we state without proof.

LEMMA (Communicated to Friedman by G. Bergman): Let F_3 be the free group on three generators, say, a,b,c. There's a recursive sequence $(w_n : n < \omega)$ of words in a,b,c which are independent in the sense that no w_n lies in the normal subgroup of F_3 generated by $\{w_m : m \neq n\}$.

Now fix such a $(w_n : n < \omega)$, and for $x \subseteq \omega$, let $G_x = F_3/\langle\{w_i : i \in x\}\rangle$, where for $X \subseteq G$, a group, $\langle X\rangle$ is the normal subgroup of G generated by X. Its then immediate by Bergman's lemma that:

(1) G_x is finitely generated (by [a], [b], [c]) and
$x = \{i : w_i([a],[b],[c]) = e \text{ in } G_x\}$.

We say that x is represented by G_x. Further $x \mapsto G_x$ is clearly Borel.

Before defining $F: \mathcal{G} \to \mathcal{G}$, we develop some machinery for associating reals and countable sets of reals to groups. Let $G \in \mathcal{G}$, and $\alpha,\beta,\gamma \in G$. We let:

$$|G|_{\alpha,\beta,\gamma} = \{i: w_i(\alpha,\beta,\gamma) = e \text{ in } G\}, \text{ and we let}$$

$$|G| = \{|G|_{\alpha,\beta,\gamma}: \alpha,\beta,\gamma \in G\}.$$

We then note:

(2) $\qquad G_1 \preccurlyeq G_2 \Rightarrow |G_1| \subseteq |G_2|$.

As in (1.3), we then define $F: \mathcal{G} \to \mathcal{G}$ by: $F(G) = G$ if $|G|$ is a model of Z_2; otherwise, we let $F(G) = \bigoplus\{G_x: x \in |G|^*\}$ = the direct sum[1] of the G_x for $x \in |G|^*$. This is easily verified to be Borel and isomorphically invariant. Further, if $(\omega,|G|,\in,\ldots)$ is not a model of Z_2, then clearly $|F(G)| \supseteq |G|^* \supsetneq |G|$, so, by (2), if $F(G) \preccurlyeq G$, then $(\omega,|G|,\in,\ldots)$ is a model of Z_2, so $T_{\mathcal{G}}$ guarantees the existence of an ω-model of Z_2.

§2. Let E be an equivalence relation on a space S. We abuse notation by setting $\vec{s}\ E\ \vec{t}$, for $\vec{s}, \vec{t} \in S^\omega$, iff $\{(\vec{s})_i/E: i < \omega\} = \{(\vec{t})_i/E: i < \omega\}$. $F: S^\omega \to S$ is E - invariant iff whenever $\vec{s}, \vec{t} \in S^\omega$, $\vec{s}\ E\ \vec{t} \Rightarrow F(\vec{s})\ E\ F(\vec{t})$.

We consider the following strengthening of Proposition D, p. 235 of [04]:

<u>PROPOSITION D'</u>: Whenever E is an analytic equivalence relation on $\mathbb{R}$ and $F: \mathbb{R}^\omega \to \mathbb{R}$ is Borel and E-invariant, there's $\vec{x} \in \mathbb{R}^\omega$ such that for some $i < \omega$, $F(\vec{x})\ E(x)_i$.

We shall prove:

<u>THEOREM 2</u>: "ω_1 iterations of the power set operation" (over ZF-P) suffice to prove Proposition D'.

Theorem 2 was claimed in [04], where it was announced as to appear in [05]. Our proof of Theorem 2 will closely follow the proof of Theorem 3.2 in [04]. We also refer to [04], §1, for precise formulations of "ω_1 iterations of the power-set operation suffice to prove ---" and "ω_1 iterations of the power-set operation

---

[1] Since we are not in the Abelian category, it is, perhaps, an abuse of notation to call this the direct sum. What we mean, of course, is the finite support subgroup of the direct product.

are necessary to prove ...", in terms of the axiom CRA, the rules R_1, R_2 and the existence of T_α-models for countable ordinals α. In the reversal of Theorem 2, below, we shall work directly with countable T_α-models. Finally, as in [04], our base theory will be ZF-P, but, as remarked in the introduction, we could replace this by ATR_0 at the price of some coding.

Rather than reversing Proposition D' (the ostensibly weaker Proposition D was reversed in [04]), we shall reverse one of its consequences.

PROPOSITION E: If $F: P(\mathbb{Q}) \to P(\mathbb{Q})$ is Borel and F maps order-isomorphic arguments to order-isomorphic values then for some $A \subseteq \mathbb{Q}$, $F(A)$ is order isomorphic to an interval in A (in fact to an interval (a,b), with $a,b \in A$).

LEMMA 2: Over ZF-P, Proposition D' => PROPOSITION E.

Theorem 2 and Lemma 2 provide one direction of Theorem 5 of [11]. The other direction is furnished by:

REVERSAL 2: The following is provable in ZF-P + Proposition E: for all countable limit ordinals λ, there's a countable T_λ-model. Thus "ω_1-iterations of the power-set operation" are necessary to prove Proposition E.

(2.1) We turn to the proof of Theorem 2. As mentioned above, we closely follow the proof of Theorem 3.2 in [04], so we shall largely confine ourselves to dealing with the differences.

So, let E be analytic over $\mathbb{R}$ and let $u \subseteq \omega$ be an analytic code for E. Let F be Borel, $F: \mathbb{R}^\omega \to \mathbb{R}$, F E-invariant and let $v \subseteq \omega$ be a Borel code for F. Towards a contradiction, suppose that the conclusion of Proposition D' fails for F. We'll say that F is an invariant Borel diagonalizer for E. Then, define, for $x,y \in \mathbb{R}$,

$$x \precsim y \quad \text{iff} \quad (\exists \vec{z} \in \mathbb{R}^\omega)\ (\exists i < \omega)\ \ xE(z)_i \wedge F(\vec{z})Ey.$$

By our hypotheses on F, $x \precsim y \Rightarrow x\not{E}y$, so E and $(\precsim \cup \succsim)$ are disjoint. Further, $(\precsim \cup \succsim)$ is clearly analytic, so there's a Borel separator, i.e. there's R with $E \subseteq R$, $R \cap (\precsim \cup \succsim) = \phi$. Fix such an R, and let $w \subseteq \omega$ be a Borel code of R.

Our "ω_1-iterations of the power-set operation" will be embodied by Borel determinacy. It is still the case that, as remarked in [04], it's unclear whether Stern's techniques, [17], can be used to replace Borel determinacy.

In one respect, our task is simpler than in [04], in that we will not require any careful computations concerning the Borel ranks of various notions arising in the proof.

(2.2) Let Y be the set of structures (ω,R) such that u,v,w are internal, (ω,R) is an ω-model and which model the following theory:

1) KP + "V = L[u,v,w]" + "every set is countable",
2) "there are arbitrarily large (u,v,w) admissible ordinals",
3) "E is an equivalence relation on $\mathbb{R}$",
4) "F is an invariant Borel diagonalizer for E",
5) "$\mathcal{R}$ is a Borel separator of E from $(\preceq \cup \succeq)$";

sentences 3), 4), 5) are written in terms of u, (u,v), (u,v,w) respectively.

We then define, by transfinite recursion in (ω,R), G_R from the ordinals of (ω,R) to the reals of (ω,R) as in paragraph 3, p. 236 of [04]:

> $G_R(0) = 0$; for $\alpha > 0$, $G_R(\alpha) = F(G_R \circ g)$, where g is first in (ω,R)'s constructible hierarchy relative to u,v,w which maps ω onto α.

We obtain the following conclusion, stronger than in [04], p. 236, paragraph 3: if $\alpha R \beta$, then $G_R(\alpha) \preceq G_R(\beta)$; the witnessing $\vec{z}$ is $G_R \circ g$, where g is as in the definition of $G_R(\beta)$. Thus, for $\alpha R\beta$, $\mathcal{R}(G_R(\alpha), G_R(\beta))$, $\mathcal{R}(G_R(\alpha), G_R(\beta))$ both fail in (ω,R) and so, by absoluteness, both, in fact do fail.

Then, H' is defined as in [04], p. 236, paragraph 4, but replacing E by $\mathcal{R}$, i.e. given (ω,R), $(\omega,S) \in Y$, we let $(a,b) \in H'$ iff a is an ordinal of (ω,R), b is an ordinal of (ω,S) and $\mathcal{R}(G_R(a), G_S(b))$.

Now H' need not be single-valued, so we let H" be the restriction of H' to those a dom H' such that H' is single-valued at a. Finally, we let H be the largest subfunction of H" which maps an initial segment of the ordinals of (ω,R) onto an initial segment of the ordinals of (ω,S) in an order-preserving way. The following information about H" and H' will be useful later on.

PROPOSITION: Let a be an ordinal of (ω,R), b an ordinal of (ω,S). Let $R(a) = \{a' : a'Ra\}$, and similarly for $S(b)$. We say a (resp. b) is well-founded if $(R(a),R|R(a))$ is (resp. $(S(b),S|S(b))$ is). If a (resp. b) is well-founded, $\alpha(a) =$ o.t. $(R(a),R|R(a))$ and similarly for $\alpha(b)$. Also, let $\alpha(R) = \sup\{\alpha(a)+1 : a$ is a well-founded ordinal of $(\omega,R)\}$, and similarly for $\alpha(S)$.

Then, if a is well-founded, and $\alpha(S) > \alpha(a)$, let b be the well-founded ordinal of (ω,S) with $\alpha(b) = \alpha(a)$. Then, b is the unique ordinal b' of (ω,S) with $(a,b') \in H'$; further, $G_R(a)\ E\ G_S(b)$.

Proof: By induction on $\alpha(a)$. By the induction hypothesis, $H|R(a)$ is the unique order isomorphism of $(R(a),R|R(a))$, $(S(b),S|S(b))$, and for each $(a',b') \in H|R(a)$, $G_R(a')\ E\ G_S(b')$. Now, let g be as in the definition of $G_R(a)$, g' be as in the definition of $G_S(b)$. The induction hypotheses then give that $G_R \circ g\ \ E\ \ G_S \circ g'$, so by the invariance of F, $G_R(a)\ E\ G_S(b)$. Thus $(a,b) \in H'$. It remains to show that if $b' \neq b$, b' an ordinal of (ω,S) then $(a,b') \notin H'$. Suppose first that $b'\ S\ b$. Then $\mathcal{R}(G_R(a), G_S(b'))$. We derive a contradiction by showing that $G_S(b') \preccurlyeq G_R(a)$. The witness, $\vec{z}$, is $G_R \circ g$, where g is as in the definition of $G_R(a)$. Thus, $F(\vec{z}) = G_R(a)$, but also, for some i, $G_S(b)\ E\ (\vec{z})_i$, since by the induction hypothesis if $a'\,R\,a$ is such that $\alpha(a') = \alpha(b')$, then $G_R(a')\ E\ G_R(b')$. On the other hand, if $b\ S\ b'$, we derive a contradiction by showing that $G_R(a)\ E\ G_S(b')$. This time, the witness, $\vec{z}$, is $G_S \circ g$ where g is as in the definition of $G_S(b)$. Thus, $F(\vec{z}) = G_S(b')$, but for some i, $G_S(b) = (\vec{z})_i$, and we've already shown that $G_R(a)\ E\ G_S(b)$.

We then say that (ω,S) is longer than (ω,R) iff H maps all the ordinals of (ω,R) onto an initial segment of the ordinals of (ω,S) determined by an ordinal of (ω,S), or the domain of H is a proper initial segment of the ordinals of (ω,R) not determined by an ordinal of (ω,R) and the range of H is either all the ordinals of (ω,S) or is an initial segment of the ordinals of (ω,S) determined by an ordinal of (ω,S). As in [04] we prove that "longer than" on Y is Borel. We have the

COROLLARY: (ω,S) longer than $(\omega,R) \Rightarrow \alpha(R) < \alpha(S)$.

Proof: The definition of "longer than" guarantees that all well-founded ordinals of (ω,R) are in the domain of H.

(2.3) Our next task is to prove two lemmas below corresponding to Lemmas 3.2.2, 3.2.3 of [04]. We shall only indicate the necessary changes in the proofs of [04].

LEMMA A: For all $\alpha < \omega_1^{L[u,v,w]}$, there's $\alpha < \gamma < \omega_1^{L[u,v,w]}$ such that $L_\gamma[u,v,w]$ models (1), (3) - (5) above and:

(2') there are arbitrarily large (u,v,w)-inaccessible ordinals

and such that (6) every element of $L_\gamma[u,v,w]$ is definable in parameters u,v,w.

The proof is as in [04] for Lemma 3.2.2, except that 3), 4), 5) now have complexity Π_2^1 instead of Π_1^1, so Π_1^1 is replaced by Π_2^1, Δ_2 is replaced by Δ_3 and "(u,v)-admissible" is replaced by "(u,v,w)-inaccessible".

LEMMA B: If $\alpha < \omega_1^{L[u,v,w]}$, there's $\alpha < \gamma < \omega_1^{L[u,v,w]}$ such that $L_\gamma[u,v,w]$ models (1), (3) - (5) above, (6) of the previous lemma and:

(2") there are arbitrarily large (u,v,w)-admissible ordinals and a largest (u,v,w) inaccessible ordinal.

The proof of Lemma 3.2.3 in [04] is modified in the same way the statement was to give the statement of the lemma.

We then let $((\omega,R), (\omega,S)) \in Z$ iff:

$((\omega,R) \in Y$ and (ω,R) models (2') above) =>
$((\omega,S) \in Y$ and (ω,S) models (2") above and (ω,S) is longer than $(\omega,R))$.

Regarding Z as a (Borel) subset of $2^\omega \times 2^\omega$, we consider the game with payoff set Z (for player II). We shall derive a contradiction to Borel determinacy. By Borel determinacy and absoluteness, there's $J \in L_{\omega_1}[u,v,w]$ (ω_1 means $\omega_1^{L[u,v,w]}$) which is a winning strategy for one of the players. We prove:

LEMMA: a) J is not a winning strategy for II,
b) J is not a winning strategy for I.

Proof: Using the Corollary of (2.2), the proof of a) is essentially the same as the proof of the corresponding statement in [04], Lemma 3.2.4. Suppose I plays (ω,R) isomorphic to an $L_\gamma[u,v,w]$ as in Lemma A, with $J \in L_\gamma[u,v,w]$ (this is possible, since, by absoluteness, $J \in L[u,v,w]$ and therefore, letting $\sigma = \omega_1^{L[u,v,w]}$, $J \in L_\sigma[u,v,w]$). We may further suppose that R is arithmetic in the theory of $(L_\gamma[u,v,w], \in u,v,w)$. Then, $(\omega,R) \in Y$ and (ω,R) models (2'), above. Thus, if J is winning for II, and we let S be the result of playing by J against R, we have $(\omega,S) \in Y$ and (ω,S) is longer than (ω,R). So $\alpha(S) > \alpha(R)$, but since the well-founded core of (ω,S) models KP, $\alpha(S)$ is in fact (u,v,w)-admissible. Then, as in [04], we obtain a contradiction by observing that this guarangees that the ωth jump of S is internal to (ω,S).

The proof that J is not a winning strategy for I is somewhat harder. We first have the following immediate consequence of the material of (2.2).

(*) If (ω,S) is well-founded and $(\alpha(R) < \alpha(S)$ or R is non-well-founded and $\alpha(R) = \alpha(S))$ then (ω,S) is longer than (ω,R).

This is clear if $\alpha(R) < \alpha(S)$, since then if R is well-founded, the first clause of the definition of "longer than" applies, while if R isn't well-founded, the second part of the second clause applies. Finally, if $\alpha(R) = \alpha(S)$ and R isn't well-founded, the first clause of the definition of "longer than" applies.

Now, suppose J is winning for I. Let II play (ω,S) isomorphic to $L_\gamma[u,v,w]$ as in Lemma B with $J \in L_\gamma[u,v,w]$ (as above, this is possible). S will be chosen arithmetic in the theory of $(L_\gamma[u,v,w],\in,u,v,w)$. Then, $(\omega,S) \in Y$ and (ω,S) models (2") above. Let R be J applied to S. Then $(\omega,R) \in Y$, and (ω,R) models (2') above, but (ω,R) is not longer than (ω,S). So, we have the following possibilities:

a) $\alpha(R) = \alpha(S)$ and R is well-founded,

b) $\alpha(R) > \alpha(S)$.

The former is impossible since then $(\omega,R) \cong (\omega,S)$, but (ω,R) models (2') and (ω,S) models (2"). If $\alpha(R) > \alpha(S)$, then, as above, we get that the ωth jump of R is internal to (ω,R), a contradiction.

(2.4) We now turn to proving Lemma 2 and Reversal 2. These proofs use the motion of the dense mix of a countable set of (countable) linear orderings.

DEFINITION: Suppose $D = \{\langle L_n, \leq_n \rangle : n \in \omega$, where each $\langle L_n, \leq_n \rangle$ is a linear ordering. Let $\varphi : \mathbb{Q} \to D$ be such that for each n, $\varphi^{-1}[\{\langle L_n, \leq_n \rangle\}]$ is dense. We define Mix (φ) to be the linear ordering $\langle M, \leq \rangle$, where $M = \bigcup \{\{q\} \times L_n : \varphi(q) = \langle L_n, \leq_n \rangle\}$ and $(q, \ell) \leq (q', \ell')$ iff $q <_{\mathbb{Q}} q'$ or $q = q'$ and $\ell \leq_n \ell'$, where $\varphi(q) = \langle L_n, \leq_n \rangle$.

REMARK: If $\varphi, \varphi' : \mathbb{Q} \to D$ are as above, then Mix $(\varphi) \cong$ Mix (φ'). This is easily proved by a variant of the Cantor back-and-forth argument, by showing that there's an automorphism of $\mathbb{Q}$ sending each D_n onto D'_n, where $D_n = \{q \in Q : \varphi(q) = \langle L_n, \leq_n \rangle\}$, $D' = \{q \in Q : \varphi'(q) = \langle L_n, \leq_n \rangle\}$. Thus, we are justified in changing notation and taking Mix (D) = Mix (φ), for any $\varphi : \mathbb{Q} \to D$ as above: we say that φ represents Mix (D). In what follows, it will often be convenient to work with a representative of Mix (D).

Proof of Lemma 2: Suppose Proposition D' holds, let $F: P(\mathbb{Q}) \to P(\mathbb{Q})$ be as in the hypothesis of Proposition E. Let E be the analytic equivalence relation of order isomorphism. Define $F^*: P(\mathbb{Q})^{\omega} \to P(\mathbb{Q})$ by: $F^*(A_n : n \in \omega) = F(\text{Mix (D)})$, where, for $n \in \omega$, $\langle L_n, \leq_n \rangle$ is the linear ordering with underlying set $A_n \cup \{-\infty_n, \infty_n\}$, $-\infty_n$ is the left endpoint, ∞_n the right endpoint, $\leq_n | A_n = \leq_{\mathbb{Q}} | A_n$. By Proposition D' let $(A_n : n \in \omega)$ be such that $F^*(A_n : n \in \omega) \cong A_i$, for some $i \in \omega$. Then, $F(\text{Mix (D)}) \cong A_i$ which is an interval of Mix (D) of the form $(-\infty_i, \infty_i)$. We have sloughed over how we embed Mix (D) isomorphically as a subset of Q, but, by the invariance of F, this is immaterial.

REMARK: In fact, the form of Proposition E which does not mention the endpoints a, b is provably equivalent to Proposition E as stated, over a weak theory, using devices analogous to those of the fake endpoints $-\infty_n, \infty_n$, above.

(2.5) To prove Reversal 2, we need the notion of a T_{α}- model.

DEFINITION: If $\alpha < \omega_1$, a T_{α}-model is a transitive set $A \in V_{\alpha+1}$ such that for all $\beta < \alpha$, all $\in$ -formulas φ with one free variable and parameters from $A, \{x \in V_{\beta} : (A, \in) \models \varphi(x)\} \in A$.

The existence of a T_{α}-model has roughly the strength of "there are α cardinals" (over $ZF + AC_{\omega} - P$, or ATR_0); a T_{ω}-model is essentially an ω-model of Z. To see that Reversal 2, as stated, gives "the necessity of ω_1-iterations of power-set", see the discussion in [04].

Proof of Reversal 2: We define $\text{Codes} \subseteq P(\omega)$ and $\Gamma: \text{Codes} \to_{\text{onto}} \text{HC} =$ the set of hereditarily countable sets. We have $\text{Codes} = \bigcup\{\text{Codes}(\alpha): \alpha < \omega_1\}$, $\Gamma = \bigcup\{\Gamma_\alpha: \alpha < \omega_1\}$, where $\Gamma_\alpha = \Gamma|\text{Codes}(\alpha): \text{Codes}(\alpha) \to \text{HC}_\alpha$, where $\text{HC}_0 = \phi$; $\text{HC}_\lambda = \bigcup\{\text{HC}_\alpha: \alpha < \lambda\}$ for countable limit λ, and $\text{HC}_{\alpha+1}$ = the set of countable subsets of HC_α. We need some preliminary notation. If $x \subseteq \omega$, $n \in \omega$, $(x)_n = \{i: \langle n,i+1\rangle \in x\}$; we set $y \mathrel{E} x$ iff $y \neq \phi$ and for some $n \in \omega$, $y = (x)_n$.

We then set: $\text{Codes}(0) = \{\{0\}\}$; $\Gamma_0(\{0\}) = \phi$; $\text{Codes}(\lambda) = \bigcup\{\text{Codes}(\alpha): \alpha < \lambda\}$; $\Gamma_\lambda = \bigcup\{\Gamma_\alpha: \alpha < \lambda\}$, and:

> $x \in \text{Codes}(\alpha+1)$ iff $x \in \text{Codes}(\alpha)$ or $x \neq \phi$ and for all $n \in \omega$, $(x)_n \neq \phi \Rightarrow (x)_n \in \text{Codes}(\alpha))$; then for $x \in \text{Codes}(\alpha+1) \setminus \text{Codes}(\alpha)$, $\Gamma_{\alpha+1}(x) = \{\Gamma_\alpha((x)_n): (x)_n \neq \phi\}$.

We also set $x \equiv_\alpha y$ iff $x,y \in \text{Codes}(\alpha)$ and $\Gamma_\alpha(x) = \Gamma_\alpha(y)$, and $y\bar{E}_\alpha x$ iff $x,y \in \text{Codes}(\alpha)$ and for some n, $y \equiv_\alpha (x)_n$. Of course, at the price of some tedium, we can define $\equiv_\alpha$ by recursion directly (i.e., without direct reference to Γ). Each $\text{Codes}(\alpha)$, Γ_α, $\equiv_\alpha$, $\bar{E}_\alpha$ is Borel, and, of course for $x,y \in \text{Codes}(\alpha)$ $y \mathrel{\bar{E}_\alpha} x$ iff $\Gamma_\alpha(y) \in \Gamma_\alpha(x)$. We let $\equiv = \bigcup\{\equiv_\alpha: \alpha < \omega_1\}$.

Our goal is to define $H: \text{Codes} \to P(\mathbb{Q})$. Of course, $H = \bigcup\{H_\alpha: \alpha < \omega_1\}$ where the $H_\alpha: \text{Codes}(\alpha) \to P(\mathbb{Q})$ are defined recursively. We need some preliminaries. We fix a partition $(D_n: n \in \omega)$ of $\mathbb{Q}$ into dense sets. We also fix an order-isomorphism σ from $\mathbb{Q} \times \mathbb{Q}$ with the lexicographic order to the interval $\mathbb{Q} \cap (0,1)$. We conventionally define $\text{Mix}(\phi)$ = some subset of $(0,1)$ order isomorphic to $\mathbb{Z}$. Then:

> $H(x) = Q \cap (-\infty,0) \cup \sigma[\text{Mix}_\varphi(\{H((x)_n): (x)_n \neq \phi\}] \cup Q \cap (1,\infty)$, where, if $x \in \text{Codes} - \{\{0\}\}$, we let $(n_k: k \in \omega)$ enumerate (possibly with repetitions) $\{n: (x)_n \neq \phi\}$ and for $k \in \omega$, $q \in D_k$, $\varphi(q) = H((x)_{n_k})$.

Clearly H is invariant, in that for $x,y \in \text{Codes}$, $x \equiv y \Rightarrow H(x)$

is order-isomorphic to $H(y)$. Also, each H_α is Borel. Now H induces a map from HC to $P(\mathbb{Q})$ mod order isomorphism. We shall abuse notation by referring to the induced map as H, and its restriction to HC_α as H_α. We summarize the remaining crucial facts about the map induced by H in the next lemma, but first we need some further notions about mixes.

(2.5.1) DEFINITION: Let $D \neq \phi$ be a set of linear orderings and let $(M,<) = \text{Mix}(D)$. Choose a representative $\varphi : \mathbb{Q} \to_{\text{onto}} D$. Let $X \subseteq M$; we let $\mathbb{Q}(X) = \{q \in \mathbb{Q}: (\exists x \in X)(\exists \ell \in \varphi(q))(x=(q,\ell))\}$, and for $q \in \mathbb{Q}(X)$, let $X(q) = \{\ell \in \varphi(q): (q,\ell) \in X\}$.

(2.5.2) LEMMA: Suppose $A, A' \in HC$. Then

a) $\text{Mix } H''A \cong \text{Mix } H''A'$ iff $A = A'$,

b) suppose further that I' is an interval of $\text{Mix } H''A'$, and that $\text{Mix } H''A \cong I'$ with the induced ordering. Then, either:

i) $A = A' = \phi$ and $I' = \text{Mix } H''A = \mathbb{Z}$, or $A' \neq \phi$, and letting φ' be a representative of $\text{Mix } H''A'$,

ii) $\mathbb{Q}(I')$ is an open interval of $\mathbb{Q}$ and $A = A'$, or

iii) $\mathbb{Q}(I')$ is a singleton, say $\{q'\}$, and letting $\varphi(q') = H(B')$, we have $A \in$ the transitive closure of $\{B'\}$ and $I' \subseteq \text{Mix } H''B'$.

The lemma is proved by induction on (rank A, rank A') in the Godel well-ordering of $OR \times OR$.

In fact, as the next lemma shows, we can do even better than this. For $\alpha < \omega_1$, we let $R_\alpha = \{a \subseteq \mathbb{Q}$: for some $x \in \text{Codes}(\alpha)$, a is order-isomorphic to $H(x)\}$.

(2.5.4) LEMMA: For all $\alpha < \omega_1$, R_α is Borel and there's a Borel function $I_\alpha: P(\mathbb{Q}) \to P(\omega)$ such that $I_\alpha(H_\alpha(x)) \equiv_\alpha x$ for all $x \in \text{Codes}(\alpha)$, and for $L, L' \in R_\alpha$ if L and L' are order-isomorphic, then $I_\alpha(L) = I_\alpha(L')$.

Proof: It will suffice, of course, to show that R_α is Borel and define I_α restricted to R_α in a Borel fashion, since, off R_α we can set $I_\alpha(L) = \phi$.

We proceed by induction on α. The limit case is trivial. The case $\alpha = 0$ requires us to recognize when a is order-isomorphic to $Q + \mathbb{Z} + Q$. So, passing from α to $\alpha + 1$, given $L \subseteq Q$ we must first recognize when L has unbounded dense end segments. This is accomplished by considering the sets $\{x: (-\infty,x] \text{ is dense}\}$, $\{x: [x,\infty) \text{ is dense}\}$, requiring that they be disjoint, non-empty, that neither has endpoints, and their union is not all of L. Then, setting $L' = L \setminus$ these end-segments, we require that what remains of L must be isomorphic to the mix of subintervals in R_α. We first generate these orderings as certain maximal intervals of L', in a Borel fashion; we then require that L' is the disjoint union of these intervals and that it's isomorphic to the mix of these intervals. If not, $I_{\alpha+1}(L) = \phi$. If so, $I_{\alpha+1}(L)$ = the set of $I_\alpha(I)$, for I one of the subintervals. The only problematical point will be to determine, in a Borel fashion, whether L' is isomorphic to the mix of these intervals; apparently this is a Σ^1_1 condition. We first show how to generate the subintervals.

If $x,y \in L'$, $x < y$ and $(x,y) \cap L' \in R_\alpha$, set $(x',y') \in Y(x,y)$ iff $x' < x$, $y < y'$, $x',y' \in L'$ and $(x',y') \cap L' \in R_\alpha$. We let $I(x,y) = \bigcup Y(x,y)$ and we say that (x,y) is <u>good</u> if $I(x,y) \in R_\alpha$ and $I(x,y)$ does not have the form (x^*,y^*) for $x^*,y^* \in L'$. The collection of intervals is the set of $I(x,y)$ for (x,y) good. A proof of (2.5.2), when written out, will show that $L \in R_{\alpha+1}$ iff L' is isomorphic to the mix of the $I(x,y)$, (x,y) good. So, it remains to show that this can be recognized in a Borel fashion. We require, in addition to the fact that the $I(x,y)$, (x,y) good, form a partition of L', that:

for all good (x,y), (x',y'), (x'',y''), all z all w in L', if $I(x,y) \neq I(x',y')$ and $z \in I(x,y)$ and $w \in I(x',y')$ and $z < w$, then there's good (x^*,y^*) with $z < t < w$ for all $t \in I(x^*,y^*)$ such that $I_\alpha(I(x'',y'')) = I_\alpha(I(x^*,y^*))$.

To conclude the proof, we remark that, starting from a fixed enumeration of $\mathbb{Q} \times \mathbb{Q}$, we obtain, in a Borel fashion from L, an enumeration $(I_n: n \in \omega)$ (possibly with repetitions) of the set of $I(x,y)$ for (x,y) good. We then let $I_{\alpha+1}(L) = x$, where, for $n \in \omega$, $(x)_n = I_\alpha(I_n)$.

Now, fix λ a countable limit ordinal. Given $L \subseteq \mathbb{Q}$, let $(x,y) \in I(L)$ iff $x,y \in L$, $x < y$ and $L \cap (x,y) \in R_\lambda$. We define:

$|L|$ = the transitive closure of the set of $I_\lambda(L \cap (x,y))$, for $(x,y) \in I(L)$.

Then, if $|L|$ is not a T_λ-model, there's a new parameter-definable (over $|L|$) subset of $|L|$ of rank less than λ. Let α be least such that there's one such of rank α. Let $|L|^*$ = the set of all parameter definable (over L) subsets of $|L|$ of rank α. Note that, in this case, we can clearly obtain, in a Borel fashion from L, an enumeration of $|L|^*$, which, in turn allows us to obtain, in a Borel fashion from L, a $u \in \text{Codes}(\lambda)$ such that $\Gamma(u) = |L|^*$. We then define:

$F(L) = \phi$, if $|L|$ is a T_λ-model; otherwise
$F(L) = H_\lambda(u)$, where u is as above.

Thus, F is Borel and clearly F maps isomorphic arguments to isomorphic values. So, let L be as guaranteed for F by Proposition E, i.e. F(A) is isomorphic to an interval $L \cap (x,y)$ of L, where $(x,y) \in L$. We claim that $F(L) = \phi$, i.e. that $|L|$ is a T_λ-model. Suppose not. Then $F(L) = H_\lambda(u)$, i.e. $F(L) \in H(|L|^*)$, viewing H_λ as a map from HC_λ to isomorphism classes. Further, $F(L) = L \cap (x,y)$, so $(x,y) \in L$. Also, for some new parameter definable (over $|L|$) $a \subseteq |L|$ of rank α, $H_\lambda(a)$ is the isomorphism class of an interval of $L \cap (x,y)$ (again, viewing H_λ as a map from HC_λ to isomorphism classes). But then, by finitely many applications of (2.5.2) we'd obtain that $a \in$ the transitive closure of $|L|$ which equals $|L|$; this is a contradiction.

(2.6) We now show that Proposition D' is essentially optimal by showing that it fails if we replace "analytic" by "coanalytic". This result was mentioned in [04]. Instead of $\mathbb{R}$, we work with a space of trees labelled by subsets of ω.

<u>DEFINITION</u>: $A \in \mathcal{A} \iff A: \omega^{<\omega} \to 2^\omega$. Nodes of $\omega^{<\omega}$ are denoted by s, t, etc.

$\mathcal{A}$ has a natural Borel isomorphism with Baire space. Our goal is to define a one-to-one Borel $F: \mathcal{A}^\omega \to \mathcal{A}$ with Borel range and a Borel subset $\mathcal{C} \subseteq \mathcal{A}$ with $F[\mathcal{C}^\omega] \subseteq \mathcal{C}$.

For $A \in \mathcal{A}$, $s \in \omega^{<\omega}$, define $A^s(t) = A(s \frown t)$. If $s = \langle n \rangle$, we shall abuse notation by letting $A^n = A^s$.

Now suppose $\vec{A} \in \mathcal{A}^\omega$. We easily obtain an enumeration $(f_n : n \in \omega)$ of $\bigcup\{\text{range } A_n : n \in \omega\}$ as a Borel function of $\vec{A}$. This permits us to define F.

DEFINITION: If $\vec{A} \in A^\omega$, $F(\vec{A})(n^\frown s) = A^n(s)$, for $n \in \omega$, $s \in \omega^{<\omega}$, and $F(\vec{A})(\phi)$ = the result of applying the Cantor diagonal procedure to $(f_n: n \in \omega)$, i.e. $F(A)(\phi)(n) = 1 - f_n(n)$, where $(f_n: n \in \omega)$ is as above.

We note that for $\vec{A} \in A^\omega$, $n \in \omega$, $F(\vec{A})^n = A_n$. We now define an equivalence relation E on A. E will be $E_{\omega_1} = \bigcup\{E_\alpha: \alpha < \omega_1\}$, where $(E_\alpha: \alpha < \omega_1)$ is the continuous tower with E_0 = the identity on A, and for $\alpha < \omega_1$, $A, A' \in A$,

$$AE_{\alpha+1}A' \iff \begin{array}{l} 1)\ \{A^n/E_\alpha: n \in \omega\} = \{(A')^n/E_\alpha: n \in \omega\} \text{ and} \\ 2)\ A(\phi) \in \bigcup\{\text{range } A^n: n \in \omega\} \text{ iff} \\ \quad A'(\phi) \in \bigcup\{\text{range } (A')^n: n \in \omega\}. \end{array}$$

Clause 1) is the substantial one; clause 2) is either satisfied or fails at level $\alpha = 0$, and then doesn't change. We easily verify, by induction that $E_\alpha \subseteq E_{\alpha+1}$. The purpose of clause 2) will become clearer when we verify that F is a diagonalizer.

We shall also verify that E is coanalytic. This hinges on the fact that, for $A \in A$, letting $T(A) = \{A^s: s \in \omega^{<\omega}\}$, we can define $E|T(A)$ as the smallest equivalence relation E' on $T(A)$ such that for $B, B' \in T(A)$, 1') and 2') imply $BE'B'$, where:

1') is: $\{B^n/E': n \in \omega\} = \{(B')^n/E': n \in \omega\}$,

2') is: $B(\phi) \in \bigcup\{\text{range } B^n: n \in \omega\}$ iff $B'(\phi) \in \bigcup\{\text{range } (B')^n: n \in \omega\}$.

Then, to determine whether $A\ E\ B$, we let $C_{2n} = A$, $C_{2n+1} = B$, $C = F(\vec{C})$ and in $T(C)$ we decide whether $A\ E\ B$, as above.

We now show that F is a diagonalizer for E. We show by induction on α that if $B \in A$ and $B(\phi) \notin \bigcup\{\text{range } B^n: n < \omega\}$, then for all $n < \omega$, $B \not E_\alpha B^n$. This clearly suffices, since if $\vec{A} \in A^\omega$ then $B = F(\vec{A})$ has the desired property and as we've already noted, in this case $B^n = A_n$.

Clearly if $\alpha = 0$, then $B \neq B^n$, since $B(\phi) \notin \text{range } B^n$. The limit case is trivial, so suppose for all $B' \in A$ with the desired property, for all $n \in \omega$, $B' \not E_\alpha (B')^n$, suppose $B \in A$ has the desired property, $n \in \omega$, and, suppose towards a contradiction that $B\ E_{\alpha+1}\ B^n$. Then, by the clause 2) of the definition of $E_{\alpha+1}$, B^n has the desired property, so, by the induction hypothesis for $B' = B^n$, for all $k \in \omega$, $B^n \not{E}_\alpha B^{nk}$. But this is a contradiction since by the clause 1) of the definition of $E_{\alpha+1}$, $\{B^k/E_\alpha : k \in \omega\} = \{(B^n)^k/E_\alpha : k \ \ \omega\}$ and, of course $B^n/E_\alpha \in \{B^k/E_\alpha : k \in \omega\}$. Let $A \in C$ iff $A(\phi) \notin \bigcup\{\text{range } A^n : n \in \omega\}$.

We now show that F is E-invariant: suppose $\vec{A},\vec{B} \in C^\omega$, and for all $n \in \omega$, $A_n\ E\ B_n$. Then for some α, for all $n \in \omega$, $A_n E_\alpha B_n$. Further, since $F(\vec{A}), F(\vec{B}) \in C$, $F(A)(\phi) \notin \bigcup\{\text{range } (F(\vec{A}))^n : n \in \omega\}$, $F(B)(\phi) \notin \bigcup\{\text{range } (F(B))^n : n \in \omega\}$. Finally, since $A_n = (F(\vec{A}))^n$, $B_n = (F(\vec{B}))^n$, we have $\{(F(\vec{A}))^n/E_\alpha : n \in \omega\} = \{(F(\vec{B}))^n/E_\alpha : n \in \omega\}$, i.e. $F(\vec{A})E_{\alpha+1}F(\vec{B})$.

(2.7) We conclude this section, by remarking that Propositions D', E are Π^1_3 statements. There are analogous Π^1_2 statements, obtaining by strengthening the notion of isomorphic invariance to be: "commutes with permutations". These forms are provable from the same theories (i.e., the analogue of Theorem 2 holds). It remains open however, whether we can obtain the analogue of Reversal 2, and therefore, a fortiori, whether the Π^1_2 versions are provably equivalent to the Π^1_3 versions over a weak base theory.

§3. In this section, G will denote a closed (in fact equationally defined) subspace of an appropriate B_s, where, as in §1, s is a finite similarity type without relation symbols. G_f will denote the subspace of G of finitely generated points (as in §1). We will be interested in three cases:

1) G = products $(G = B_2)$
2) G = semigroups with unit $(G \subseteq B_2 \times \omega)$
3) G = groups $(G \subseteq B_2 \times B \times \omega)$.

We shall be considering four kinds of statements for each of our spaces G.

<u>PROPOSITION F</u>(G): Whenever $F: G_f^\omega \to G_f$ is Borel and isomorphically invariant (as usual, this means that if for all n $x_n \cong y_n$ then $F(\vec{x}) \cong F(\vec{y})$), then for some $\vec{x} \in G_f^\omega$ and some $n \in \omega$, $F(\vec{x}) \preccurlyeq x_n$(as in §1).

Thus, Proposition F is an infinite-dimensional version of statements considered in §1, but cutting down from G to G_f. This has surprising consequences for the consistency strength, as we shall see. The next statement involves blending Proposition F with a Ramsey property of the type considered by Galvin-Prikry, [07], and Silver, [13].

PROPOSITION G(G): Whenever F: $G_f^\omega \to G_f$ is Borel and isomorphically invariant, there's $\vec{x} \in G_f^\omega$ such that for all subsequences $\vec{y}$ of $\vec{x}$, there's k such that $F(\vec{y}) \preccurlyeq y_k$.

We can obtain weakenings and strengthenings of Proposition G, by changing the punch line to:

a) there's k such that $F(\vec{y}) \prec x_k$,

b) there's k such that for all subsequences $\vec{y}$ of $\vec{x}$, $F(\vec{y}) \preccurlyeq y_k$ (of course a subsequence of $\vec{x}$ is, as usual $\vec{x} \circ f$, where $f: \omega \to \omega$ is order-preserving).

Proposition G will turn out to be very strong, reversing to the existence of $x^{\#}$ for all sets x. Propositions F', G', below are analogues of Propositions F, G, where we suppress mention of coordinates of sequences in favor of direct limits. We shall have that Proposition F' => Proposition F, Proposition G' => Proposition G. Before stating Propositions F', G', we should explicate the notion:

"z is embeddable in a direct limit of $\vec{x}$."

This means that there is a sequence of embeddings $j_i: x_i \to x_{i+1}$ such that z is embeddable in the direct limit of $((x_k, j_{ik}): i \leq k \in \omega)$, where the j_{ik} $(i < k)$ are obtained from the $j_i = j_{i,i+1}$ by composition, to make the system commutative. Now, we can state Propositions F', G'.

PROPOSITION F'(G): Whenever F: $G_f^\omega \to G$ is Borel and isomorphically invariant, there's $\vec{x} \in G_f^\omega$ such that $F(\vec{x})$ is embeddable in a direct limit of $\vec{x}$.

PROPOSITION G'(G): Whenever F: $G_f^\omega \to G$ is Borel and isomorphically invariant, there's $\vec{x} \in G_f^\omega$ such that for all subsequences $\vec{y}$ of $\vec{x}$, $F(\vec{y})$ is embeddable in a direct limit of $\vec{y}$.

This time we have only one variant:

a) $F(\vec{y})$ is embeddable in a direct limit of $\vec{x}$.

We shall prove:

<u>THEOREM 3</u>: For each of the spaces G mentioned above:

a) Proposition $F'(G)$ is provable in $ZF + AC_\omega - P +$ "$V_{\omega+\omega}$ exists",

b) Proposition $G'(G)$ is provable in ZFC + "there exists a measurable cardinal".

<u>REVERSAL 3</u>: For each of the spaces G mentioned above:

a) $ZF + AC_\omega - P +$ Proposition $F(G)$ proves: "there's an ω-model of $ZC + V = L$",

b) $ZF + AC_\omega - P +$ Proposition $G(G)$ proves: "there's an ω-model of $Z_2 - AC$ (the scheme of countable second-order choice) and an M-definable predicate U on the "ordinals" of M, such that L[U] can be constructed in M and satisfies $ZFC + \forall x(x^{\#}$ exists).

<u>COROLLARY 3</u>: For each of the spaces G mentioned above

a) Propositions $F(G)$, $F'(G)$ are not provable in $ZC + V = L$,

b) Propositions $G(G)$, $G'(G)$ are not provable in $ZFC + \forall x(x^{\#}$ exists).

Corollary 3 is immediate from Reversal 3 and the fact, already mentioned, that Proposition $F'(G)$ => Proposition $F(G)$ and Proposition $G'(G)$ => Proposition $G(G)$.

We shall proceed as follows. We first introduce some "Turing degree" analogues of Propositions F,F' and Propositions G,G'. We shall prove Theorem 3' and Reversal 3', with these Turing degree analogues replacing the appropriate Propositions. We then develop some abstract sufficient conditions on a space in order for Theorem 3' and Reversal 3' to transfer from Turing degrees to G. This involves our ability to transform elements of G_f, and G into Turing degrees in a Borel fashion, while preserving certain information. We obtain Theorem 3 for the statements involving $F(G)$ and $G(G)$ and Reversal 3 by showing, for each of our spaces G, that these sufficient conditions are met. This last argument is particularly interesting when G = groups, as it involves some non-trivial group theory. We need a special argument for Theorem 3 as stated, which we develop in parallel with our proofs that our sufficient conditions are met.

In (3.1), we introduce the Turing degree statements and we prove Theorem 3' and Reversal 3' for the Turing-degree analogue of Propositions F,F'. In (3.2) we prove Theorem 3' and in (3.3) we prove Reversal 3' for the Turing-degree analogue of Propositions G,G'. The argument of (3.3) is by far the most subtle of this paper. In (3.4) we obtain our necessary and sufficient conditions for transfer to $\mathcal{G}$, and in (3.5) we establish these conditions for each of our three spaces, $\mathcal{G}$, above, and do the extra work involved in proving Theorem 3 for the statements F'($\mathcal{G}$), G'($\mathcal{G}$).

Before beginning our development, we should note that Theorem 3a and Corollary 3a (together with the fact that Proposition F' => Proposition F yield Theorem 7 of [11]. Concerning Theorem 8 of [11], we should remark that there appears to have been some imprecision concerning the precise sense of "Ramsey cardinal"; had the statement of Theorem 8 of [11] said "Ramsey-type cardinals", it would have been entirely accurate, since when we interviewed Friedman, he said that, at the time, he was confident that the Propositions G($\mathcal{G}$) reversed only to the Ramsey-type cardinals $\kappa \to (\alpha)^{<\omega}$ for all $\alpha < \omega_1$. Of course Corollary 3b) is much stronger.

(3.1) Without further ado, we state the Turing-degree analogues of Propositions F,F', G,G'.

<u>PROPOSITION F(T)</u>: Whenever $F\colon B^\omega \to B$ is Borel and Turing-invariant (i.e., if for all n, $x_n \equiv_T y_n$, then $F(\vec{x}) \equiv_T F(\vec{y})$) there is $\vec{x} \in B^\omega$ and $n \in \omega$ such that $F(\vec{x}) \leq_T x_n$.

<u>PROPOSITION G(T)</u>: Whenever $F\colon B^\omega \to B$ is Borel and Turing-invariant, there is $\vec{x} \in B^\omega$ such that for all subsequences $\vec{y}$ of $\vec{x}$, there's $n \in \omega$ such that $F(\vec{y}) \leq_T y_n$.

As before we have variants (two, this time):

a) there's $n \in \omega$ such that $F(\vec{y}) \leq_T x_n$

b) there's $n \in \omega$ such that for all subsequences $\vec{y}$ of $\vec{x}$, $F(\vec{y}) \leq_T y_n$.

We now obtain Theorem 3' and Reversal 3' by replacing Proposition F'($\mathcal{G}$) (respectively Proposition F($\mathcal{G}$)) by Proposition F(T), and replacing Proposition G'($\mathcal{G}$) (respectively Proposition G($\mathcal{G}$)) by Proposition G(T).

We turn now to proving Theorem 3'a). We start with $u \in B$, a Borel code of a Borel, Turing invariant $F: B^\omega \to B$, and $M = L_\theta[u]$, a model of a sufficiently rich fragment of $ZF - P +$ "$V_{\omega+\omega}$ exists", where $\theta < \omega_1$. We shall force over M, producing, in $M[G]$, a sequence $(\tau_n : n < \omega) \in B^\omega$. We shall show that $F(\vec{\tau}) \leq_T \tau_n$ for some $n \in \omega$. Our treatment is close in spirit to portions of §5 of [04], and we shall refer to there for the detailed proofs of certain statements.

$\vec{\tau}$ will be τ_G for a canonical term τ. G will be a union of "layers" G_n. τ will be defined in such a way that if G' is also M-generic and for each n, G_n and G'_n differ by a finite amount of information (G and G' will be called similar), then τ_G and $\tau_{G'}$ are termwise Turing equivalent, and therefore $F(\tau_G) \equiv_T F(\tau_{G'})$. Therefore $F(\tau_G) \in M[G']$ and vice-versa. We shall then argue that there is n such that G_n determines $F(\tau_G)$; if not, we shall obtain a perfect set of $F(\tau_{G'})$, where the G' are pairwise similar, which contradicts that the $F(\tau_{G_i})$ are pairwise Turing equivalent. Since this last argument is rather standard, we shall not develop it any further. For such an n, we shall be able to verify that $F(\vec{\tau}) \leq_T \tau_{n+1}$.

From the above discussion, it's clear that the crucial property of τ is that similar generic sets produce pairwise Turing-equivalent $\vec{\tau}$. There is a rather simple-minded way of producing a τ' which <u>fails</u> to have this property. For expository purposes, we shall, nevertheless, first show how to produce such a τ', and then say how to modify the construction to produce τ with the desired property.

(3.1.1) For the remainder of this section, for $1 \leq n < \omega$, $\aleph_n$ means $\aleph_n^M$. Working in M, let P be the set of finite partial functions $f: \aleph_\omega \times \omega \to M$ such that for all (α, n), $f(\alpha, n) \in L_\alpha[u]$, ordered by reverse inclusion. Thus if G is M-generic/$\mathbb{P}$, $\bigcup G: \aleph_\omega \times \omega \to_{onto} L_{\aleph_\omega}[u]$, and for each $\alpha < \aleph_\omega$, $f_\alpha = \bigcup G(\alpha, \cdot)$: $\omega \to_{onto} L_\alpha[u]$. The $(f_\alpha : \alpha < \aleph_\omega)$ are mutually generic. The elements of $L_{\aleph_\omega}[u]$ are terms in the forcing language for $\mathbb{P}$. We can regard them as terms for subsets of ω in the following way. If G is M-generic/$\mathbb{P}$ and $x \in L_{\aleph_\omega}[u]$, define:

$$e_G(x) = \{k: (\exists f \in G)\ (k, f) \in x\}.$$

For $\alpha < \aleph_\omega$, let $G_\alpha = G \cap \mathbb{P}_\alpha$, where $\mathbb{P}_\alpha$ is $\mathbb{P}$ restricted to $\{f \in P: \mathrm{dom}\, f \subseteq \alpha \times \omega\}$. For limit ordinals $\lambda < \aleph_\omega$, f_λ gives an enumeration of $L_\lambda[u]$ in type ω, and $(e_G \circ f_\lambda(n): n \in \omega)$ enumerates in type ω all the subsets of ω in $M[G_\lambda]$ which are named by terms in $L_\lambda[u]$. Regard this enumeration as an element $y_\lambda \in B$. Note that $\lambda < \lambda' \Rightarrow y_\lambda \leq_T y_{\lambda'}$. Further, the y_λ give us all the reals of $M[G]$ constructible from u and a bounded segment of $(f_\alpha: \alpha < \aleph_\omega)$. Now define $\tau' = \tau'_G$ by letting $\tau'_n = y_{\aleph_n}$ and we let τ' be the canonical term in M for this sequence.

(3.1.2) We now describe how the construction of τ' can be modified. The key idea is that of superimposing a condition on another, or on a generic G. The notion appears already in [04] on p. 254, the paragraph immediately preceding Lemma 5.1.1; incidentally the development in (3.1.1) is essentially that of the three paragraphs immediately preceding the introduction of the superimposition.

So, for $f, g \in P$, $g|f$ is defined by: $\mathrm{dom}\, g|f = \mathrm{dom}\, g \cup \mathrm{dom}\, f$; $(g|f)|\mathrm{dom}\, f = f$; $(g|f)|\mathrm{dom}\, g \setminus \mathrm{dom}\, f = g|\mathrm{dom}\, g \setminus \mathrm{dom}\, f$. We also define $G|f = \{g|f: g \in G\}$ and we observe that $G|f$ is generic.

Then, for limit λ, instead of constructing our enumeration $UG(\lambda, \cdot)$ as before, we arrange instead, an enumeration (p_n, t_n) of a subset of $P_{\lambda+1} \times L_\lambda[u]$ such that $(\mathrm{dom}\, p_n: n \in \omega)$ is $\subseteq$-non-decreasing, such that $L_\lambda[u] = \{t_n: n \in \omega\}$ and such that $\{p_n: n \in \omega\}$ is dense in $\mathbb{P}_{\lambda+1}$. This can be accomplished using the genericity of G. Then, the nth subset of ω in the ω-sequence associated with λ and G will be the interpretation of t_n by $G|p_n$.

Now suppose G' is M-generic/$\mathbb{P}$ and that for some $q \in \mathbb{P}_{\lambda+1}$, $G'_{\lambda+1} = G_{\lambda+1}|q$. Of course $G_{\lambda+1} \in M[G'_{\lambda+1}]$ and conversely, so, the range of the ω-sequence of subsets of ω associated with λ and G = the range of the ω-sequence of subsets of ω associated with λ and G'. Further, the sequence of (p_n, t_n) produced by G is tail-equal to the sequence of (p_n, t_n) produced by G'. Finally if n is sufficiently large to get us into the tail and to guarantee that $\mathrm{dom}\, q \subseteq \mathrm{dom}\, p_n$, then the nth subset of ω in the λ, G-sequence equals the nth subset of ω in the λ, G'-sequence. Thus, the two sequences are tail equal with equal ranges so the Turing degree of the y_λ

produced by G equals the Turing degree of the y_λ produced by G'.

The reader will remark that in all of the preceding, well-foundedness was never used in an essential way; an ω-model would have sufficed as ground model. In fact, Proposition F(T) is equivalent over ATR_0 to: for all countable admissible A, there's an ω-model of Zermelo set theory with comprehension for L_A-formulas. This remark is important, since it shows that cutting down the set-theoretic universe will not permit us to settle the status of Proposition F(T) negatively, as long as our universe is closed under hyperjump. F(T) is also Π^1_2. Later, we'll see that G(T), which is Π^1_3, exhibits similar behavior, more typical of Π^1_2.

(3.1.3) To obtain a) of Reversal 3', we shall define an invariant Borel $F: B^\omega \to B$ by cases. The fourth case will be the only case in which we can have $F(\vec{x}) \leq_T x_n$ for some n, and this case will yield an ω-model of Z_2 + "only countably many reals are constructible from any given real". Standard methods then permit us to get an ω-model of ZC + V = L. Therefore, we limit ourselves to the definition of F, since it will be obvious that in cases 1-3, we can never have $F(\vec{x})$ reducible to any co-ordinate of $\vec{x}$. In §3.3 we shall build upon this F, by subdividing Case 4 into further cases.

Given $\vec{x} \in B^\omega$, we work with $\vec{d}$, where d_n is the degree of x_n. This is simply to make it crystal clear that the F we define is Turing-invariant, but it will be clear that everything could just as easily be done with $\vec{x}$, so there will be no problem lifting the definition to obtain elements of B (rather than degrees) as values of F.

So, given $\vec{d}$, we let M = {r:r is recursive in some finite join of the d_n's}. We define the degree $d^* = F(\vec{d})$ by cases:

Case 1: M fails to model Z_2. Then, for some φ, say with k real variables, and one integer variable and some $x_1,\dots,x_k \in M$, the set $\{k: (k,x_1,\dots,x_k)$ satisfies φ in $M\}$ is a real outside of M. Let i be the least Godel number such that φ_i has this property, and, for the k appropriate for φ_i, let $x_1,\dots,x_k \in M$ be witnessing parameters. Let m be least such that $x_1,\dots,x_k$ are all recursive in the join of $d_0,\dots,d_m$.

Now, letting $(y_1,\ldots,y_k) \in M^k$ vary over sequences whose terms are all recursive in the join of $d_0,\ldots,d_m$, consider all the reals $\{n: (n,y_1,\ldots,y_k)$ satisfies φ_i in $M\}$, at least one of which lies outside M. Thus, we have the map:

$$(e_1,\ldots,e_k) \longmapsto \begin{pmatrix} \text{the set of } n \text{ such that } \varphi_i \text{ holds} \\ \text{of } n \text{ and the parameters} \\ \{e_1\}^{d_0\vee\ldots\vee d_m},\ldots,\ \{e_k\}^{d_0\vee\ldots\vee d_m}, \end{pmatrix}$$

and we define $F(\vec{d})$ = (the degree of) this map coded as a real. Different representatives of the d_n result in a primitive recursive permutation of the indices $(e_1,\ldots e_k)$. Clearly $F(\vec{d})$ can't be recursive in any d_n, lest it lie in M, which is impossible, since a counterexample to M's being a model of Z_2 is recursive in $F(\vec{d})$.

So, in Cases 2)-4) we shall know that M models Z_2.

<u>Case 2</u>: For some m, the jump of $d_0\vee\ldots\vee d_m$ is not reducible to any d_n. We then let $F(\vec{d})$ = the jump of $d_0\vee\ldots\vee d_m$.

<u>Case 3</u>: For some $x \in M$, there's no enumeration in M of the set of reals constructible from x in M. Let s be least such that there's no emuneration in M of the set of reals constructible from d_s in M. Thus, for each t there's a real constructible from d_s which is not recursive in d_t. Letting t vary through natural numbers and e vary through indices, we have the enumeration:

$$(t,e) \mapsto x_t(e) = \text{the first real constructible from } d_s \text{ in } M \text{ which is non-recursive in } \{e\}^{d_t}.$$

We then let $F(\vec{d}) = d^*$ = the degree of this enumeration coded as a real. Clearly this is Borel, invariant and can be viewed as taking values in B rather than in degrees. And, of course, $F(\vec{d})$ cannot be recursive in d_t, since $x_t(e)$ is recursive in $F(\vec{d})$ but not in d_t (for any e).

<u>Case 4</u>: Otherwise, i.e. M models Z_2 and for all reals r the set of reals constructible in r is countable. We take $F(\vec{d}) = d_0$. But this case must arise, by Proposition F(T).

(3.2) Our approach to obtaining b) of Theorem 3' is very close to our method, above for obtaining a), except now we start from a countable transitive model M of a suitably large fragment of ZFC + "there's a measurable cardinal". Once again, $u \in M$ is a Borel code for Turing invariant Borel F. Let κ be a measurable cardinal of M and $U \in M$ be a normal ultrafilter on κ. This time we force with finite partial functions, ordered by reverse inclusion to add a generic sequence of collapses $f_\alpha : \omega \to_{onto} V_\alpha^M$, for $\alpha < \kappa$. As in (3.1.1), for each limit λ, we obtain an ω-sequence of subsets of ω, which, as before, we code as a member of B, b_λ. Now, return to M. Force over M to add a Prikry sequence of limit ordinals $\vec{\alpha} = (\alpha_n : n \in \omega)$ to κ, using U. Now, using <u>both</u> $\vec{\alpha}$ and $(f_\alpha : \alpha < \kappa)$, we let $x_n = b_{\alpha_n}$. We'd like to repeat the argument of (3.1.2) to show that for some k, $F(\vec{x}) \leq_T x_k$ <u>and</u> that this holds <u>regardless</u> <u>of</u> <u>the</u> Prikry sequence $\vec{\alpha}$. This would yield Proposition G(T), since any subsequence of a Prikry sequence is a Prikry sequence, which means the same conclusion holds for any subsequence $\vec{y}$ of $\vec{x}$ in the place of $\vec{x}$.

Unfortunately, $(f_\alpha : \alpha < \kappa)$ is not $M[\vec{\alpha}]$-generic; if it were, the above sketch would be a complete proof. However, we'll show that $(f_\alpha : \alpha < \kappa)$ can be "adjusted" (from outside) to produce a new sequence $(f'_\alpha : \alpha < \kappa)$ generic over $M[\vec{\alpha}]$, <u>and</u> such that $(f'_\alpha : \alpha < \kappa)$ is similar to $(f_\alpha : \alpha < \kappa)$, as in (3.1.2), i.e., for all $\beta < \kappa$, $(f'_\alpha : \alpha < \beta)$ is $(f_\alpha : \alpha < \beta)|p$ for some p. This accomplished, by essentially the same arguments as in (3.1.2), we can argue that the sequence $(x'_n : n \in \omega)$ produced from $(f'_\alpha : \alpha < \kappa)$ is termwise Turing-equivalent to $(x_n : n \in \omega)$, and that for some $k < \omega$, $F(\vec{x}') \leq_T x'_k$.

Thus, we limit ourselves to showing how to produce the desired $(f'_\alpha : \alpha < \kappa)$. The crucial observation is that, since Prikry forcing adds no new bounded subsets of κ, all bounded segments $(f_\alpha : \alpha < \beta)$, where $\beta < \kappa$, really are $M[\vec{\alpha}]$-generic/$\mathbb{P}$. Then, from outside $M[\vec{\alpha}]$, enumerate in type ω, the dense subsets of $\mathbb{P}$ lying in $M[\vec{\alpha}]$, say as $(D_n : n \in \omega)$. At stage n, we've already constructed $(f'_\alpha : \alpha < \alpha_{m(n)})$ which is similar to $(f_\alpha : \alpha < \alpha_{m(n)})$. We seek a condition $p \in D_n$ such that $p|\alpha_{m(n)} \times \omega \subseteq \bigcup\{f'_\alpha : \alpha < \alpha_{m(n)}\}$. But, since $(f'_\alpha : \alpha < \alpha_{m(n)})$ is $M[\vec{\alpha}]$-generic/$\mathbb{P}_{\alpha_{m(n)}}$, if there were no such p, it would be forced that there's no such p, which is absurd. We then superimpose $p|(\mathrm{dom}\ p \setminus \alpha_{m(n)} \times \omega)$ on $(f_\beta : \alpha_{m(n)} \leq \beta < \alpha_{m(n+1)})$, where $\alpha_{m(n+1)}$ is least such that $\mathrm{dom}\ p \subseteq \alpha_{m(n+1)} \times \omega$.

In fact, we can even obtain the variant a) of Proposition G(T). The difficulty is that the k such that membership in $F(\vec{x}(\vec{\alpha}))$ as determined by $(f'_\alpha : \alpha < \alpha_\kappa)$ a priori depends on the Prikry sequence $\vec{\alpha}$. The least such k is definable from $\vec{\alpha}$, which gives us a natural number-valued term, $\bar{k}$, for Prikry forcing over M.

A basic property of Prikry forcing is that any forcing statement is decided by a Prikry condition of the form (ϕ, S), where $S \in U$. Thus, each of the statements "$\bar{k} = k$" is so decided, say by (ϕ, S_k). But $(S_k : k \in \omega) \in M$ so $\bigcap \{S_k : k \in \omega\} \in U$. Clearly there's a single $k_0 \in \omega$ such that "$\bar{k} = k_0$" is forced by $(\phi, \bigcap\{S_k : k \in \omega\})$. Now, taking only Prikry sequences from $\bigcap\{S_k : k \in \omega\}$, k_0 is the fixed natural number which works for all subsequences $\vec{y}$ of $\vec{x}$.

As noted at the close of (3.1.2) for F(T), well-foundedness was never used; an ω-model M of ZFC + "there's a measurable cardinal" with $u \in M$ suffices, provided we still have that every subsequence of a Prikry sequence is a Prikry sequence. This is not apparent in this context, but is true in virtue of Mathias' characterization of Prikry sequences as those which have a tail in every set in the ultrafilter, and this characterization does carry over to this setting. Thus, though G(T) is Π^1_3, it is absolute in the sense that, if true, its truth relativizes down to any model closed under hyperjump, since in such models all ω-models which are supposed to exist, in fact, do. Thus, e.g., if true, G(T) is true in L, in the minimal model of ZF, in L_θ, where θ is the supremum of the first ω admissible ordinals; in particular, though Π^1_3, G(T) exhibits "Π^1_2 - behavior".

(3.3) We shall now obtain b) of Reversal 3', in fact with the "weaker" variant b) of G(T). Our first step will be to show that variant b) isn't weaker after all. This is done in (3.3.1). In (3.3.2) we show that G(T) implies a slightly stronger statement. These preliminaries out of the way, we pick up the thread of (3.1.3), by assuming that Cases 1-3 of the definition of the Turing-invariant, Borel F of (3.1.3) all fail.

(3.3.1) LEMMA: Variant b) of G(T) implies G(T).

Proof: We show that variant b) of G(T) implies that if $F: B^{\omega} \to B$ is Turing-invariant and Borel then there's $\vec{x}$ such that for all subsequences $\vec{y}$ of $\vec{x}$ there's k such that $F(\vec{y}) \leq_T x_k$ and such that:

(*) there's n_0 such that $m < n_0 \Rightarrow x_m \leq_T x_{n_0}$ and $n_0 \leq m < n \Rightarrow x_m \leq_T x_n$.

But such an $\vec{x}$ of course satisfies the conclusion of G(T).

Given Turing-invariant Borel $F: B^{\omega} \to B$, we define Turing-invariant Borel F' by cases. Let $\vec{x} \in B^{\omega}$:

Case 1: for some i, x_i is $\leq_T$-maximal among the x_j. Let i_0 be the least such. Let $F'(\vec{x})$ = the jump of the join of the x_j, $j \leq i_0$. Clearly, in this case, $\vec{x}$ fails to satisfy the conclusion of variant b) of G(T).

Case 2: Case 1 fails, and the x_i's are not eventually strictly increasing for $\leq_T$. We can then produce, in a Turing-invariant, Borel fashion, a strictly increasing sequence $(i_n : n \in \omega)$ such that for all n:

$$x_{i_{4n+1}} \not\leq_T x_{i_{4n+2}}$$

$$n \text{ odd} \Rightarrow x_{i_{4n+3}} <_T x_{i_{4n+4}},$$

namely by taking $i_0 = 0$ and i_{n+i} = the least $j > i_n$ satisfying the two above conditions.

We then let $F'(\vec{x})$ = (the characteristic function of) $\{i: x_{2i+1} <_T x_{2i+2}\}$. But given a Turing degree d we can find a subsequence $\vec{y}$ of $\vec{x}$ for which the definition of F' comes under Case 2, and for which $d \leq_T [F'(\vec{y})]$. Clearly then we cannot have $\vec{x}$ satisfying the conclusion of variant b) of G(T).

Case 3: Cases 1, 2 fail. Note that $\vec{x}$ then satisfies (*) since, if not, there's i such that $x_i \not\leq_T x_k$ for any x_k in the strictly-increasing tail. But then, either this x_i is $\leq_T$-maximal or has a $\leq_T$-maximal successor in the finite non-strictly increasing initial segment. We let $F'(\vec{x}) = F(\vec{x})$, and only an $\vec{x}$ falling under this case, i.e. satisfying (*), can satisfy the conclusion of variant b) of G(T). This concludes the proof. In what follows, we let G*(T) denote the statement of the lemma.

(3.3.2) Henceforth, we work with G*(T). We first show that we can obtain a slightly stronger conclusion.

LEMMA: Let F be Turing-invariant and Borel, and let H be Borel and strongly Turing-invariant, i.e. if $\vec{x},\vec{y}$ are termwise Turing-equivalent then $H(\vec{x}) = H(\vec{y})$. Then, there's $\vec{x}$ satisfying the conclusion of G*(T) and such that for all subsequences $\vec{y}$ of $\vec{x}$, $H(\vec{y}) = H(\vec{x})$.

Proof: Let F,H be as in the statement of the lemma. We may regard H as taking values in 2^{ω}. For $\vec{x} \in B^{\omega}$ and $n \in \omega$ let ${}_n(\vec{x}) = (\lambda k)x_{n+k}$.

Let $H'(\vec{x})(n) = 0$ if $H({}_n(\vec{x})) = H({}_{n+1}(\vec{x}))$; otherwise, let $H'(\vec{x})(n) = k^*+1$, where k^* = the least k such that $H({}_n(\vec{x}))(k) \neq H({}_{n+1}(\vec{x}))(k)$. Finally, let $Q(\vec{x}) = \langle F(\vec{x}),H'(\vec{x})\rangle$, where $\langle,\rangle$ is a homeomorphism of B^2 with B. Now, apply G(T) to Q and let $\vec{x}$ be s.t. for all subsequences $\vec{y}$ of $\vec{x}$, there's k such that $Q(\vec{y}) \leq_T y_k$. Now any subsequence of $\vec{x}$ also satisfies the conclusion of G(T) for Q so, towards a contradiction, assume there's no subsequence, $\vec{x}'$ of $\vec{x}$ such that H is constant on the subsequences of $\vec{x}'$.

For $\vec{y} \in B^{\omega}$, let $S(\vec{y})$ = the set of subsequences of $\vec{y}$. Let $\vec{y} \in A$ iff $\vec{y} \in S(\vec{x})$ and $H(\vec{y}) = H({}_1(\vec{y}))$. Thus, by Galvin-Prikry, [07], there's $\vec{y} \in S(\vec{x})$ such that $S(\vec{y}) \subseteq A$ or $S(\vec{y}) \cap A = \phi$. If $S(\vec{y}) \subseteq A$, then, since H is continuous on a dense G_{δ}, there's $\vec{z} \in S(\vec{y})$ such that H is continuous on $S(\vec{z})$, so assume, without loss of generality that H is continuous on $S(\vec{y})$. Thus, H is continuous and tail-invariant on $S(\vec{y})$ and strongly Turing-invariant. We shall show that H is constant on $S(\vec{y})$, a contradiction.

So, towards a contradiction, suppose $\vec{z}, \vec{z}' \in S(\vec{y})$ and $H(\vec{z}) \neq H(\vec{z}')$, say $H(\vec{z})(a) \neq H(\vec{z}')(a)$. Since H is continuous there's k such that if $z''_i = z_i$ for all $i \leq k$, then $H(\vec{z}'') \mid a+1 = H(\vec{z}) \mid a+1$ (actually, we don't need that $z_i = z''_i$ for all $i \leq k$, only that for all such i, z_i, z''_i have a sufficiently long common initial segment). Then let $\vec{z}''$ be defined by: $z''_i = z_i$ $(i \leq k)$, and for $j > k$, $z''_j = z'_j$. Then, $H(\vec{z})(a) = H(\vec{z}'')(a)$, as above, but also, by tail-invariance, $H(\vec{z}'') =$

$H(_{k+1}(\vec{z}'')) = H(_{k+1}(\vec{z}')) = H(\vec{z}')$, but $H(\vec{z})(a) \neq H(\vec{z}')(a)$, contradiction!

So, suppose that $S(\vec{y}) \cap A = \phi$, i.e., for all $\vec{z} \in S(\vec{y})$ $H(\vec{z}) \neq H(_1(\vec{z}))$. Let $\vec{a} \in B$ grow faster than all sets recursive in the x_i's. As before, we can assume H is continuous on $S(\vec{y})$. Recall that, since H takes values in 2^ω, for any $\vec{y}' \in S(\vec{y})$ and for any k there are at most 2^k possibilities for $H(\vec{z})|k$ $(\vec{z} \in S(\vec{y}'))$, namely the set of length k binary sequences. Thus, in at most 2^k applications of Galvin-Prikry, we obtain $\vec{z} \in S(\vec{y}')$ such that for all $\vec{z}' \in S(\vec{z})$, $H(\vec{z}')|k = H(\vec{z})|k$. Let $\vec{y}^0 = \vec{y}$, and obtain $\vec{y}^{j+1}$ as the $\vec{z}$ which results when $\vec{y}' = \vec{y}^j$ and $k = a_j$. Then, set $u_j = (\vec{y}^{j+1})_j$. Clearly u has the property:

(**): for all j, if $\vec{t}, \vec{v} \in S(\vec{u})$ and neither begins earlier than u_j then $H(\vec{t})|a_j = H(\vec{v})|a_j$.

But then $H'(\vec{u})$ grows at least as fast as $\vec{a}$, which is impossible since $H'(\vec{u})$ is recursive in $Q(\vec{u})$ which is recursive in some u_k, which is one of the x_i, while $\vec{a}$ grows faster than any set recursive in the x_i's. Henceforth, we let $G^{\#}(T)$ denote the statement of the lemma.

(3.3.3) We now pick up the thread of (3.1.3). We subdivide Case 4 of the definition of F, there, i.e. we suppose Cases 1-3 fail. Our first goal is to show that an $\vec{x}$ satisfying the conclusion of $G^{\#}(T)$ yields a model of ZF (actually, we use "only" $G(T)$). Recall that, given $\vec{x}$, we let $\vec{d}$ be the corresponding sequence of Turing degrees, and we defined M to be the set of reals recursive in some finite join of the d_n's. We've already shown in (3.1.3), using just $F(T)$, that for some $\vec{x}$, M is a model of Z_2 in which the set of reals constructible from any given real is countable, and that the d_n's are eventually strictly increasing for $<_T$.

Now the existence of such an M which is also a model of second-order (scheme form) countable choice is equiconsistent with the existence of a model of ZF, this uses the methods of [14]. We now show there's such an M by showing that if the M defined from $\vec{x}$ fails to model the scheme form of second-order countable choice then $\vec{x}$ cannot satisfy the conclusion of $G(T)$.

Case 4: M fails to satisfy the scheme-form of second-order choice. So, fix φ and a parameter $a \in M$ such that choice fails in M for φ and a, i.e., in M, i) is true and ii) is false, where:

i) is: $\forall i \, \exists y \, \varphi(i,y,a)$,
ii) is: $\exists y \, \forall i \, \varphi(i,(y)_i,a)$.

We first note that by the definition of M, and the failure of Case 2 (in 3.1.3) for any $\bar{x}$ satisfying the conclusion of $G(T)$, by rewriting φ, we can take $a = x_m$ for some m. To guarantee that our definition of F is Turing invariant, we give an explicit rewriting of φ:

$$\exists j \, \forall i \, \exists y \, \varphi(i,y,\{j\}^{x_m}).$$

Our second observation is that, for the same reasons as in dealing with i), we can take $y = \{e\}^{x_n}$, for some e,n. Thus, i) can be rewritten:

i') $\exists j \, \forall i \, \exists n \, \exists e \quad (i,\{e\}^{x_n},\{j\}^{x_m})$, though this is no longer a statement over M due to quantifying over n.

Exploiting the failure of Case 2 for a third time, the following function g is totally defined, i.e. $g \in B$:

$g(\langle j,k\rangle)$ = the least n such that for all $i \leq k$ if, in M, there's y such that $\varphi(i,y,\{j\}^{x_m})$ then, in M, there's e such that $\varphi(i,\{e\}^{x_n},\{j\}^{x_m})$.

By ii), range g is unbounded, in fact there's a fixed j such that $\{g(\langle j,k\rangle): k \in \omega\}$ is unbounded, and for all k, all $i \leq k$,

$$\exists e \, \varphi(i,\{e\}^{x_{g(k)}}, \{j\}^{x_m}).$$

The j will vary according to which representative $\vec{x}$ of $\vec{d}$ we're dealing with, and, for Turing-invariance, we must in fact treat the j as a variable, rather than a fixed parameter. That, for such a j, $\{g(\langle j,k\rangle): k \in \omega\}$ must be unbounded is by ii), since if $g(<j,k>) \le n$ for all k, then by comprehension, $h \in M$, where $h(i) =$ the least e such that $\varphi(i,\{e\}^{x_n},\{j\}^{x_m})$, and $p \in M$, where $p(<i,\ell>) = \{h(i)\}^{x_n}(\ell)$; but p is a choice function, i.e., a counterexample to ii).

We let $F(\vec{d}) =$ (the Turing degree of) g. We now show that, in this case, we cannot have $\vec{x}$ satisfying the conclusion of $G(T)$. Let $h \in B$ eventually dominate all elements of M and fix $j \in \omega$ which yields our parameter a. We define a monotone sequence $(q_n : n \in \omega) \in B$: $q_0 = m$, $q_{n+1} =$ the least $q > q_n$ such that $d_1 \vee \ldots \vee d_{q_n} \le_T d_q$ and such that for all $i \le h(n)$, $\exists e\, \varphi(i,\{e\}^{x_q},\{j\}^{x_m})$. This yields a subsequence $(x_{q_n} : n \in \omega)$. But then h is dominated by a function recursive in $F((d_{q_n} : n \in \omega))$, namely $\sigma(\ell) =$ the least k such that $F((d_{q_n} : n \in \omega))(<j,k>) > \ell$. Thus, $F((d_{q_n} : n \in \omega))$ cannot be (the Turing degree of a set) recursive in one of the d_k.

(3.3.4) Now let $\omega \in A$ be the least admissible set and let $A \in B$ be the next admissible set. Let $\vec{x} \in B^\omega$ satisfy the conclusion of $G^\#(T)$. Thus, Cases 1)-4) in the definition of F fail. We consider $M^+ = (M,d_n)_{n\in\omega}$, where each d_n is the subset of M consisting of all elements $r \in M$ with $[r] = d_n$; thus, M^+ is really $(\omega,+,\cdot,M,d_n)_{n\in\omega}$. We summarize the properties of M^+ as follows:

(1) $(d_n : n \in \omega)$ is strictly $<_T$ - increasing,
(2) M satisfies comprehension,
(3) M satisfies the scheme-form of second-order choice,
(4) in M, each $\omega_1^{L[r]}$ is countable.

We now point out that the arguments for Cases 1, 4, respectively can be easily extended to obtain the following strengthenings of 2), 3).

2^+): M^+ satisfies comprehension for L_B-formulas in the additional unary predicates interpreted by the d_n's, and with finitely many parameters; the same holds for any subsequence of $\vec{d}$ in place of $\vec{d}$.

3^+): M satisfies the scheme form of countable second order choice for L_B - formulas with finitely many parameters, <u>this time not allowing the additional unary predicates</u> interpreted by the d_n's.

Further, using the full force of $G^{\#}(T)$, and the auxiliary function $H(\vec{d})$ = the L_B theory of M^+, we obtain:

5) the d_n's are strongly indiscernible in the sense that for any subsequences $\vec{d}'$, $\vec{d}''$, $(M,\vec{d}')$ is L_B - elementarily equivalent to $(M,\vec{d}'')$.

We've now obtained what we need from $G^{\#}(T)$. Henceforth, we work with an M^+ satisfying 1), 2^+), 3^+), 4), 5) and show how to construct, within such an M^+, a predicate U on the ordinals of M such that, in M, L[U] can be constructed and satisfies ZFC + $(\forall X)X^{\#}$ exists.

(3.3.5) We should, at this point, be somewhat more specific about how to associate to M an "HC-type" model, i.e., a model $(\tilde{M},\tilde{\in})$ of ZF - P + "all sets are countable", with standard integers and which extends M (to be precise, includes an isomorphic copy of M). $\tilde{M}$ consists of isomorphism classes of certain (codes of) trees, which, in M, are well founded. The class T of trees is Π^1_1 over M; isomorphism of two elements of T is Δ^1_1 (in T) over M. $\tilde{\in}$ is arithmetic (in the isomorphism relation on T) over M. See [14] for details, where it's shown that all this can be carried out assuming only ATR_0 in M. Adding full Z_2 plus the scheme of countable second-order choice to M yields ZF-P for $\tilde{M}$. Further, adding (4) of (3.3.4) to M will then yield the power-set axiom for all the L[x] of $\tilde{M}$ $(x \in M)$.

In order to exploit the fact that 2^+), 3^+), 5) hold for M^+, we let $\alpha_i = \omega_1^{L[x_i]}$ in $\tilde{M}$, where $x_i \in d_i$ is an element of M. We let $\tilde{M}^+ = (\tilde{M},\tilde{\in},\alpha_i)_{i\in\omega}$, where each α_i interprets a constant symbol. We then have the following analogues of 2^+), 3^+), 5):

$\tilde{2}^+$): $\tilde{M}^+$ satisfies the Separation-Scheme for L_B - formulas in the additional constant symbols interpreted by the α_i, and with finitely many parameters; the same holds with any subsequence of $\vec{\alpha}$ in place of $\vec{\alpha}$.

$\tilde{3}^+$): $\tilde{M}$ satisfies the scheme of countable choice for L_B-formulas with finitely many parameters (not allowing the new constant symbols interpreted by the α_i's).

$\tilde{5}$): the α_i's are strongly indiscernible in the sense that for any subsequences $\vec{\alpha}', \vec{\alpha}''$, $(\tilde{M}, \tilde{\epsilon}, \vec{\alpha}')$ is L_B-elementarily equivalent to $(\tilde{M}, \tilde{\epsilon}, \vec{\alpha}'')$.

$\tilde{3}^+)$ is immediate from $3^+)$, $\tilde{2}^+)$ and $\tilde{5})$ result from $2^+)$, $5)$ respectively by re-interpreting L_B-formulas involving the new constant symbols as formulas over M^+; this is done in [14] for finitary formulas without the new constant symbols. We need only say how to reinterpret atomic formulas involving the constant symbols:

interpret $v \in c_i$ as: $T(v) \wedge \exists x(U_i(x) \wedge$ "x codes a well-ordering" $\wedge T(x) \wedge v \cong x)$ (here and below, U_i is the unary predicate interpreted by d_i in M^+);

interpret $c_i \in v$ as: $T(v) \wedge \exists n[\forall x(U_i(x) \wedge$ "x codes a well-ordering" $\wedge T(x)) \Rightarrow \exists m(x \cong v_{<n,m>})]$ (i.e. x is the code of a tree isomorphic to the part of T below $<n,m>$, where T is the tree coded by v);

interpret $c_i = v$ as: $T(v) \wedge$ "v codes a well-ordering" $\wedge \neg(v \in c_i) \wedge \neg(c_i \in v)$ (i.e., we write out the above formulae);

interpret $c_i \in c_j$ as true if $i < j$ and false otherwise;

interpret $c_i = c_j$ as true if $i = j$ and false otherwise.

(3.3.6) Given a sequence $\vec{\beta}$ of strong L_B-indiscernibles over $\tilde{M}$ (i.e. one which satisfies $\tilde{5})$ with $\vec{\beta}$ replacing $\vec{\alpha}$), we shall want to "normalize" $\vec{\beta}$, i.e. guarantee that any "pressing-down" function (this will be made precise in the next lemma) is constant on a set associated with $\vec{\beta}$. This can be done (as the next lemma shows) but at the price of some loss of indiscernibility. Thus, we introduce a weaker notion called medium L_B-indiscernibility over $\tilde{M}$ which is preserved under our normalizing procedures (as the next lemma shows).

If $\vec{\gamma}$ is a subsequence of $\vec{\beta}$, we say $\vec{\gamma}$ is a finite deletion subsequence of $\vec{\beta}$ if $\vec{\gamma}$ results from $\vec{\beta}$ by deleting finitely many terms of $\vec{\beta}$, i.e. if $\vec{\beta},\vec{\gamma}$ have a common tail. In what follows, we often write $\tilde{M}$, when we mean $(\tilde{M},\tilde{\in})$. We then define:

> $\vec{\beta}$ is a sequence of medium L_B-indiscernibles over $\tilde{M}$ iff $\vec{\beta}$ is strictly increasing and for any finite deletion subsequences $\vec{\gamma},\vec{\delta}$ of $\vec{\beta}$, $(\tilde{M},\vec{\gamma})$ and $(\tilde{M},\vec{\delta})$ are elementarily equivalent for L_B-sentences without parameters; further $(\tilde{M},\vec{\beta})$ satisfies comprehension for L_B-formulas with finitely many parameters and additional constant symbols interpreted by the β_n's.

(3.3.7) We now state our "normalizing" lemma, whose proof we shall defer, and its Corollary; we shall work with a normalized sequence of medium L_B-indiscernibles over $\tilde{M}$ to finish b) of Reversal 3', and then return to prove the normalizing lemma.

LEMMA: Let $\vec{\beta}$ be a sequence of medium L_B-indiscernibles over $\tilde{M}$.

a) Suppose $\theta = \theta(\vec{c},v)$ is a parameter-free L_B-formula with only v free, where $(c_n : n \in \omega)$ is a sequence of new constant symbols and that, over $(\tilde{M},\vec{\beta})$ the following holds:

$$(1) \quad \exists! v\theta(\vec{c},v) \wedge \forall v(\theta(\vec{c},v) \Rightarrow v \in c_0).$$

Then, by medium L_B-indiscernibility, (1) holds in $(M,\vec{\gamma})$ for any finite deletion subsequence of $\vec{\beta}$, so θ defines a pressing-down function $F(\vec{\gamma})$ on finite deletion subsequences of $\vec{\beta}$, by $F(\vec{\gamma})$ = the unique $a \in \tilde{M}$ such that $\theta(\vec{c},a)$ holds in $(\tilde{M},\vec{\gamma})$.

Let $1 \leq k \in \omega$. We say that F is k-supported if whenever $\vec{\gamma},\vec{\delta}$ are finite deletion subsequences of $\vec{\beta}$ such that $\gamma_j = \beta_j$ for all $j < k$ then $F(\vec{\gamma}) = F(\vec{\delta})$. We sum this up by writing:

> F: FDEL $(\vec{\beta}) \to OR_{\tilde{M}}$, is an L_B-definable, k-supported pressing-down function.

LEMMA: If $1 \leq k \in \omega$ and $F: \mathrm{FDEL}(\vec{\beta}) \to \mathrm{OR}_{\tilde{M}}$ is an L_B-definable, k-supported pressing-down function, and F is non-constant, then $\vec{\gamma}$ is a sequence of medium L_B-discernibles over $\tilde{M}$, where, for $n \in \omega$, $\gamma_n = F(\vec{\delta})$, where $\vec{\delta}$ is any finite deletion subsequence of $\vec{\beta}$ such that for $j < k$, $\delta_j = \beta_{nk+j}$.

This lemma will be proved in (3.3.11). We obtain a

COROLLARY: There's a sequence $\vec{\beta}$ of medium L_B-indiscernibles over $\tilde{M}$ such that for any $F: \mathrm{FDEL}(\vec{\beta}) \to \mathrm{OR}_{\tilde{M}}$, if F is L_A-definable, pressing-down and k-supported for some $1 \leq k \in \omega$, then F is constant.

Proof: We use a truth definition for L_A (which is available in L_B) and we iterate the lemma obtaining sequences $\vec{\delta}^n$ of medium L_B-indiscernibles over $\tilde{M}$, with $\vec{\delta}^0 = \vec{\alpha}$; δ^{n+1} = the result of applying the lemma to $\vec{\beta} = \vec{\delta}^n$ and θ = the nth L_A-formula of the appropriate type, if θ defines a pressing-down function which is k-supported for some $1 \leq k \in \omega$ (by indiscernibility, this can be recognized from a truth-definition for L_A), otherwise $\vec{\delta}^{n+1} = \vec{\delta}^n$. We claim that for some $N \in \omega$, whenever $N \leq m \in \omega$, $\vec{\delta}^m = \vec{\delta}^N$. This is because otherwise $(\delta_0^n: n \in \omega)$ has an easily definable subsequence which is strictly decreasing below α_0 an ordinal of $\tilde{M}$. But this contradicts that we can actually get $(\delta_0^n: n \in \omega) \in \tilde{M}$, by $(\tilde{3}^+)$ for $(\tilde{M}, \vec{\alpha})$, using the L_B-truth definition for L_A. We take $\vec{\beta} = \vec{\delta}^N$.

In (3.3.8) - (3.3.10) we fix $\vec{\beta}$ a normalized (i.e. satisfying the Corollary) sequence of medium L_B-indiscernibles. In (3.3.8) we show that we may assume, without loss of generality, that the β_n's are cofinal in $\tilde{M}$. In (3.3.10) we obtain, working in $(\tilde{M}, \vec{\beta})$, an inner model which satisfies for all sets X, $X^\#$ exists.

(3.3.8) We work in $(\tilde{M}, \vec{\beta})$.

LEMMA: If the β_n's are bounded, then ZFC+ "there exists a measurable cardinal" is consistent.

Proof: First note that, by separation, if the β_n's are bounded then $(\beta_n: n \in \omega) \in \tilde{M}$. Thus, again by separation, $\sup_n \beta_n$ exists in $\tilde{M}$; call it γ. By replacement in $\tilde{M}$, A = the set of tails of $(\beta_n: n \in \omega)$ exists in $\tilde{M}$; so, in $\tilde{M}$ we can construct $L[A]$. Also, pulling back to M, we see that for some real $x \in \tilde{M}$, $A \in L[x]$, so $L[A]$ is an inner model of $L[x]$, so, in $\tilde{M}$, $L[A]$ is satisfied to model ZFC. Now, in $L[A]$, let F = the filter on γ generated by A. We can easily verify that $L[A] = L[F]$ but then, as usual F is a γ-complete ultrafilter on γ in $L[F]$. We proceeded in this roundabout manner, since, lacking power-set in $\tilde{M}$, it is not clear how to define F first and obtain a member of $\tilde{M}$.

(3.3.9) We write $x \tilde{\subseteq} OR$ if $x \in \tilde{M}$ and x is a subset of an ordinal of $\tilde{M}$. As in (3.3.8), sup x exists in $\tilde{M}$; call if ν. Also, by Simpson's methods, [14] (see also (3.3.5), above), in $\tilde{M}$, we can construct $L[x]$, for such x, and since $x \in L[y]$ holds in $\tilde{M}$ for some $y \subseteq \omega$, $L[x]$ is an inner model of $L[y]$ which models ZFC, so $L[x]$ models ZFC.

For $x \tilde{\subseteq} OR$, "$x^{\#}$ exists in $\tilde{M}$" means that, in $\tilde{M}$, the tidy Ehrenfeucht-Mostowski Theory of $L[x]$ with constants for the elements of $\nu \cup \{x\}$ produces well-founded (in the sense of $\tilde{M}$) models when applied to any countable ordinal. Apparently we cannot exclude the possibility that for some $x \in \tilde{M}$, $x \subseteq \omega$, "$x^{\#}$ exists in $\tilde{M}$" fails. However, in a positive direction we have:

LEMMA: Let $x \tilde{\subseteq} OR$ be first-order definable over $\tilde{M}$ without parameters. Then "$x^{\#}$ exists in $\tilde{M}$".

Proof: Since x is light-face first-order definable, so is ν; thus $\nu < \beta_0$. We define a class I of ordinals of $\tilde{M}$, containing all the β_n's, by an L_A-formula over $\tilde{M}^+$. For $n \in \omega$, let Seq(n) be the set of finite sequences whose terms are natural numbers greater than n. Let L be the finitary language of set theory, let L^+ be L augmented by the constant symbols c_n, $n \in \omega$, and for $s \in$ Seq (= Seq(-1)), letting ℓ = length s, we let Fm(s) = the set of L^+ formulas with exactly v_0, v_1 free and all of whose constant symbols are among $c_{s(0)}, \ldots, c_{s(\ell-1)}$. Then define:

$$(*) \qquad x \in I \text{ iff } \forall y \in \nu[\bigwedge_{n<\omega} x \in c_n + 1 \Rightarrow \bigwedge_{s \in Seq(m)} \bigwedge_{\varphi \in Fm(s)} (\varphi(y,x) \Leftrightarrow \varphi(y,c_n))].$$

In (*), x, ν are not parameters; we write out their definitions over $\tilde{M}$. In other words, letting $\tilde{M}[\nu] = (\tilde{M},y)_{y\in\nu}$, $x \in I$ iff whenever $x \tilde{\in} \beta_n$, $(\tilde{M}[\nu],x,_{n+1}(\vec{\beta}))$ is elementarily equivalent to $(M[\nu],\beta_{n,n+1}(\vec{\beta}))$, for the finitary language $L^+[\nu]$ consisting of L^+, above, augmented by new constant symbols for $y \tilde{\in} \nu$.

Clearly, each $\beta_n \in I$ and, applying separation to β_n and the L_A-definition of I, each $I \cap \beta_n \in \tilde{M}$. We also claim:

(1) I is a set of $L[\nu]$ indiscernibles over $\tilde{M}[\nu]$.

(1) is immediate from (2), below, and the fact that, in $\tilde{M}$, each β_n is closed for Godel's pairing function π (if not, let $F(\beta_n)$ = the least $x < \beta_n$ such that for some $y < \beta_n$, $\pi(x,y) \geq \beta_n$; then F is a non-constant, L^+-definable, 1-supported pressing-down function).

(2) Whenever $\vec{\gamma}, \vec{\delta}$ are increasing sequences of finite length ℓ from $\vec{\beta}$, $(\tilde{M}[\nu], \vec{\gamma})$ and $(\tilde{M}[\nu], \vec{\delta})$ are elementarily equivalent for $L[\nu]$ augmented by ℓ new constant symbols interpreted by $\vec{\gamma}, \vec{\delta}$ respectively.

To prove (2), we again use the normality of $\vec{\beta}$. Let $\gamma_i = \beta_{j_i}, \delta_i = \beta_{m_i}$. Thus, if (2) fails, recalling that $x \subseteq \nu < \beta_0 \leq \min(\gamma_0, \delta_0)$ and using the closure of β_0 under Godel's pairing function to code finitely many $y \tilde{\in} \nu$ into a single $z \tilde{\in} \beta_0$, we have, for some φ, in $(\tilde{M}, \vec{\beta})$:

$$(**): \quad \exists z(z \in c_{j_0} \wedge z \in c_{m_0} \wedge \varphi(z, c_{j_0}, \ldots, c_{j_{\ell-1}}) \wedge \neg \varphi(z, c_{m_0}, \ldots, c_{m_{\ell-1}})),$$

so this holds for $(\tilde{M}, \vec{\tau})$, for all finite deletion subsequences $\vec{\tau}$ of $\vec{\beta}$. Then, letting $F(\vec{\tau})$ = the least z witnessing (**) in $(\tilde{M}, \vec{\tau})$, F is a $\leq$ - 2ℓ-supported pressing-down function on FDEL($\vec{\beta}$) which is L^+ definable. To see that F is non-constant, a contradiction, let $\vec{\eta} = (\beta_{n_k} : k \in \omega)$ be a finite deletion subsequence, such that, letting $n_{-1} = 0$, we have, for $-1 \leq k < \ell-1$, $n_{k+1} - n_k \geq \ell$. Let p = the number of distinct elements in range $\vec{\gamma} \cup$ range $\vec{\delta}$. Then, adding $p-\ell$ coordinates to $\vec{\eta}$ in two different ways, we easily obtain finite deletion subsequences $\vec{\eta}'$, $\vec{\eta}''$ of $\vec{\beta}$ such that for $i < \ell$, $\eta'_{m_i} = \eta''_{j_i}$. Then, clearly $F(\vec{\eta}') \neq F(\vec{\eta}'')$, since by indiscernibility and (*), in M, $\neg\varphi(F(\vec{\eta}'), \eta'_{m_0}, \ldots, \eta'_{m_{\ell-1}})$ holds and $\varphi(F(\vec{\eta}''), \eta''_{j_0}, \ldots, \eta''_{m_{\ell-1}})$ holds. This completes the proof of (2) and therefore of (1). Of course, the argument for (2) could have been carried out for L_A-formulas, but this does not seem to gain us anything, here.

We need a final observation:

(3) In $\tilde{M}$, each $I \cap \beta_n$ has order-type β_n.

To prove (3), we use the normality of $\vec{\beta}$ one more time. Note that each $I \cap \beta_n$ is uniformly definable over $(\tilde{M}, \vec{\gamma})$ for any finite deletion sequence $\vec{\gamma}$ with $\gamma_0 = \beta_n$; thus, the same is true of each o.t. $(I \cap \beta_n)$. But then, if any $I \cap \beta_n$ has order-type less than β_n, then all do, and this gives a 1-supported L_A-definable pressing-down function. Further, this function is non-constant (which is impossible), since $\beta_n \in I \cap \beta_{n+1}$.

Using separation in $(\tilde{M}, \vec{\beta})$, let $u \in \tilde{M}$ be the $L^+[\nu]$-theory of $(L[x], \tilde{\in}, y, \vec{\beta})_{y \tilde{\in} \nu}$. Then, by (1) and (3), and the fact that, in $L[u]$ (as constructed in $\tilde{M}$), some β_n is uncountable (since in $\tilde{M}$, $L[u]$ models ZFC), we immediately have that $x^\#$ exists in $\tilde{M}$ (and $u = x^\#$).

(3.3.10) We now complete the proof of Reversal 3', modulo the proof of the Lemma in (3.3. 7), which we give in (3.3.11), below. We define in $\tilde{M}$ sequences $(x_\alpha : \alpha \in \mathrm{OR})$, $(\#_\alpha : \alpha \in \mathrm{OR})$ where each x_α, $\#_\alpha \subseteq \mathrm{OR}$, by recursion:

$$x_0 = \phi;\ \#_\alpha = \text{the sharp of } x_\alpha;\ \text{for } \alpha > 0,$$

$$x_\alpha = \{(\beta,\gamma) : \beta < \alpha \wedge \gamma \in \#_\alpha\},\ \text{coded as a set of ordinals.}$$

Clearly each $\#_\alpha$ is defined, using (3.3.9), since otherwise the least α such that $\#_\alpha$ is undefined is first-order definable as a point in $\tilde{M}$, and therefore x_α is definable but then by (3.3.9), the sharp of x_α is defined and therefore so is $\#_\alpha$.

Let $\# = \{(\beta,\gamma) : \beta \tilde{\in} \mathrm{OR}^{\tilde{M}} \wedge \gamma \tilde{\in} \#_\beta\}$ coded as a class of ordinals of $\tilde{M}$; thus $\#$ is $\tilde{M}$-definable. Then, in $\tilde{M}$, we can carry out the construction of $L[\#]$ all the way through the ordinals of $\tilde{M}$, i.e., for each α

$$(L_\beta[\#] : \beta < \alpha),\ (\tilde{\in} \mid L_\beta[\#] : \beta < \alpha),\ L_\alpha[\#]$$

are members of $\tilde{M}$ and enjoy the usual uniform definability properties (in the $\tilde{M}$-definition of $\#$) over $\tilde{M}$, as do the classes

$(L_\beta[\#]: \beta \in OR)$, $L[\#]$.

This immediately yields that, in $\tilde{M}$, $L[\#]$ satisfies extensionality, foundation, pairing, union, infinity, collection, Δ_0-separation, every set is well-ordered, and there's a definable (over $\tilde{M}$) one-one map of OR onto $L[\#]$. Also, $L[\#]$ clearly satisfies $\forall x(x \subseteq OR \Rightarrow x^\#$ exists). We need only verify the power-set axiom for $L[\#]$, in $\tilde{M}$. We exploit the normality of $\vec{\beta}$ one last time.

Suppose, towards a contradiction, that, in $\tilde{M}$, for some $\alpha \in OR^{\tilde{M}}$, $P(\alpha) \cap L[\#] \notin L[\#]$. Then, there's a least such α in $\tilde{M}$ and this α is first-order definable in $\tilde{M}$, and therefore $\alpha < \beta_0$. Further, we have an $L[\#]$-definable and therefore first-order-definable-over-$\tilde{M}$ sequence of <u>distinct</u> subsets $(y_\xi : \xi \in OR^{\tilde{M}})$ of $L[\#]$-subsets of α. In particular, for $i < j \in \omega$, $y_{\beta_i} \neq y_{\beta_j}$, so for some $\gamma \tilde{\in} \alpha$ $(\gamma \tilde{\in} y_{\beta_i} \Leftrightarrow \gamma \tilde{\notin} y_{\beta_j})$. The least such γ then yields a 2-supported-first-order definable pressing down function, which cannot be constant by an argument similar to that for (2) in (3.3.9), but easier. This contradicts the normality of $\vec{\beta}$. We should remark that this last argument could be carried out in a somewhat more general context.

(3.3.11) We conclude §3.3 by providing a proof of the lemma of (3.3.7). We first show that $\vec{\gamma}$ is strictly increasing. If not, then since F is non-constant, by indiscernibility, $\vec{\gamma}$ would be strictly decreasing. But $\vec{\gamma}$ is nicely definable over $\tilde{M}^+$, so applying comprehension, we would obtain, in $\tilde{M}$, a decreasing sequence of $\tilde{M}$-ordinals which is impossible. Now let $\vec{\gamma}'$ be any finite deletion subsequence of $\vec{\gamma}$, say $\gamma'_n = \gamma_{i_n}$. Then $\vec{\gamma}'$ is obtained from $\vec{\beta}'$ exactly as $\vec{\gamma}$ was obtained from $\vec{\beta}$, where for all n, and all $j < k$, $\beta'_{n+j} = \beta_{i_n+j}$ (note: $\vec{\beta}'$ is finite deletion, since $\vec{\gamma}'$ is !). Thus, in a uniform way, $(\tilde{M},\vec{\gamma}')$ models φ can be rewritten as $(\tilde{M},\vec{\beta}')$ models φ'. Clearly then $\vec{\gamma}$ is a sequence of medium L_B-indiscernibles over $\tilde{M}$.

(3.4) We now develop a sufficient condition on a space G to have $F(G) \Leftrightarrow F(T)$ and $G(G) \Leftrightarrow G(T)$. For convenience, our condition will be stated for $G \subseteq B$, but can clearly be readily transferred to the spaces of concern ($B_2, B_2\times\omega, B_2\times B\times\omega$, for products, semigroups, groups respectively). The general setting is as follows. Suppose $G \subseteq B$ is a Borel space. As usual, G_f will consist of the finitely generated elements of G. Suppose further that $\leq$ is a Borel pre-order on B and $\sim$ is a Borel equivalence relation on B such that $a \sim b \Rightarrow a \leq b$ and G_f is closed for $\sim$-equivalence. We have $\leq$ playing the role of isomorphic embeddability and $\sim$ playing the role of isomorphism, respectively, between finitely generated structures (which is why these relations can be taken to be Borel rather than analytic). We seek a Borel transformation of Turing degrees into elements of G_f (actually elements of B into elements of G_f in an invariant fashion); this will be our σ, below. We also need a Borel map back, τ below. In practice, this will turn out to be the equational theory of a set of generators. We require σ,τ to preserve enough information; these are the properties a)-c), in the lemma below. In this abstract context, F(G), G(G) are the propositions involving $\leq$ in place of isomorphic embeddability and $\sim$ in place of isomorphism. We use $a,b,\vec{a},\vec{b}$ as variables over B, B^ω, and $x,y,\vec{x},\vec{y}$ as variables over G_f, G_f^ω. As usual $\vec{x} \sim \vec{y}$ means for all n, $x_n \sim y_n$. Similarly, $\sigma(\vec{a})$ means $(\sigma(a_n): n \in \omega)$, $\tau(\vec{x})$ means $(\tau(x_n): n \in \omega)$.

LEMMA: Let $\leq$, $\sim$ be as above. Suppose further that there are Borel functions $\sigma,\tau: B \to B$ with range $\sigma \subseteq G_f$ such that:

a) $a \equiv_T b \Rightarrow \sigma(a) \sim \sigma(b)$

b) τ is monotone, i.e. $x \leq y \Rightarrow \tau(x) \leq_T \tau(y)$

c) for all $x \in G_f$, $b \in B$:
  i) $\tau(x) \leq_T b \Rightarrow x \leq \sigma(b)$,
  ii) $\sigma(b) \leq x \Rightarrow b \leq_T \tau(x)$.

Then: F(G) <=> F(T) and G(G) <=> G(T).

Proof: Suppose the hypotheses of the lemma hold and suppose that $F: B^\omega \to B$ is Borel and Turing-invariant, $H: G_f^\omega \to G_f$ is Borel and $\sim$-invariant. We let

$\tilde{F}: G_f^\omega \to G_f$, $\hat{H}: B^\omega \to B$ be defined by:

$$\tilde{F}(\vec{x}) = \sigma F \tau(\vec{x})$$
$$\hat{H}(\vec{a}) = \tau H \sigma(\vec{a}).$$

Clearly $\tilde{F},\hat{H}$ are Borel. Also, by a) (for σ) and two applications of b) (for τ), σ,τ are both invariant, so $\tilde{F},\hat{H}$ are

$\sim$, $\equiv_T$ invariant respectively.

Now let $\vec{a},\vec{b} \in B^{\omega}$, $\vec{x},\vec{y} \in G_f^{\omega}$, $m,n \in \omega$ and suppose that $\hat{H}(\vec{a}) \leq_T b_m$ and $\tilde{F}(\vec{x}) \leq y_n$, i.e. $\tau H\sigma(\vec{a}) \leq_T b_m$ and $\sigma F\tau(\vec{x}) \leq y_n$. By c), $H\sigma(\vec{a}) \leq \sigma(b_m)$ (by i)) and $F\tau(\vec{x}) \leq_T \tau(y_n)$ (using ii)). We conclude easily that $F(G) \Leftrightarrow F(T)$, since if $\hat{H}(\vec{b}) \leq_T b_m$, then, taking $\vec{a} = \vec{b}$, $H\sigma(\vec{b}) \leq \sigma(b_m)$, so $F(T) \Rightarrow F(G)$. Similarly, if $\tilde{F}(\vec{y}) \leq y_n$, then, taking $\vec{x} = \vec{y}$, $F\tau(\vec{y}) \leq \tau(y_n)$ so $F(G) \Rightarrow F(T)$. The proof that $G(T) \Leftrightarrow G(G)$ is similar, once we remark that if $\vec{b} \in B^{\omega}$, then $\{\sigma(\vec{a})$: $\vec{a}$ is a subsequence of $\vec{b}\}$ is the set of all subsequences of $\sigma(\vec{b})$, and that if $\vec{y} \in G_f^{\omega}$, then $\{\tau(\vec{x})$: $\vec{x}$ is a subsequence of $\vec{y}\}$ is the set of all subsequences of $\tau(\vec{y})$. Now we apply the above with $\vec{a}$ an arbitrary subsequence of $\vec{b}$, $\vec{x}$ an arbitrary subsequence of $\vec{y}$.

We make several remarks. First, c) implies:

c') for all $x \in G_f$, for all $b \in B$, $x \leq \sigma\tau(x)$ and $b \leq_T \tau\sigma(b)$; this is clear.

Further, b) plus c') for $\tau\sigma$ implies c),i), since if $\sigma(b) \leq x$, then, by b), $\tau\sigma(b) \leq_T \tau(x)$, while by c') for $\tau\sigma$, $b \leq_T \tau\sigma(b)$, yielding $b \leq_T \tau(x)$. Similarly, if instead of b), we have:

b') τ and σ are both monotone,

then b') for σ plus c') for $\sigma\tau$ implies c),ii). We shall verify a), b'), c') for G = semigroups, products; for G = groups, we verify a), b), i) of c) and c') for $\tau\sigma$. This suffices by the above.

Secondly, we note that our machinery will actually show:

(*): if $x \in G$, x is recursive in b, then $x \leq \sigma(b)$.

Note that $x \in G$, perhaps $x \notin G_f$. By "x is recursive in b", we mean that the product operation of x is recursive in b (in the case of G = groups, it's enough to show this since, if so, the inverse automatically is, because $z^{-1} = \mu y[zy=e]$). This will be done in (3.5.9).

(3.5) We proceed as follows. In (3.5.2) we prove the sufficient condition of (3.4) for semigroups. In (3.5.3) we do the same for products. In (3.5.4), (3.5.5) we prove two group-theoretic lemmas, prior to proving the sufficient condition of (3.4) for groups, in (3.5.6) - (3.5.8). This will complete the proof of Reversal 3, and will provide a proof of the analogue of Theorem 3 for the Propositions $F(G)$, $G(G)$. In (3.9), we show how the proofs of Theorem 3', in (3.1), (3.2), can be modified to prove Theorem 3 as stated. The group-theoretic arguments of

(3.5.4) - (3.5.8) seem interesting in their own right. In particular, the theorem of (3.5.8) strengthens a theorem of Higman, Chapter IV, Theorem 7.3 of [09].

(3.5.1) Before embarking on this program, we give a unified treatment, for all of our spaces, G, of the map τ of the lemma of (3.4). Recall that $\leq$ is isomorphic embeddability and $\sim$ is isomorphism. Since G_f is a Borel subset of the closed subspace G of the appropriate product space, we restrict to defining τ on G_f, filling in arbitrarily off G_f.

Actually, we shall define τ on finitely generated structures of the appropriate variety, even for those which have not been isomorphed onto ω. So, let x be such a structure, let $a_1,\ldots,a_p \in x$ and, as in (1.2), let $E_{a_1\ldots a_p} = E^x_{a_1\ldots a_p}$ be the set of (natural number codes of) equations in $v_1,\ldots,v_p$ which are satisfied by $a_1,\ldots,a_p$ in x. If the language for our variety has a symbol for the identity, this symbol is allowed as a term in our equations. Our main observation, which is well-known in the folklore, is:

LEMMA: Suppose $h: y \to x$ is an isomorphic embedding, $c_1,\ldots,c_m \in y$ and $h(c_1),\ldots,h(c_m)$ are members of the substructure of x generated by $a_1,\ldots,a_p$. Then $E^y_{c_1\ldots c_m} \leq_T E^x_{a_1\ldots a_p}$.

Proof: For $1 \leq i \leq m$, let $t_i(v_1,\ldots,v_p)$ be such that $h(c_i) = t^x_i(a_1,\ldots,a_p)$. If e is an equation in $v_1,\ldots,v_m$, rewrite e as e' by simultaneously substituting t_i for each occurrence of v_i, for $1 \leq i \leq m$. Clearly, the map $e \mapsto e'$ is primitive recursive; let g be the associated function on codes. Then

$$i \in E^y_{c_1\ldots c_m} \iff g(i) \in E^x_{a_1\ldots a_p},$$

since h is an isomorphic embedding. But then $E^y_{c_1\ldots c_m}$ is recursive in $E^x_{a_1\ldots a_p}$, as required.

COROLLARY 1: If $\{a_1,\ldots,a_p\}$ and $\{c_1,\ldots,c_m\}$ both generate x, then $E^x_{a_1\ldots a_p} \equiv_T E^x_{c_1\ldots c_m}$.

Thus, we define τ from finitely generated structures to $P(\omega)$ (and thence to B in a standard fashion) by

$\tau(x) = E^x_{a_1 \ldots a_p}$ where $(a_1, \ldots, a_p)$ is the first finite sequence (relative to a fixed enumeration of $\omega^{<\omega}$ in type ω) such that $\{a_1, \ldots, a_p\}$ generates x.

COROLLARY 2: If x, y are finitely generated structures and $y \preccurlyeq x$, then $\tau(y) \leq_T \tau(x)$.

This gives b) for τ of the lemma of (3.4).

(3.5.2) Let $b \in B$. Consider $\sigma'(b)$, the semigroup of partial b-recursive functions under composition, with the identity function as identity element. We shall show that this is finitely generated, and we shall take $\sigma(b)$ to be some isomorph of $\sigma'(b)$ onto ω, uniformly obtained in a Borel fashion from b (for the record one such isomorph is obtained by first identifying a partial b-recursive function φ with its least b-index, and then bijecting the resulting set of indices with ω; this can be done effectively in some finite number of Turing jumps of b). However, in proving a), c') and b') for σ of the Lemma of (3.4) for semigroups, we shall work directly with σ'.

LEMMA A: $\sigma'(b)$ is finitely generated.

Proof: We let φ be the partial b-recursive function defined by: $\varphi(n) = m$ iff n is of the form $2^r 3^s$ (otherwise φ is undefined) and $\varphi^b_r(s) = m$. Then, for each r, $\varphi^b_r = \varphi \circ (\lambda s(2^r 3^s))$. But each $\lambda s(2^r 3^s)$ is generated from $d(n) = 2n$ and $h(s) = 3^s$ as h followed by the r-fold iteration of d. Thus, $\{\varphi, d, h\}$ generates $\sigma'(b)$.

LEMMA B: i) $b \leq_T b' \Rightarrow \sigma'(b) \preccurlyeq \sigma'(b')$,

ii) $b \equiv_T b' \Rightarrow \sigma'(b) \cong \sigma'(b')$.

Proof: Both statements are immediate, since if $b \leq_T b'$ then $\sigma'(b)$ is literally a subsemigroup of $\sigma'(b')$.

LEMMA C: i) $b \leq_T \tau\sigma'(b)$, for all $b \in B$,

ii) $x \preccurlyeq \sigma'\tau(x)$ for all finitely-generated semigroups x.

Proof: For i), our main observation is that $b(n) = m$ iff $b \circ \gamma_n = \gamma_m$, where, for $k \in \omega$, γ_k is the constant function on ω with value k. It's easily seen that there's total recursive $c\colon \omega \to \omega$ such that for all $k \in \omega$, $\gamma_k = \varphi^b_{c(k)}$. Unpacking the argument of Lemma A, we define by recursion on $k \in \omega$: $t'_0 = v_3$; $t'_{k+1} = v_2 \circ v_3$. Then, letting t_k be $v_1 \circ t'_k$, for all $k \in \omega$,

$t_k^{\sigma'(b)}(\varphi,d,h) = \varphi_k^b$. Let t^* be a term in v_1,v_2,v_3 such that $(t^*)^{\sigma'(b)}(\varphi,d,h) = b$. Finally, we clearly have a total recursive $p\colon \omega \times \omega \to \omega$ such that for all $n,m \in \omega$, $p(n,m)$ codes the equation: $t_{c(m)} = t^* \circ t_{c(n)}$. But then $b(n) = m$ iff $p(n,m) \in \tau\sigma'(b)$, so b is recursive in $\tau\sigma'(b)$.

For ii), let $x = (|x|,g,e)$ be a finitely generated semigroup (we shall abuse notation by writing x for $|x|$, as usual) and suppose the sequence of generators $(a_1,\ldots,a_p)$ was used to define $\tau(x)$. Let T be the set of terms in $v_1,\ldots,v_p$ in the language of x. We shall abuse notation by regarding $t \in T$ both as a term and as an integer code for the term. For $t,t' \in T$, let $\tilde{g}(t,t') = t'$, if the equation "$t = \bar{e}$" is a member of $\tau(x)$; otherwise, let $\tilde{g}(t,t')$ = the least t'' such that the equation "$t'' = \bar{g}(t,t')$" is a member of $\tau(x)$. Clearly $\tilde{g}$ is a total $\tau(x)$-recursive function and therefore so is each of its sections, $(\lambda t')\tilde{g}(t,t')$, for $t \in T$.

Note further that by construction, $(\lambda t')\tilde{g}(\bar{e},t')$ is the identity function and that:

(*) if $t_1,t_2 \in T$ and $t_1^x(a_1,\ldots,a_p) = t_2^x(a_1,\ldots,a_p)$, then $(\lambda t')\tilde{g}(t_1,t') = (\lambda t')\tilde{g}(t_2,t')$.

This is because if $t_1^x(a_1,\ldots,a_p) = t_2^x(a_1,\ldots,a_p)$, then the equation "$t_1 = t_2$" $\in \tau(x)$. Thus, "$t_1 = \bar{e}$" $\in \tau(x)$ iff "$t_2 = \bar{e}$" $\in \tau(x)$ and for all $t',t'' \in T$, "$t'' = \bar{g}(t_1,t')$" $\in \tau(x)$ iff "$t'' = \bar{g}(t_2,t')$" $\in \tau(x)$.

We then define $\pi\colon x \to \sigma'(\tau(x))$ by:

$$\pi(t^x(a_1,\ldots,a_p)) = (\lambda t')\tilde{g}(t,t').$$

The presence of the identity, e, guarantees that π is one-to-one since for all $t \in T$, the equation "$\bar{g}(t,\bar{e}) = t$" $\in \tau(x)$, and thus, if $t_1^x(a_1,\ldots,a_p) \neq t_2^x(a_1,\ldots,a_p)$, then the equation "$t_1 = t_2$" $\notin \tau(x)$, and therefore for no $t'' \in T$ are both of the following equations in $\tau(x)$: "$t'' = \bar{g}(t_1,\bar{e})$", "$t'' = \bar{g}(t_2,\bar{e})$". This guarantees that $(\lambda t_1)\tilde{g}(t_1,t') \neq (\lambda t_2)\tilde{g}(t_2,t')$. Similarly, using the fact that for all $t_1, t_2, t' \in T$, the equation expressing the instance of the associative law for $(\bar{g}, t_1, t_2 t')$ is a member of $\tau(x)$ guarantees that for all $t_1, t_2, t', t'' \in T$, $e_1 \in \tau(x)$ iff $e_2 \in \tau(x)$, where e_1 is "$t'' = \bar{g}(\bar{g}(t_1, t_2), t'))$" and e_2 is "$t'' = \bar{g}(t_1, \bar{g}(t_2, t'))$". But this means that for all $t_1, t_2 \in T$, $(\lambda t')\tilde{g}(\bar{g}(t_1, t_2), t') = (\lambda t')\tilde{g}(t_1, t') \circ (\lambda t')\tilde{g}(t_2, t')$, i.e. π is a homomorphism. But then π is an isomorphic embedding of x in $\sigma'\tau(x)$.

Lemma B establishes a), and b') for σ' of the lemma of (3.4); Lemma C establishes c').

(3.5.3) We now work with G = products = B_2. If $b \in B$, once again, let $(\varphi_e^b : e \in \omega)$ be the natural enumeration of the unary partial b-recursive functions. We define $\sigma(b)$ to be the product on ω given by: $\sigma(b)(0,n) = \sigma(b)(n,0) = 0$; $\sigma(b)(e+1, n+1) = \varphi_e^b(n)$, if $\varphi_e^b(n)$ is defined; otherwise $\sigma(b)(e+1, n+1) = 0$. The intuition is that 0 is a special element for "undefined".

<u>LEMMA A</u>: $\sigma(b)$ is finitely generated.

<u>Proof</u>: If i is a b-index of the successor function then $\{0,1,i\}$ is a set of generators.

Our main tool in proving Lemmas B, C, analogous to those of (3.5.2), will be Friedman's analysis, [02], of BRFT's (basic recursive function theories). We refer the reader to [02] for details; we sketch only that portion of the material needed for our purposes, and only in the special cases of interest to us.

An ω-BRFT is a system (ω, F, φ_n) such that $F = \bigcup\{F_n : 1 \leq n \in \omega\}$, each F_n a countable collection of n-ary partial functions $\varphi : \omega^n \to \omega$ with $\varphi_n \in F_{n+1}$, φ_n an enumeration of F_n, such that the successor function is a member of F_1. In addition, F must contain all the constant functions, projections, a four-place function $\lambda abcx(b \text{ if } x=a;\ c \text{ if } x \neq a)$; F must be closed under generalized composition, and have abstract versions of the functions of the s-m-n theorem, i.e., for $m, n > 0$, functions $S_n^m \in F_{m+1}$ such that for all $x, x_1, \ldots, x_m$, we have $\lambda y_1 \ldots y_n(\varphi_{m+n}(x, x_1, \ldots, x_m, y_1, \ldots, y_n)) =$
$\lambda y_1 \ldots y_n(\varphi_n(S_n^m(x, x_1, \ldots, x_m), y_1, \ldots, y_n))$. Naturally, the paradigms are all BRFT's, that is, the structures where for some $\alpha \in B$, F consists of all the partial α-recursive functions (of any number

of arguments), and the φ_n's are the natural enumerating functions. We refer to this as the standard α-recursive BRFT. One of the main results of [02] is that, up to isomorphism, a BRFT is determined by its F; in other words, the choice of enumerating functions is immaterial, as long as they have the S_n^m-property. This result, Theorem 2.5 of [02], will be our main tool; we shall state it more precisely below, and specify the notion of isomorphism involved.

Closely related to the notion of BFRT is that of an enumerative system (ES), a system (ω, F) where F is a countable collection of unary and binary partial functions, closed under generalized composition, containing all the constant functions, the identity function, the successor function, a binary enumerating function for the set of unary functions in F, all unary constant functions, a binary pairing function with its unary co-ordinate inverses, and each of the binary functions $E_{z,w}$ for $z,w \in \omega$, $z \neq w$, where $E_{z,w} = \lambda xy(z \text{ if } x=y;\ w \text{ if } x \neq y)$. Friedman also proves, Theorem 1.1 of [02], that if (ω, G) is an ES, then (ω, G) can be extended to a BRFT (ω, H, φ_n) with $G_i = H_i$, $i = 1,2$. Further, the proof shows that, due to the presence of the pairing apparatus in an ES, if $a \in B$ and G_i = the set of all i-place partial a-recursive functions, $i = 1,2$, then for all $1 \leq k \in \omega$, H_k = the set of all k-place partial a-recursive functions.

Now let γ be a binary partial α-recursive function. We describe an embedding, π, from (ω, γ) to (ω, ψ), where ψ is the binary partial (α-recursive) enumerating function of an ES (ω, G), with G_i = the set of all i-place partial α-recursive functions, $i = 1,2$. Let $\pi(n) = 2n$, let φ be the standard binary enumerating partial α-recursive function for the unary partial α-recursive functions. Let $\psi(2n+1,m) \simeq \varphi(n,m)$; let $\psi(2n,2m+1)$ be undefined and let $\psi(2n,2m) \simeq \gamma(n,m)$. Clearly π, ψ are as required.

We now specify Friedman's notion of isomorphism of BRFT's, which he called piecewise isomorphism (Definitions 1.7, 1.8 of [02]), and state the above cited isomorphism theorem.

<u>DEFINITION</u>: If (ω, F, φ_n), (ω, F, ψ_n) are BRFT's, then $(G_m : 1 \leq m \in \omega)$ is an isomorphism of (ω, F, φ_n) and (ω, F, ψ_n) iff for each m, G_m is a permutation of ω lying in F_1 such

that for all $x, x_1, \ldots, x_m \in \omega$,
$G_m(\varphi_m(x, x_1, \ldots, x_m)) \simeq \psi_m(G_m(x), G_m(x_1), \ldots, G_m(x_m))$.

THEOREM (Friedman, Theorem 2.5 of [02]): Whenever (ω, F, φ_n), (ω, F, ψ_n) are BRFT's, there's $(G_m : 1 \leq m \in \omega)$ which is an isomorphism of (ω, F, φ_n) and (ω, F, ψ_n).

Now, combining Friedman's theorems, 1.1 and 2.5 of [02], with our embedding, above, of g into an ES, we have:

LEMMA A: Suppose $\alpha \in B$, g is a binary partial α-recursive function. Then, there's an embedding $\pi' : (\omega, g) \to (\omega, \varphi^{\alpha,1})$, where $\varphi^{\alpha,1}$ is the standard binary partial α-recursive enumerating function for the unary partial α-recursive functions.

Proof: First, embed (ω, g) into (ω, ψ), ψ the binary partial enumerating function for the unary partial α-recursive functions of an ES, as above. Next, extend this ES to a BRFT, (ω, H, ψ_n), where, for $1 \leq k \in \omega$, H_k is the set of k-place partial α-recursive functions, $\psi_1 = \psi$ (by Theorem 1.1 of [02]). Finally, let $(G_m : 1 \leq m \in \omega)$ be an isomorphism of (ω, H, ψ_n) with the standard α-recursive BRFT. Then $\pi' = G_1 \circ \pi$ is as required.

We now define $\sigma(b)$. Let $\varphi = \varphi^{b,1}$, and define $\theta = \theta(b)$, a binary partial α-recursive function, by $\theta(m,n)$ is undefined, if $m = 0$ or $n = 0$; $\theta(m+1, n+1) \simeq \varphi^{b,1}(m,n)$. We then define $\sigma = \sigma(b)$ by: $\sigma(m,n) = \theta(m,n)+1$, if $\theta(m,n)$ is defined; otherwise $\sigma(m,n) = 0$.

REMARK: $\sigma(b)$ is Turing-equivalent to the Turing jump of b.

LEMMA B: $\sigma(b)$ is finitely-generated.

Proof: $\{0, 1, i+1\}$ is a generating set, where i is a b-index of the successor function.

LEMMA C: i) $b \leq_T b' \Rightarrow \sigma(b) \preccurlyeq \sigma(b')$

ii) $b \equiv_T b' \Rightarrow \sigma(b) \cong \sigma(b')$

Proof: ii) is by Friedman's Theorem 2.5 of [02]. For i), we use Lemma A, above, with $\gamma = \varphi^{b,1}$, $\alpha = b'$. Then, letting π' be as guaranteed by Lemma A, let $\pi''(0) = 0$, $\pi''(n+1) = \pi'(n)+1$. Then π'' is as required.

LEMMA D: i) $b \leq_T \tau\sigma(b)$, for all $b \in B$,

ii) $x \preccurlyeq \sigma\tau(x)$, for all $x \in B_2$.

Proof: For i), we use the remark that $\sigma(b)$ is Turing equivalent to the Turing-jump of b, so we let j be a $\sigma(b)$-index of b. Using the generating set of Lemma A, and the approach of (3.5.2) to obtain, for $k \in \omega$, canonical terms $\overline{k}$ such that $\overline{k}^{\sigma(b)}(0,1,i) = k$, and a total recursive $h: \omega \times \omega \to \omega$, such that for all m,k, h(m,k) is a code for the equation: $\overline{k+1} = \overline{\sigma}(\overline{j+1},\overline{m+1})$. So $k = b(m)$ iff $h(m,k) \in \tau\sigma(b)$.

For ii), we let $\tilde{g}$ be the binary function $\tilde{g}(t,t') =$ the least t" such that the equation "t" = $\overline{x}(t,t')$", where t,t',t" are (codes of) x-terms in $v_1,\ldots,v_p$. As before, $\tilde{g}$ is $\tau(x)$-recursive. Now (ω,x) embeds in $(\omega,\tilde{g})$, by $\pi(k) = \mu t(k=t^x(a_1,\ldots,a_p))$ (here, ω is identified with the set of x-terms in $v_1,\ldots,v_p$). Since $\tilde{g}$ is recursive in $\tau(x)$, by Lemma A with $\gamma = \tilde{g}$, $\alpha = \tau(x)$, $(\omega,\tilde{g})$ embeds in $\varphi^{\tau(x),1}$ and then (by the successor function) into $\sigma\tau(x)$. So (ω,x) embeds in $(\omega,\sigma\tau(x))$ as required.

(3.5.4) In this subsection and in (3.5.5), we prove two group-theoretic lemmas which will permit us to define, in (3.5.6), a Borel function $H: G^\omega \to G_f$, with nice properties. In (3.5.7) we use H to define our $\sigma: B \to G_f$ and in (3.5.8) we verify it satisfies the hypotheses of the lemma of (3.4). This will complete the proofs of transfer, and therefore of Reversal 3 and Theorem 3 for the propositions F(G), G(G). In (3.5.9), we finish the proof of Theorem 3 for the propositions F'(G), G'(G); when G = groups, we draw upon the material of this subsection and (3.5.6), (3.5.7).

We write X! for the group of permutations of a (non-empty) set X. (X!)* denotes the set of permutations of X which have infinitely many orbits. $f \in X!$ is regular if $f \in (X!)^*$ and each orbit of f is infinite.

LEMMA: If $f \in (\omega!)^*$ then there are regular $g,h \in \omega!$ such that $f = g \circ h$. Further, $f \mapsto (g,h)$ can be taken to be a Borel function.

Proof: The last assertion will be clear from our construction of g,h. We start from a fixed partition $(E_n: n \in \omega)$ of ω into infinite subsets, such that for all n, $f[E_n] = E_n$ (each E_n will be a union of orbits). We construct g,h through finite approximations, α,β, one-to-one partial functions from ω to ω, which satisfy:

a) for all n, $\alpha[E_n]$, $\beta[E_n] \subseteq E_n$,

b) α, β have no cycles,

c) $\alpha(\beta(n)) = m \Rightarrow f(n) = m$,

d) dom α = range β.

It will suffice to show that if α, β satisfy a) - d) (from now on, we'll say (α, β) is nice), then for all n, there's nice (α', β'), $\alpha' \supseteq \alpha$, $\beta' \supseteq \beta$ with $n \in \text{dom } \alpha' \cap \text{dom } \beta' \cap \text{range } \alpha'$ (if we can find such α', β', then $n \in \text{range } \beta'$ since (α', β') is nice, and by d)).

Given n, let $n \in E_q$. If $n \notin \text{dom } \alpha$, by d), $n \notin$ range β; let $m \in E_q \setminus (\text{dom } \alpha \cup \text{dom } \beta \cup \{n\}$; let $\beta'(m) = n$, $\alpha'(n) = f(m)$. Note that $f(m) \notin \text{range } \alpha$, since if $f(m) = \alpha(k)$, then, for some ℓ, $k = \beta(\ell)$, but then by c), $\alpha(\beta(\ell)) = f(\ell)$, so $m = \ell$, contradiction. This guarantees that b) is preserved, so this procedure preserves niceness. Thus, we may assume that $n \in \text{dom } \alpha$.

If $n \notin \text{dom } \beta$, then, by an argument similar to that above, $f(n) \notin \text{range } \alpha$. Choose $m \in E_q \setminus (\text{dom } \alpha \cup \text{dom } \beta \cup \{n\})$, and set $\beta'(n) = m$, $\alpha'(m) = f(n)$. This preserves niceness, so we may assume $n \in \text{dom } \beta$.

Finally, if $n \notin \text{range } \alpha$, as above, $f^{-1}(n) \notin \text{dom } \beta$. Choose $m \in E_q \setminus (\text{dom } \alpha \cup \text{dom } \beta \cup \{n\})$. Set $\beta'(f^{-1}(n)) = m, \alpha'(m) = n$.

(3.5.5) As a first step towards obtaining a strong, uniform, effective version of a theorem of Higman in (3.5.8), below, we show how a theorem of Higman, Neumann and Neumann, Chapter IV, Theorem 3.1 of [09], can be strengthened.

<u>LEMMA</u>: Every $G \in \mathcal{G}$ embeds in a 2-generated $G' \in \mathcal{G}_f$ with generators α, β, $\alpha \neq \beta$ such that $\lambda x(\alpha x)$, $\lambda x(\beta x) \in (\omega!)^*$. Further, (G', α, β) can be taken to be a Borel function of G.

<u>Proof</u>: The theorem of [09], mentioned above, gives that any $H \in \mathcal{G}$ can be embedded in a 2-generated $G' \in \mathcal{G}_f$. Take H isomorphic to the direct product of our given G with the direct sum of infinitely many copies of $\mathbb{Z}_2$. Embed this H in $G' \in \mathcal{G}_f$ with generators α, β.

Thus, G' has infinitely many elements $g_n (n \in \omega)$, each self-inverse and such that for all n, m, $g_n g_m = g_m g_n$. Now, if α is of finite order, each of the orbits of $\lambda x(\alpha x)$ are finite, and so, in this case α has infinitely many orbits. If α has infinite order, we show that if $m \neq n$ then g_m, g_n are in distinct orbits of $\lambda x(\alpha x)$. If not, say $m \neq n$ and g_m, g_n are in the same orbit. But then $g_m g_n^{-1} = g_m g_n$ is a power of α. But $g_m g_n$ has order two and so α has finite order, contradiction. The argument for β is similar.

(3.5.6) We work in the following context, until (3.5.9). Our aim is to prove:

<u>THEOREM</u>: There's a Borel function $\mathcal{H}: \mathcal{G}^{\omega} \to \mathcal{G}_f$ such that for all $\vec{G} \in \mathcal{G}^{\omega}$, all $n \in \omega$, $G_n \preccurlyeq \mathcal{H}(\vec{G})$. Further, if H is any finitely generated subgroup of $\omega!$, then $\mathcal{H}$ can be taken to be invariant under the action of elements of H, i.e., if $f \in H$ then $\mathcal{H}(\vec{G}) \cong \mathcal{H}(\vec{G} \circ f)$.

Note that by the theorem of [09], $\mathcal{H}$ can be taken to be invariant under the action of elements of H where H is any fixed countable subgroup of $\omega!$. The theorem, as stated, is clearly equivalent to:

<u>THEOREM'</u>: There's a Borel function $\mathcal{H}: \mathcal{G}^{\mathbb{Z}} \to \mathcal{G}_f$ such that for all $\vec{G} \in \mathcal{G}^{\mathbb{Z}}$, all $i \in \mathbb{Z}, G_i \preccurlyeq \mathcal{H}(\vec{G})$. Further, if H is any countable subgroup of $\mathbb{Z}!$ then $\mathcal{H}$ can be taken to be invariant under the action of elements of H.

It will be somewhat more convenient to prove the Theorem'. In (3.5.8) we shall apply it with H a countable subgroup of $\mathbb{Z}!$ given by recursion theory, cf. the Proposition of (3.5.7)

In proving the Theorem', given $\vec{G} \in \mathcal{G}^{\mathbb{Z}}$, we shall abuse notation by writing simply xy, x^{-1} for the group operations of G_i, when i is clear from context. Also, we shall get $\mathcal{H}(\vec{G})$ as a group of permutations of $\Omega = \mathbb{Z} \times \omega$; accordingly, we shall prefer to regard G_i as a group on $\omega_i = \{i\} \times \omega$ via the identification of n and (i,n). Also, we shall sometimes regard G_i as a group of permutations of ω_i via the identification of n and $\lambda(i,m)((i,n)(i,m))$. We let $B(i) = \omega_i^{\omega_i}$

<u>Proof of Theorem'</u>: We first note that we may assume that $\mathcal{H}$ is defined on sequences of the form $\vec{G}' =$ $((G_i', \alpha_{i,1}, \alpha_{i,2}, \beta_{i,1}, \beta_{i,2}): i \in \mathbb{Z})$, where, letting $\alpha_i = \alpha_{i,1}\alpha_{i,2}$, $\beta_i = \beta_{i,1}\beta_{i,2}$, G_i' is generated by $\{\alpha_i, \beta_i\}$, $\lambda x(\alpha_i x), \lambda x(\beta_i x) \in (\omega!)^*$ and $\lambda x(\alpha_{i,j} x)$, $\lambda x(\beta_{i,j} x)$ are regular. This is by (3.5.4),

(3.5.5), since given $\vec{G}$ we can obtain such a $\vec{G}'$ as a Borel function of $\vec{G}$, where, for $i \in \mathbb{Z}$, $G_i \preccurlyeq G_i'$. But then, if each $G_i' \preccurlyeq H(\vec{G}')$, we also have each $G_i \preccurlyeq H(\vec{G}')$, so we can take $H(\vec{G}) = H(\vec{G}')$. Further, if $f \in H$ then the $\vec{G}'$ obtained from $\vec{G} \circ f$ is $\vec{G}' \circ f$, so invariance for $H(\vec{G}')$ guarantees invariance for $H(G)$ (under the action of elements of H, in both cases).

We next note that we can assume that $\vec{G}'$ comes equipped with sequences $s_{i,j}, t_{i,j} (i \in \mathbb{Z}, j=1,2)$ such that $s_{i,j}$ (respectively $t_{i,j}$) is a choice function on the orbits of $\alpha_{i,j}$ (respectively $\beta_{i,j}$). Formally, $s_{i,j}$, for example, will lie in B(i) and for $m,n \in \omega$, if m,n are in the same orbit of $\lambda x(\alpha_{i,j} x)$, then $s_{i,j}(m) = s_{i,j}(n)$ is a member of this orbit. Once again, we can make this assumption because we can obtain such $s_{i,j}, t_{i,j}$ as a Borel function of $\vec{G}'$.

Now the $s_{i,j}, t_{i,j}$ give us canonical maps $\lambda_{i,j}, \mu_{i,j}$ such that (regarding the $\alpha_{i,j}, \beta_{i,j}$ as permutations of ω_i), $\lambda_{i,j} \circ \alpha_{i,j} \circ (\lambda_{i,j})^{-1} = \alpha_{i+1,j}$ and $\mu_{i,j} \circ \beta_{i,j} \circ (\mu_{i,j})^{-1} = \beta_{i+1,j}$; $\lambda_{i,j}, \mu_{i,j}$ are induced by the maps: $s_{i,j}(i,n) \mapsto s_{i+1,j}(i+1,n)$, $t_{i,j}(i,n) \mapsto t_{i+1,j}(i+1,n)$. More generally, $\lambda_{i,k,j}, \mu_{i,k,j}$ are the canonical conjugations of $\alpha_{i,j}, \beta_{i,j}$ into $\alpha_{k,j}, \beta_{k,j}$, induced by $s_{i,j}(i,n) \mapsto s_{k,j}(k,n)$, $t_{i,j}(i,n) \mapsto t_{k,j}(k,n)$. Note that these $\lambda_{i,k,j}, \mu_{i,k,j}$ form a commutative system.

We also let $\sigma_i : \omega_i \to \omega_{i+1}$ be given by $\sigma_i(i,n) = (i+1,n)$, and $\sigma_{i,k}(i,n) = (k,n)$. We let $\sigma \in \Omega!$ be given by $\sigma = \bigcup_i \sigma_i$. We let $\lambda^j, \mu^j \in \Omega!$ be given by $\bigcup_i \lambda_{i,j}, \bigcup_i \mu_{i,j}$. Let $\alpha^j, \beta^j \in \Omega!$ be given by $\alpha^j | \omega_0 = \alpha_{0,j}, \beta^j | \Omega \setminus \omega_0$ = the identity, $\beta^j | \omega_0 = \beta_{0,j}, \beta^j | \Omega \setminus \omega_0$ = the identity (regarding $\alpha_{0,j}$ as $\lambda(0,n)(\alpha_{0,j}(0,n))$.

Finally, let $f_1, \ldots, f_p$ generate H. We let $\tilde{f}_k \in \Omega!$ be given by: $f_k | \omega_i = \sigma_{if_k(i)}$, (where $\sigma_{\ell\ell}$ is the identity on ω_ℓ).

We let $H(\vec{G}')$ = the subgroup of $\Omega!$ generated by $\sigma, \lambda^1, \lambda^2, \mu^1, \mu^2, \alpha^1, \alpha^2, \beta^1, \beta^2, \tilde{f}_1, \ldots, \tilde{f}_p$.

We first show that each G_i' embeds in $H = H(\vec{G}')$. Note that for each $i \in \mathbb{Z}$ there are permutations $\lambda^*_{i,j}, \mu^*_{i,j}$ in the subgroup of $\Omega!$ generated by $\{\sigma, \lambda^j, \mu^j\}$ $(j=1,2)$, such that $\lambda^*_{i,j}(0,n) = \lambda_{i,j}(i,n), \mu^*_{i,j}(0,n) = \mu_{i,j}(i,n)$ for each $n \in \omega$. Define $\xi_{k,j}, \zeta_{k,j}$ by recursion on $k \in \omega$, $k \geq 1$: $\xi_{1,j} = \lambda^*_{0,j}, \zeta_{1,j} = \mu^*_{0,j}$; $\xi_{k+1,j} = \lambda^*_{k+1,j} \circ \xi_{k,j}, \zeta_{k+1,j} = \mu^*_{k+1,j} \circ \zeta_{k,j}$, and let $\gamma_{k,j}, \delta_{k,j}$ be $\xi_{k,j}\alpha^j(\xi_{k,j})^{-1}, \zeta_{k,j}\beta^j(\zeta_{k,j})^{-1}$. For $k < 0$, we let $\gamma_{k,j}, \delta_{k,j}$ be $(\xi_{\ell,j})^{-1}\alpha^j\xi_{\ell,j}, (\zeta_{\ell,j})^{-1}\beta^j\zeta_{\ell,j}$, where $\ell = |k|$. Clearly then, for all $k \neq 0$, all $n \in \omega$, $\gamma_{k,j}(0,n) = \alpha_{k,j}(k,n), \delta_{k,j}(0,n) = \beta_{k,j}(k,n), \gamma_{k,j}|\Omega \setminus \omega_0 = \delta_{k,j}|\Omega \setminus \omega_0 =$ the identity. Thus, G_i' embeds in H by the embedding induced by: $\alpha_{i,j} \mapsto \gamma_{i,j}, \beta_{i,j} \mapsto \delta_{i,j}$ (for $i \neq 0$), $\alpha_{0,j} \mapsto \alpha^j, \beta_{0,j} \mapsto \beta^j$.

We now show that H is invariant under the actions of elements of H. Clearly it suffices to show that H is invariant under the actions of $\tilde{f}_1, \ldots \tilde{f}_p$. This is essentially because we've added each $\tilde{f}_1, \ldots, \tilde{f}_p$ to the set of generators of $H(\vec{G}')$. So, suppose $1 \leq e \leq p$, let $f = f_e$ and let $\vec{G}'' = \vec{G}' \circ f$. We mean, by this, that not only is $G_i'' = G'_{f(i)}$, but also that $\alpha''_{i,j} = (i,n)$, where $\alpha_{f(i),j} = (f(i),n)$, $\beta''_{i,j} = (i,m)$, where $\beta_{f(i),j} = (f(i),m)$, $s''_{i,j} = \sigma^{-1}_{i,f(i)} \circ s_{f(i),j} \circ \sigma_{i,f(i)}$, $t''_{i,j} = \tau^{-1}_{i,f(i)} \circ t_{f(i),j} \circ \tau_{i,f(i)}$. Note that our last equations can be rewritten as $s''_{i,j} = \tilde{f}^{-1} \circ s_{f(i),j} \circ \tilde{f}$, $t''_{i,j} = \tilde{f}^{-1} \circ t_{f(i),j} \circ \tilde{f}$. This immediately yields that $(\lambda^j)'' = \tilde{f}^{-1} \circ \lambda^j \circ \tilde{f}$, $(\mu^j)'' = \tilde{f}^{-1} \circ \mu^j \circ \tilde{f}$. By construction, $(\alpha^j)'' = \tilde{f}^{-1} \circ \alpha^j \circ \tilde{f}$, $(\beta^j)'' = \tilde{f}^{-1} \circ \beta^j \circ \tilde{f}$. Finally, for $1 \leq \ell \leq p$, the generators $\tilde{f}_\ell$ are common to $H(\vec{G}')$, $H(\vec{G}'')$ since obviously $\sigma_{i,k} = \sigma''_{i,k} = \tilde{f}^{-1} \circ \sigma_{f(i),f(k)} \circ \tilde{f}$. Thus the generators of $H(\vec{G}'')$ all lie in $H(\vec{G}')$. The converse is proved by interchanging the roles of $\vec{G}', \vec{G}''$ and noting that $\vec{G}' = \vec{G}'' \circ f^{-1}$. This proves our Theorem'. Our assumptions, above, on $s''_{i,j}$, $t''_{i,j}$ are justified since they are obtained from $\vec{G}''$ via the same Borel function used to obtain $s_{i,j}, t_{i,j}$ from $\vec{G}'$. Note that we've shown not only isomorphism, but <u>equality</u> of $H(\vec{G}')$, $H(\vec{G}'')$.

(3.5.7) We now avail ourselves of a standard recursion theoretic fact.

PROPOSITION: There's a fixed sequence $(f_n : n \in \omega) \in (\omega!)^\omega$ such that for all $x,y \in B$ if $x \equiv_T y$ then for some $n \in \omega$, for all $e \in \omega$, $\varphi_e^x = \varphi_{f_n(e)}^y$.

If $n = \langle p,q \rangle$, then f_n is constructed via a back-and-forth argument so that for all x,y, if $y = \varphi_p^x$, $x = \varphi_q^y$, then $\varphi_e^x = \varphi_{f_n(e)}^y$, for all e.

We now define σ. We let H be a finitely generated subgroup of $\mathbb{Z}!$ containing all the $\hat{f}_n$, where $(f_n : n \in \omega)$ is as in the Proposition and $\hat{f}_n$ is the permutation of $\mathbb{Z}$ which fixes the negative integers and is f_n on ω. Let $b \in B$. Let $(x_n^b : n \in \omega)$ enumerate, in a standard b-effective way all structures $x = (\omega, \varphi, \theta, k)$, where φ is a binary, θ a unary partial b-recursive function and $k \in \omega$. Let $G_n^b = x_n^b$, if x_n^b is a group; if not, let G_n^b be a fixed, canonically chosen isomorph of $\mathbb{Z}$ onto a group on ω. If $i \in \mathbb{Z} \setminus \omega$, let G_i^b be the same isomorph of $\mathbb{Z}$. Let $\sigma(b) = H(\vec{G}^b)$.

(3.5.8) We now show how σ provides a strengthened, uniform effective version of Higman's Theorem, [09]. This will involve verifying (*) of (3.4) for G = groups. We also verify a), b), c) of (3.4).

LEMMA A: If $b \equiv_T b'$ then $\sigma(b) \cong \sigma(b')$.

Proof: This is by the Proposition of (3.5.7), the choice of H in the definition of σ and by the Theorem' of (3.5.6).

We now state Friedman's strong uniform effective version of Higman's Theorem, using σ. Given a group $G \in G$, we obtain a $G' \in G_f$ in which G embeds. Further, up to isomorphism, G' depends not on G but only on the Turing-degree of the group operation of G, i.e. any $H \in G$ whose group operation is recursive in G's will also embed in G'. Further, $G \mapsto G'$ is Borel, since it's obtained as $\sigma(b)$, where $b \in B$ codes the group operation of G, via a pairing function. This theorem also establishes (*) of (3.4) when G = groups.

THEOREM B: For all $b \in B$, every group recursive in b embeds in $\sigma(b)$.

Proof: Immediate from the definition of $\sigma(b)$ and the Theorem' of (3.5.6).

LEMMA C: i) of c) of (3.4) holds; i.e. if $x \in G_f$, $b \in B$ and $\tau(x) \leq_T b$ then $x \preccurlyeq \sigma(b)$.

Proof: The argument of Lemma D, ii) of (3.5.3), above, can be trivially modified to work for G= groups, thereby proving that x embeds in a group recursive in $\tau(x)$, and therefore, since $\tau(x) \leq_T b$, in a group recursive in b. But, by Theorem B, such a group embeds in $\sigma(b)$ and therefore so does x.

We now complete the proof of Reversal 3 and of the version of Theorem 3 that replaces $F'(G)$, $G'(G)$ by $F(G)$, $G(G)$, by showing that c') of (3.4) holds for $\tau\sigma$. This completes the proof of transfer for G = groups.

LEMMA D: For all $b \in B$, $b \leq_T \tau\sigma(b)$.

Proof: This is just like the proof of Lemma C, i), in (3.5.2).

(3.5.9) We now complete the proof of Theorem 3 by showing how to modify the proof of Theorem 3' in (3.1), (3.2) to yield a proof of the Propositions $F'(G)$,$G'(G)$ from the appropriate theories. The basic idea is to apply the transfer map σ pointwise to the special sequences $\vec{\tau} \in B^\omega$ produced in (3.1.2), (3.2). For technical reasons, a somewhat more elaborate construction will be necessary, to obtain (1), (*), below. Note that our $\vec{\tau}$ notation from (3.1), (3.2) conflicts with our τ notation for the map back from G_f to B, of (3.5.1). We resolve this conflict in favor of the latter: henceforth $\vec{b}$ denotes the special sequences from B^ω produced in (3.1.2), (3.2).

Before doing so, we need some preliminaries. First, we should note that when G = semigroups, products, σ has the same properties as guaranteed for G = groups, by Theorem B of (3.5.8), i.e. (*) of (3.4) holds for all of our spaces G. For semigroups, this is because if the semigroup operation, $*$, of $x = (\omega,*)$ is recursive in b, then so is each of its cross-sections, $\lambda n(m*n)$, for $m \in \omega$. Now recall we're using "semigroup" to mean "semigroup with identity" (what is sometimes called monoid). Thus, $m_1 \neq m_2 \Rightarrow \lambda n(m_1 * n) \neq \lambda n(m_2 * n)$. But then $m \mapsto \lambda n(m * n)$ is an embedding of $(\omega,*,e)$ to $\sigma'(b)$, the semigroup of partial b-recursive functions under composition, with the identity function as identity element. If G = products, then this is by Lemma A of

(3.5.3) and the definition of σ. Summing up, we have:

LEMMA A: (*) of (3.4) holds for all our spaces G.

Thus, we have strong, uniform effective Higman style theorems for G = products, semigroups, as well as G = groups.

Second, we shall need some amalgamation machinery for G = semigroups, similar to that developed for G = products, groups in (1.1), (1.4), above.

LEMMA B: Let $(\omega, *_n, e)$ be semigroups, for $n \in \omega$, and let $H \subseteq \omega$ be such that $e \in H$, $(H, *_n, e)$ is a subsemigroup of $(\omega, *_n, e)$, and such that for all $m, n \in \omega$, $*_m | H^2 = *_n | H^2$. Then, there's a semigroup $(\omega, *, e)$ with $(H, *, e)$ a subsemigroup, $* | H^2 = *_n | H^2$ for all (some) $n \in \omega$, and in which each $(\omega, *_n, e)$ embeds by an embedding extending the identity on H.

Proof: Isomorph $(\omega, *_n, e)$ so that its underlying set is $H \cup (\{n\} \times \omega \setminus H)$. Add a new element α, to H, where $\alpha \notin \omega \cup \bigcup_n (\{n\} \times (\omega \setminus H))$. Let $S = H \cup \{\alpha\} \cup \bigcup_n (\{n\} \times (\omega \setminus H))$. Define $\otimes$ on S^2 by: $\otimes | H^2 = *_n | H^2$ for all (some) $n \in \omega$; if $n \in \omega$, $k \in H$, $i \in \omega \setminus H$, then $k \otimes (n,i) = k *_n i$, if $k *_n i \in H$, otherwise $k \otimes (n,i) = (n, k *_n i)$, and similarly for $(n,i) \otimes k$. If $n \in \omega$, $i,j \in \omega \setminus H$, then define $(n,i) \otimes (n,j) = i *_n j$, if $i *_n j \in H$; otherwise $(n,i) \otimes (n,j) = (n, i *_n j)$. If $m, n \in \omega$, $i,j \in \omega \setminus H$ and $m \neq n$, define $(m,i) \otimes (n,j) = \alpha$. Finally, define α to be a "zero" for $(S, \otimes, e)$, i.e., for all $x \in S$, $\alpha \otimes x = x \otimes \alpha = \alpha$. Clearly $(S, \otimes, e)$ is a semigroup and each $(\omega, *_n, e)$ embeds in $(S, \otimes, e)$ by the identity on H together with $i \mapsto (n,i)$ for $i \in \omega \setminus H$. Now isomorph $(S, \otimes, e)$ to a semigroup $(\omega, *, e)$ by a bijection π which is the identity on H.

We turn now to modifying the proof of Theorem 3'. Recall that we work in the following context. We are given $F: G_f^\omega \to G$, Borel and isomorphically invariant, with $u \in B$, a Borel code of F. We start from M, a countable transitive (but recall that as remarked at the end of (3.1), (3.2), ω-models would suffice) model of a sufficiently rich fragment of ZFC-P + "$V_{\omega+\omega}$ exists" (for a) of Theorem 3; here M will have the form $L_\theta[u]$), or of ZFC + "there's a measurable cardinal" (for b) of Theorem 3; here we fix $U \in M$, a normal ultrafilter in the sense of M on κ, a measurable cardinal of M). We forced over M adjoining a sequence of collapsing functions $(f_\alpha : \alpha < \aleph_\omega^M)$, each $f_\alpha : \omega \to_{\text{onto}} L_\alpha[u]$ (for a)) or $(f_\alpha : \alpha < \kappa)$, each $f_\alpha : \omega \to_{\text{onto}} V_\alpha^M$ (for b)). For b) we also arranged that $(f_\alpha : \alpha < \kappa)$ was actually generic over $M[\vec{\alpha}]$, where $\alpha = (\alpha_n : n \in \omega)$ is a Prikry sequence through κ relative to U and M.

For technical reasons which we sketch below, we shall not be able to work directly with the transferred sequence $(\sigma(b_n) : n \in \omega)$. Rather, from $\vec{b}$ we shall obtain, using σ, a sequence $\vec{x} \in G_f^\omega$ satisfying:

(1) $(x_n : n \in \omega)$ is a <u>strong</u> directed system, i.e. in addition to having embeddings $j_n : x_n \to x_{n+1}$ (which we extend to a commutative system $(j_{n,m} : n \leq m)$ in the obvious way), the x_n's also have the following property:

(*): suppose $H, H' \in G_f$, H a substructure of H' and $i : H \to x_n$ is an isomorphic embedding. Suppose further that for some $n < m$, $H' \preccurlyeq x_m$. <u>Then</u> there's $\bar{i} : H' \to x_{m+1}$, an isomorphic embedding, such that $\bar{i} \supseteq j_{n,m+1} \circ i$.

(2) $F(\vec{x})$ has the property that any finitely generated substructure of $F(\vec{x})$ lies in some $M[(f_\alpha : \alpha < \beta)]$, where, for a), $\beta < \aleph_\omega^M$, for b), $\beta < \kappa$.

(3) Whenever $H \in G_f$ lies in some $M[(f_\alpha : \alpha < \beta)]$ as in (2), then for some n, $H \preccurlyeq x_n$.

We first argue that this suffices, and then show how to produce such an $\vec{x}$. We claim that $F(\vec{x})$ is embeddable in the direct limit of the $((x_n, j_{n,m}) : n \leq m)$ provided that every finitely generated substructure of $F(\vec{x})$ is embeddable in some x_n. This is clear, using (*) of (1) repeatedly, starting from a decomposition $F(\vec{x}) = \bigcup_n H_n$ into a tower of finitely-generated substructures. But (2) and (3) provide just this sufficient condition.

Now the proof of (2) will be just like the argument, sketched before the start of (3.1.1), that $F(\vec{b})$ is determined by some G_n, lest we obtain a perfect set of $F(\vec{b}_{G'})$ corresponding to generic G' which are similar to G, contradicting that such $F(\vec{b}_{G'})$ are Turing-equivalent to $F(\vec{b})$. Here, however, we run this argument relative to a term, t, in the forcing language, forced by some condition to be a real, not lying in any $M[(f_\alpha : \alpha < \beta)]$, and which codes a complete equational theory of a finitely-generated substructure of $F(\mathring{x})$ ($\mathring{x}$ is the canonical term for the sequence $\vec{x}$). In this setting, we produce a perfect tree of G' similar to G, yielding distinct interpretations of t, all of which, however, are realized in $F(\vec{x})$ (this last is by similarity and corresponds to the fact that for G' similar to G, $F(\vec{b}) \equiv_T F(\vec{b}_{G'})$). But this is absurd, since then the countable $F(\vec{b})$ would have uncountably many finitely-generated substructures.

So, we turn now to (1) and (3), revealing at last how to produce $\vec{x}$. The difficulty with using $\vec{x} = (\sigma(b_n): n \in \omega)$ is (1). Notice, however, that we do have (3) for $(\sigma(b_n): n \in \omega)$; this is by the construction of the special sequence $\vec{b}$, and the properties of σ, as expressed in the Theorem' of (3.5.6) and Lemma B of (3.5.8), when $\mathcal{G}$ = groups, and as expressed in c) of the Lemma of (3.4), for $\sigma\tau$, when $\mathcal{G}$ = semigroups or products. Thus, we need only define x_n so that x_n satisfies (1), $\sigma(b_n) \preccurlyeq x_n$, and the construction of x_n can be carried out canonically within $M[G_{n+1}]$, where $G_n = G_{\prec_n}$, for a), $G_n = G_{\alpha_n}$, for b), where $\vec{\alpha}$ is the Prikry sequence. This last requirement is to guarantee that the argument above for (2) really does work.

We should note that for all n, $\sigma(b_n) \preccurlyeq \sigma(b_{n+1})$. For semigroups and products, this is because $b_n \leq_T b_{n+1}$, and in virtue of b') of (3.4) for σ. For groups, this needs a slightly more delicate argument. We use Theorem B, the fact (easily verified) that $\sigma(b)$ is recursive in a small finite number of jumps of b, and the fact (again, easily verified) that if b' results from applying the jump operation a finite number of times to b_n, then $b' \leq_T b_{n+1}$; in fact, much more is true, the b_n's are really quite far apart in $<_T$. We should note that this last argument remains valid for semigroups and products, even though it was not needed here. However, we shall need it below. In any case, this does give that $(\sigma(b_n): n \in \omega)$ is directed, since the choice of the embedding of $\sigma(b_n)$ in $\sigma(b_{n+1})$ can be made canonically, giving us the embedding j_n.

However, in order to obtain the strong directedness property, (*), of (1) above, we shall need to "homogenize" $\sigma(b_n)$, using the amalgamation machinery developed in (1.1) for products, (1.4) for groups and above, for semigroups. Our construction of x_n is recursive. Our induction hypotheses are:

a)$_k$: if $n < \ell < k$, if $i: H \to x_n$ is an embedding, if $H \subseteq H'$, where $H, H' \in \mathcal{G}_f$ and if $H' \preccurlyeq x$, then there's

$\bar{i}$: $H' \to x_{\ell+1}$ such that $\bar{i}|H = j_{n,\ell+1} \circ i$,

b)$_k$: for all $\ell \leq k$, there's a canonical j'_ℓ: $x_\ell \to \sigma(b_{\ell+1})$. Below, when we verify that the induction hypothesis b)$_k$ is preserved, we will use the fact (which will be clear from the definition of x_n) that x_m is recursive in a finite number of jumps of b_m. The canonical embedding j'_ℓ will come from the proof that (*) of (3.4) holds. The fact that x_m is recursive in a finite number of jumps of b_m will also guarantee that x_n can be constructed canonically in $M[G_{m+1}]$; this was needed in the argument for (2), above.

So, we let $x_0 = \sigma(b_0)$, $x_1 = \sigma(b_1)$. Clearly the induction hypothesis a)$_1$ holds trivially; b)$_1$ holds using the fact that $\sigma(b_0)$ is recursive in b_1, $\sigma(b_1)$ is recursive in b_2, as argued above. To define x_{m+2}, we assume a)$_{m+1}$, b)$_{m+1}$ hold. In particular, we have j'_{m+1}: $x_{m+1} \to \sigma(b_{m+2})$. We shall let x_{m+2} be the result of amalgamating ω copies of $\sigma(b_{m+2})$ over $A = j'_{m+1} \circ j_m[x_m]$. One of these copies is the "standard" copy, and we take j_{m+1}: $x_{m+1} \to x_{m+2}$ to be j'_{m+1} with target the standard copy of $\sigma(b_{m+2})$. The other copies correspond to the following countably many situations which code up all of the systems (H,H',i,p) where $H,H' \in \mathcal{G}_f$, $H \subseteq H'$, i: $H \to x_m$ is an embedding, p: $H' \to x_{m+1}$ is an embedding:

(a,a',q,b'), where a is a finite subset of A, b' is a finite subset of range j'_{m+1}, $a' \subseteq b'$ and q: $a \to a'$ is a bijection which extends to an isomorphism of the substructures of $\sigma(b_{m+2})$ generated by a,a', respectively.

The (a,a',q,b') - copy of $\sigma(b_{m+2})$ is an isomorph resulting from interchanging the underlying sets of the substructures of $\sigma(b_{m+2})$ generated by a and a' respectively (via the isomorphism extending q); all other points are left fixed. Now, we've already argued that this construction of x_{m+2} makes b)$_{m+2}$ true. For a)$_{m+2}$, we need only argue when $\ell = m+1$, $n = m$. Now, if (H,H',i,p) is as above, let $\bar{a}$ be a set of generators of H, let $\bar{b} \supseteq \bar{a}$ be a set of generators of H', let $a = i[\bar{a}]$, let $b' = p[\bar{b}]$, $a' = p[\bar{a}]$. Let $q = p \circ i^{-1}$. The desired embedding, $\bar{i}$, is just the result of following $j'_{m+1} \circ p$ by the isomorphism between $\sigma(b_{m+2})$ and the (a,a',q,b') - copy of $\sigma(b_{m+2})$. This completes the proof.

(3.6) We conclude §3 with some remarks. First, Friedman can obtain the Reversals 3, 3' with Propositions $F(\mathcal{G})$, $G(\mathcal{G})$,

F(T), G(T) restricted to Borel functions, F, of low Borel rank. However, he has not yet calculated precisely how low the bound on these ranks can be taken. Second, it remains open whether the weaker versions of Propositions G(G), G(T), which replace arbitrary subsequences by finite deletion subsequences, reverse to the existence of sharps, as Propositions G(G), G(T) do.

§4. In this section we consider anti-diagonalization statements for Turing degrees, Propositions H, H', respectively, which we prove from AD(L[ℝ]) and reverse to projective determinacy. We use unpublished theorems of Woodin, [18], for the reversal, since we shall prove a reversal to Z_2 + the countable second order-scheme form of choice + Turing degree determinacy for projective sets, formulated as a scheme. Woodin's theorem states that this theory is equiconsistent to ZFC + PD, where PD is formulated as a scheme. Simple modifications of Woodin's proof yield ZFC + PD, where PD is formulated as an axiom.

(4.1) Before stating Propositions H, H', we need a

<u>DEFINITION</u>: Suppose $\vec{x} = (x_n : n \in \omega) \in B^\omega$ and for all n, $x_n \leq_T x_{n+1}$ (we'll say x is increasing). If $\vec{y} \in B^\omega$, then $\vec{y}$ is a <u>separating sequence</u> for $\vec{x}$ if there's a strictly increasing $(n_k : k \in \omega)$ such that for all k, $x_{n_{2k}} \leq_T y_k \leq_T x_{n_{2k+1}}$.

<u>PROPOSITION H</u>: Whenever $F: B^\omega \to B$ is Turing-invariant and Borel, there's <u>increasing</u> $\vec{x} \in B^\omega$ and $n \in \omega$ such that whenever $\vec{y}$ is a separating sequence for $\vec{x}$, $F(\vec{y}) \leq_T y_n$.

We have a weakening of Proposition H, where we don't require any uniformity on the term of $\vec{y}$ in which $F(\vec{y})$ is recursive. However, since any separating sequence of an increasing sequence is, itself, increasing, this version is clearly equivalent to:

<u>PROPOSITION H'</u>: Whenever $F: B^\omega \to B$ is Turing-invariant and Borel, there's increasing $\vec{x} \in B^\omega$ such that whenever $\vec{y}$ is a separating sequence for $\vec{x}$, there's $n \in \omega$ such that $F(\vec{y}) \leq_T x_n$.

Now, since any subsequence of an increasing sequence is a separating sequence, we clearly have that Proposition H' implies variant a) of Proposition G(T) and Proposition H implies variant b) of Proposition G(T). Clearly, too, Proposition H => Proposition H'.

We shall prove:

THEOREM 4: Proposition H is provable in ZFC + AD(L[$\mathbb{R}$]).

REVERSAL 4: ZF + AC_ω - P + Proposition H' proves "there's an ω-model of Z_2 + the second-order-scheme form of countable choice + Turing-degree determinacy for projective sets formulated as a scheme".

COROLLARY 4: Proposition H' is not provable in ZFC + PD.

As mentioned above, we shall get Corollary 4 from Reversal 4, using:

THEOREM (Woodin, unpublished): The theory of Reversal 4 is equiconsistent to ZFC + PD.

Actually, as noted above, Woodin proved this with PD as a scheme in the place of PD, but easy modifications of his proof allow us to get PD as an axiom.

In (4.2) we prove Theorem 4. In (4.3) we prove Reversal 4. In (4.4), we comment on some refinements of the arguments of (4.2), (4.3). In (4.5), we close by presenting a more geometric statement which Friedman, in very recent work, has related to determinacy.

(4.2) We start from a countable transitive model, M, of a sufficiently large fragment of ZFC + AD(L[$\mathbb{R}$]), with $u \in M$, where u is a Borel code for Borel, Turing-invariant $F: B^\omega \to B$. We force over M' = the L[$\mathbb{R}$] of M. Before defining our forcing conditions, we need some preliminaries. We work in M' until (4.2.2). Let D = the set of Turing degrees. If $R \subseteq D \times D$, R is full if, for each $d \in D$, $R_d = \{d' : dRd'\}$ is non-empty and upward closed, i.e. R is transitive and for all d there's d' with dRd'.

DEFINITION: $p \in P$ iff $p = ((d_1,\ldots,d_n),R)$, where $R \subseteq D \times D$ is full, $n \in \omega$ (so if $n = 0$, p is (ϕ,R)), and $d_1 < \ldots < d_n$, for the ordering of degrees; $((d_1,\ldots,d_n),R) \leq ((d_1^*,\ldots,d_m^*),S)$ (by which we mean that $((d_1,\ldots,d_n),R)$ gives less information) iff $n \leq m$, for all $1 \leq i \leq n$, $d_i = d_i^*$, $S \subseteq R$, and for all $n \leq i < m$, $R(d_i^*,d_{i+1}^*)$. We let $\mathbb{P} = (P,\leq)$.

(4.2.1) We state two lemmas, which we prove below in (4.2.3).

LEMMA A: If $E \subseteq D^n \times D^m$, there's full $R \subseteq D \times D$ such that for degrees $\alpha_1 < \ldots < \alpha_n < \beta_1 < \ldots < \beta_m$, $\alpha_1 < \ldots < \alpha_n < \gamma_1 < \ldots < \gamma_m$, if $\beta_1 = \gamma_1$ and for all $1 \leq i < m$, $\beta_i R \beta_{i+1}$ and $\gamma_i R \gamma_{i+1}$ then $(\vec{\alpha}, \vec{\beta}) \in E$ iff $(\vec{\alpha}, \vec{\gamma}) \in E$.

LEMMA B: If φ is a statement in the forcing language for $\mathbb{P}$, there's full $R \subseteq D \times D$ such that for any non-empty, finite, strictly increasing $\vec{\alpha}$, $(\vec{\alpha}, R)$ decides φ.

(4.2.2) We prove two lemmas about forcing with $\mathbb{P}$, and then fix $\vec{\alpha}$, generic over M' to finish the proof of Theorem 4.

LEMMA A: Forcing with $\mathbb{P}$ adds no new reals.

Proof: Let $\varphi(v)$ be a forcing statement with free variable v, such that it's forced that $\forall v(\varphi(v) \Rightarrow v \in \omega)$. For each $n \in \omega$, choose R_n according to Lemma B of (4.2.1) for $\varphi(\bar{n})$ (where $\bar{n}$ is the canonical term for n). Let $R = \bigcap_n R_n$. Then R simultaneously decides each $\varphi(\bar{n})$ (in the sense of Lemma B of (4.2.1)).

LEMMA B: A sequence $\vec{\alpha} = (\alpha_n : 1 \leq n \in \omega)$ is M'-generic/$\mathbb{P}$ iff for all full $R \subseteq D \times D$ from $L[\mathbb{R}]$, for all sufficiently large i, $R(\alpha_i, \alpha_{i+1})$.

Proof: First suppose $\vec{\alpha}$ is M'-generic/$\mathbb{P}$, and, towards a contradiction, suppose R is a counterexample. Then, for some $n \geq 1$, some S, $((\alpha_1, \ldots, \alpha_n), S)$ forces that R is a counterexample. Then, we can find increasing $(\beta_k : 1 \leq k \in \omega)$ with $\alpha_n < \beta_1$, such that $\alpha_1, \alpha_2, \ldots, \alpha_n, \beta_1, \ldots, \beta_k, \ldots$ is generic and compatible with $((\alpha_1, \ldots, \alpha_n), R \cap S)$. But R is not a counterexample for $\alpha_1, \ldots, \alpha_n, \beta_1, \ldots, \beta_k, \ldots$, which contradicts that this is forced by $((\alpha_1, \ldots, \alpha_n), S)$.

Now suppose $(\alpha_n : 1 \leq n \in \omega)$ has the desired property. Let φ be a forcing statement. Choose R for φ, according to Lemma B of (4.2.1). Let k be such that $i \geq k \Rightarrow R(\alpha_i, \alpha_{i+1})$. Then $((\alpha_1, \ldots, \alpha_k), R)$ decides φ and is compatible with $(\alpha_n : 1 \leq n \in \omega)$. This proves Lemma B.

Now, let $\vec{\alpha} = (\alpha_n : 1 \leq n \in \omega)$ be M'-generic/$\mathbb{P}$, and let $M^* = M'[\vec{\alpha}]$. We now have an important lemma which firms up the analogy between $\mathbb{P}$ and Prikry forcing, and the Propositions H and G(T).

LEMMA C: If $\vec{\beta}$ is any separating sequence for $\vec{\alpha}$, then $\vec{\beta}$ is also M'-generic/$\mathbb{P}$.

Proof: Let $R \in M'$ be full. Let $R^* = \{(\alpha, \beta): (\forall\gamma \leq \alpha)(\gamma, \beta) \in R\}$. Since R^* is full, by Lemma B of (4.2.1), for all sufficiently large i, $R^*(\alpha_i, \alpha_{i+1})$. But then, by definition of R^*, and since $\vec{\beta}$ is separating for $\vec{\alpha}$, for all sufficiently large i, $R(\beta_i, \beta_{i+1})$. Again, by Lemma B of (4.2.1), $\vec{\beta}$ is M'-generic/$\mathbb{P}$.

Our next two lemmas involve an auxiliary forcing over M^*. Let $\mathbb{P}' \in M^*$ be the partial order of finite sequences of reals ordered by extension. So forcing with $\mathbb{P}'$ adds a generic enumeration of $\mathbb{R}$ in type ω. We let $N = M^*[H]$, where H is M^*-generic/$\mathbb{P}'$.

LEMMA D: Let $\vec{b}$ be any sequence of representatives of $\vec{\alpha}$. Then $F(\vec{b}) \in M^*$.

Proof: In N, we have some sequence, $\vec{b}'$ of representatives of $\vec{\alpha}$, so $F(\vec{b}') \in N$. But $F(\vec{b}')$ depends only on $\vec{\alpha}$, up to Turing degrees. Thus, there's a countable set $X \subseteq B$ (X is just the Turing degree of $F(\vec{b}')$) such that whenever $N' = M^*[H']$, H' is M^*-generic/$\mathbb{P}'$, $X \in N'$. Then, as usual, $X \in M^*$.

LEMMA E: There's a parameter-free formula $\varphi(v, \vec{c})$, where $\vec{c}$ is a sequence of new constant symbols, such that whenever $\vec{\beta}$ is M'-generic/$\mathbb{P}$, $n \in \omega$, n satisfies $\varphi(v, \vec{c})$ in $(M'[\vec{\beta}], \in, \vec{\beta})$ (i.e. β_n interprets c_n) iff for all (some) sequences $\vec{b}$ of representatives of $\vec{\beta}$, $F(\vec{b}) \leq_T b_n$.

Proof: Clear, from the proof of Lemma D.

LEMMA F: For some n and some tail segment $\vec{\beta}$ of $\vec{\alpha}$, whenever $\vec{\gamma}$ is a separating sequence for $\vec{\beta}$ then for all (some) sequences $\vec{b}$, of representatives of $\vec{\gamma}$, $F(\vec{b}) \leq_T b_n$.

Proof: By Lemma A and genericity, all generic sequences are cofinal in M''s and therefore M^*'s ordering of Turing degrees. By Lemma B of (4.2.1), let R be full such that for all $\delta \in D$, there's n such (δ, R) forces "$\bar{n}$ satisfies $\varphi(v, \vec{c})$ in $M'[\mathring{G}]$", where φ is as in Lemma E. Fix n and $\delta \in D$ such that for all $\delta' \geq \delta$, (δ', R) forces this. Let R^* be as in Lemma C for this R; and, as in Lemma C, for sufficiently large i, $\delta' < \alpha_i$, and for all $j \geq i$, $R^*(\alpha_j, \alpha_{j+1})$. Let $\beta_j = \alpha_{i+j-1}$, for $1 \leq j \in \omega$. Further, if $\vec{\gamma}$ is any separating sequence for $\vec{\beta}$, then $\vec{\gamma}$ (which is generic, by Lemma C) is compatible with (γ_1, R). Finally, (γ_1, R) forces "$\bar{n}$ satisfies $\varphi(v, \vec{c})$ in $M'[\mathring{G}]$", by the property of δ. This completes the proof of Lemma F.

The proof of Theorem 4 is now clear, modulo the proofs, below, of Lemmas A, B of (4.2.1): let $\vec{x}$ be a sequence of representatives of $\vec{\beta}$.

(4.2.3) We now prove Lemmas A, B of (4.2.1). For Lemma A, we work by induction on m, using Turing-degree determinacy. If $m = 1$, the statement is trivially true. So, suppose $m \geq 1$ and $E \subseteq D^n \times D^{m+1}$. Let $(\vec{\alpha},(\beta_1,\ldots,\beta_m)) \in E^*$ iff $\{\beta_{m+1}: (\vec{\alpha},(\beta_1,\ldots,\beta_m,\beta_{m+1})) \in E\}$ contains a cone of Turing degrees. Let R^* be as guarangeed by the induction hypothesis for E^*. Let $(\beta_m,\beta_{m+1}) \in R^{**}$ iff whenever $\alpha_n < \beta_1 < \ldots < \beta_m$, β_{m+1} is the base of a cone of Turing-degrees which is homogeneous for $\{\gamma_{m+1}: (\vec{\alpha},(\beta_1,\ldots,\beta_m,\gamma_{m+1})) \in E\}$. Then, by Turing-degree determinacy, $R = R^* \cap R^{**}$ is full and is as required for Lemma A.

For Lemma B, if $k \in \omega$, let φ_k be the kth parameter-free sentence of the forcing language (in Lemma A of (4.2.2) we apply this Lemma where φ contains $\bar{n}$ as a parameter, but this is clearly a convenience; $\bar{n}$ can be eliminated in favor of its canonical definition). If $m,n \geq 1$, let $((\alpha_1,\ldots,\alpha_n),(\alpha_{n+1},\ldots,\alpha_{n+m})) \in E_{k,m,n} \subseteq D^n \times D^m$ iff for some S $(\vec{\alpha},S)$ forces φ_k (we say, in this case, that $\vec{\alpha}$ <u>locally forces</u> φ_k). Let $R_{k,m,n}$ be as guaranteed by Lemma A for $E_{k,m,n}$ and let R be the intersection of all the $R_{k,m,n}$. We claim that R is as required.

Clearly R is full. Now fix $\alpha_1 < \ldots < \alpha_n \in D$. Let $\vec{\alpha} \frown \vec{\gamma} \in K_m$ iff $\alpha_n < \gamma_1$, $R(\alpha_n,\gamma_1)$ and for $1 \leq i < m$, $R(\gamma_i,\gamma_{i+1})$. By construction of R and the property of the $R_{k,m,n}$, for all k, either all elements of K_m locally force φ_k, or none do. If all do, let $\vec{\alpha} \frown \vec{\gamma} \in K_m$, and let S be such that $(\vec{\alpha} \frown \vec{\gamma},S)$ forces φ_k. Choose γ_{m+1} such that $R(\gamma_m,\gamma_{m+1})$, $S(\gamma_m,\gamma_{m+1})$. Then $\vec{\alpha} \frown \vec{\gamma} \frown \gamma_{m+1} \in K_{m+1}$ and $\vec{\alpha} \frown \vec{\gamma} \frown \gamma_{m+1}$ locally forces φ_k. Thus, every member of K_{m+1} locally forces φ_k, and we've shown that either, for all sufficiently large m, all elements of K_m locally force φ_k, or for all m, no elements of K_m locally force φ_k. If the latter, then clearly $(\vec{\alpha},R)$ forces $\neg\varphi_k$. So, it suffices to show that if the former, then $(\vec{\alpha},R)$ forces φ_k. But this is clear, since if $(\vec{\alpha},R) \leq (\vec{\alpha} \frown \vec{\gamma},S)$ and $(\vec{\alpha} \frown \vec{\gamma},S)$ forces $\neg\varphi_k$, then, by extending $(\vec{\alpha} \frown \vec{\gamma},S)$, if necessary, we may take $\vec{\gamma}$ to be sufficiently long so that $\vec{\alpha} \frown \vec{\gamma}$ locally forces φ_k, which is a contradiction.

(4.3) We prove Reversal 4. We pick up the thread of the proof of Reversal 3', b) at the end of Case 4 in (3.3.3). We can do this since, as remarked in (4.1), before the statement of Theorem 4, Proposition H' implies variant a) of the Proposition G(T). Thus, given $\vec{x} \subset B^{\omega}$, letting M = the set of reals recursive in some finite join of the x_n's, we have that M models Z_2 + the scheme form of countable second-order choice (in fact, more holds, corresponding to the failure of cases 1) - 3) of (3.1.3)), whenever $\vec{x}$ is as guaranteed by Proposition H' for any F defined as in Cases 1 - 4.

We add a new case:

Case 5: The scheme of Turing-degree determinacy for projective sets fails for M.

Then, for some second-order φ, and some finite $\vec{\beta}$ from M, $\{\alpha\colon \alpha$ satisfies $\varphi(v,\vec{\beta})$ in $M\}$ forms a set, S, of Turing degrees such that neither S nor its complement contains a cone. Let $\vec{b} \in B^{\omega}$ be given (so M was obtained from $\vec{b}$). Let k be least such that for some $\vec{\beta}$, the S obtained from $\vec{\beta}$ is a counterexample and each term of $\vec{\beta}$ is recursive in b_k (we can do this, in virtue of the failure of Case 2 of (3.1.3)). So, as in §3, we obtain an enumeration, in type ω, of reals, $g(\vec{\beta}) = \{i\colon b_i$ satisfies $\varphi(v,\vec{\beta})$ in $M\}$, as $\vec{\beta}$ varies over finite sequences of the right length, all of whose terms are recursive in b_k. Of course, we let $F(\vec{b})$ be the effective join of this enumeration.

Now suppose that $\vec{b}$ satisfies the conclusion of Proposition H'. For some $\vec{\beta}$, as above, the S obtained from $\vec{\beta}$ is a counterexample to Turing-degree determinacy. Therefore, letting $\vec{b}'$ vary over separating sequences for $\vec{b}$, there are continuum many $g_{\vec{b}'}(\vec{\beta})$. This, however, contradicts that all the $g_{\vec{b}'}(\vec{\beta})$ are supposed to be recursive in one of the b_n. Thus, if $\vec{b}$ satisfies the conclusion of Proposition H', Case 5 must fail.

(4.4) There's an obvious generalization of this last argument and of Woodin's Theorem, above, which yields:

Over ATR_0, Proposition H' implies that:

(*) for all well-orderings x of ω, there's an ω-model of $ZFC + AD(L_x[\mathbb{R}])$.

Friedman has shown, recently, that, over ATR_0, Proposition H' is equivalent to (*). The proof of Proposition H' from (*) is not entirely straightforward, even if we replace (*) by ZFC + $AD(L_{\omega_1}[\mathbb{R}])$. Here we force over $L_{\omega_1}[\mathbb{R}]$ (of a countable transitive model of a sufficiently rich fragment) using the conditions of (4.2). The forcing is viewed as adding no sets, just a generic predicate to $L_{\omega_1}[\mathbb{R}]$. The forcing language consists of infinitary formulas in $\in$, = and $\overset{\circ}{G}$, which are hereditarily countable, in $L_{\omega_1}[\mathbb{R}]$, and with arbitary quantification over $\mathbb{R}$ but only bounded quantification over the universe. Constants for integers and reals are available.

The truth lemma goes through, but verifying the definability of forcing requires a fair amount of rather subtle work. This done, Lemma B of (4.2.2) can be reproved. Given a generic predicate, the generic extension is defined as an interpretation of third order arithmetic as follows: integers are standard. Reals are as in $L_{\omega_1}[\mathbb{R}]$. Sets of reals are the denotations of forcing formulas on the reals of $L_{\omega_1}[\mathbb{R}]$, where the generic predicate is interpreted by the degree of the join of the α_n's. We can then prove a restricted form of second-order comprehension for formulas involving the generic predicate. The formulas which are allowed are those coded in the hereditarily countable sets of $L_{\omega_1}[\mathbb{R}]$, with parameters available for reals and sets of reals. The proof is like that of Lemma A of (4.2.2). The remainder of the proof is as in (4.2.2). Again, this sketch merely illustrates the ideas in the more delicate proof of Proposition H' from (*).

As above, and indeed, as is typical of Friedman's work in this vein since [04], while quite strong, Propositions H, H', if true, remain true even when the set-theoretic universe is cut down fairly drastically. Once again, this is because the use of well-founded models is inessential. This is not difficult to see in the setting of (4.2), but requires additional care for the proof of Proposition H' from (*).

(4.5) Very recently Friedman has found a family of statements which are, level-by-level, equivalent to determinacy. The statements are similar to the statements of Turing-degree determinacy, but lines in $\mathbb{R}^2$ replace cones of Turing degrees. To give a precise statement, we must introduce some notions.

DEFINITION: Let E be an arbitrary set and let V be a set of functions $F: E \to E$ and let $A \subseteq E$. We write $c\ell_V(A)$ for the least $A \subseteq S \subseteq E$ with $F[S], F^{-1}[S] \subseteq S$, for all $F \in V$.

Lines in $\mathbb{R}^2$ are infinite in both directions. F is countable-to-one if each $F^{-1}[\{a\}]$ is countable (i.e. finite or countably infinite). Let $X \subseteq P(\mathbb{R}^2)$ be any of the following: Borel sets, $\underset{\sim}{\Delta}^1_n$ sets, $\underset{\sim}{\Sigma}^1_n$ sets $(1 \leq n \in \omega)$, sets constructible from a real, sets ordinal-definable from a real, arbitrary sets. We can now state, for each of these X:

PROPOSITION I(X): There's a finite set V of countable-to-one, continuous functions on $\mathbb{R}^2$ into $\mathbb{R}^2$ such that for any $E \in X$, $c\ell_V(E)$ contains or is disjoint from a line in $\mathbb{R}^2$. In fact, either $c\ell_V(E)$ contains lines of every rational slope (including ∞) or for all rational r (including ∞) there's a line of slope r disjoint from $c\ell_V(E)$.

THEOREM (Friedman, to appear): For all of the above X, Proposition I(X) is equivalent to determinacy for sets in X.

Two of the ingredients in Friedman's proof are a lemma of R. H. Bing on the extendability from closed subsets to all of $\mathbb{R}^2$ of countable-to-one continuous functions into $\mathbb{R}^2$, and the equivalence (due to Friedman (for Borel), Harrington (for $\underset{\sim}{\Sigma}^1_1$), and Woodin) of determinacy for sets in X and Turing-degree determinacy for sets in X.

REFERENCES

[01] Apt, K. R. and Marek, W., Second order arithmetic and related topics, Annals of Math. Logic 6 (1974) 177-230.

[02] Friedman, H., Axiomatic recursive function theory, in: Gandy, R. and Yates, C.M.E. (eds.), Logic Colloquium '69 (North-Holland, Amsterdam, 1971).

[03] Friedman, H., Higher set theory and mathematical practice, Annals of Math. Logic 2 (1971) 326-357.

[04] Friedman, H., On the necessary use of abstract set theory, Advances in Math. 41 (1981) 209-280.

[05] Friedman, H., Some applications of Borel determinacy to analytic relations, unpublished manuscript.

[06] Friedman, H., Necessary uses of abstract set theory in finite mathematics, to appear in Advances in Math.

[07] Galvin, F. and Prikry, K., Borel sets and Ramsey's theorem, J. of Symbolic Logic 38 (1973) 193-198.

[08] Kreisel, G., A survey of proof theory, J. of Symbolic Logic 33 (1968) 321-388.

[09] Lyndon, R. and Schupp, P., Combinatorial Group Theory, (Springer-Verlag, Berlin, 1977).

[10] Mansfield, R. and Weitkamp, G., Recursive Aspects of Descriptive Set Theory, (Oxford University Press, Oxford, 1985).

[11] Nerode, A. and Harrington, L., The work of Harvey Friedman, this volume.

[12] Schmerl, J., Hyper-Ramsey logic, preprint.

[13] Silver, J., Every analytic set is Ramsey, J. of Symbolic Logic 35 (1970) 60-64.

[14] Simpson, S., Set-theoretic aspects of ATR_0, in: van Dalen, D. et al. (eds.), Logic Colloquium '80 (North-Holland, Amsterdam, 1982).

[15] Simpson, S., Σ^1_1 and Π^1_1 transfinite induction, in: van Dalen, D. et al. (eds.), Logic Colloquium '80 (North-Holland, Amsterdam, 1982).

[16] Simpson, S., Subsystems of second-order arithmetic, this volume.

[17] Stern, J., Effective partitions of the real line into Borel sets of bounded rank, Annals of Math. Logic 18 (1980) 29-60.

[18] Woodin, H., to appear.

HARVEY FRIEDMAN'S RESEARCH ON THE FOUNDATIONS
OF MATHEMATICS, L.A. Harrington et al. (editors)
Elsevier Science Publishers B.V. (North-Holland), 1985

NONPROVABILITY OF CERTAIN COMBINATORIAL PROPERTIES OF FINITE TREES*

Stephen G. Simpson

Department of Mathematics
Pennsylvania State University
University Park, PA 16802

Abstract. In this paper we exposit some as yet unpublished results of Harvey Friedman. These results provide the most dramatic examples so far known of mathematically meaningful theorems of finite combinatorics which are unprovable in certain logical systems. The relevant logical systems, ATR_0 and Π^1_1-CA_0, are well known as relatively strong fragments of second order arithmetic. The unprovable combinatorial theorems are concerned with embeddability properties of finite trees. Friedman's methods are based in part on the existence of a close relationship between finite trees on the one hand, and systems of ordinal notations which occur in Gentzen-style proof theory on the other.

Contents

---

* This is a translation of the author's paper "Nichtbeweisbarkeit von gewissen kombinatorischen Eigenschaften endlicher Bäume", to appear in Archiv für mathematische Logik und Grundlagenforschung 25, No. 1-2 (1985) by permission of Verlag W. Kohlhammer GmbH. Preparation of this paper was partially supported by NSF grant 8107867 and by a grant from the Deutsche Forschungsgemeinschaft while the author was on sabbatical leave at the University of Munich.

§0. Introduction.

Gödel's 1931 Incompleteness Theorem established that in any formal system for mathematics there exist undecidable finite combinatorial statements. In other words, in any formal system for mathematics there are unprovable finite combinatorial theorems.

Although this result of Gödel is astonishing and remarkable, the criticism has often been raised that Gödel's concrete examples of unprovable finite combinatorial theorems are mathematically rather artificial, since they involve coding of logical syntax. This objection is to be taken seriously, for it is just conceivable that all mathematically meaningful finite combinatorial theorems are provable.

The situation has become much clearer as a result of the 1977 work of Paris and Harrington [23]. Namely, this work furnished a completely transparent theorem of finite combinatorics, actually a very simple strengthening of the finite Ramsey theorem, which is not provable in Z_1. (Here Z_1 is the well known formal system of first order arithmetic, i.e. so-called Peano arithmetic.) Subsequent to this breakthrough of Paris and Harrington, several other results of the same type appeared. In this connection, the 1981 results of Kirby and Paris [16] were especially notable. Here again some combinatorial theorems were obtained which are mathematically very appealing yet not provable in Z_1.

Unfortunately, the above mentioned results apply only to the particular formal system Z_1. As far as the formal system Z_2 of second order arithmetic is concerned, the situation is still completely unclear. For there are still no published examples of mathematically meaningful theorems of finite combinatorics which are not provable in Z_2.

Instead of the full system Z_2, it is natural to consider subsystems of Z_2. As regards such subsystems, the present situation is a little more complicated. Firstly, in the last few years, a small number of subsystems of Z_2 have been isolated which are especially well adapted to the development of ordinary mathematical practice. (This isolation of a few mathematically significant subsystems of Z_2 was accomplished with the help of the ongoing research program of so-called "reverse mathematics". See [6], [7], [10], [11], [34], [35], [36], [37].) Secondly, it has turned out that only two of these mathematically significant subsystems of Z_2 are genuinely stronger than Z_1. These two subsystems are called ATR_0 and $\Pi^1_1\text{-}CA_0$. (See [7], [10], [11], [33], [35], [36] and also the discussions at the beginning of §2 and §5 of the present paper.) Thirdly, in 1979 [10] a mathematically meaningful finite combinatorial theorem was offered which is not provable in ATR_0. This combinatorial theorem belongs to the domain of finite Ramsey theory, and its statement is easily understandable although perhaps a little inelegant. See also Shelah [30].

But until now, except for [10] and [30], there have been no other mathematically meaningful, finite combinatorial theorems which are not provable in ATR_0 or $\Pi^1_1\text{-}CA_0$. The purpose of the present paper is to present some new examples of such theorems. These examples, and indeed all of the new results of the present paper, are due to Harvey Friedman [8], [9].

Our combinatorial considerations have their origin is certain embeddability questions in the theory of finite graphs. A conjecture of Wagner assets that for every 2-dimensional manifold M there are only finitely many minimal graphs which are not embeddable into M. (Here minimality is to be understood with respect to the relation of homeomorphic embeddability. Recently Robertson and Seymour [25a] have announced a proof of this conjecture.) A famous conjecture of Vászonyi asserts that every set of finite graphs in which each node is of degree ≤ 3 has only finitely many minimal elements. A famous theorem of Kruskal asserts that every set of finite trees has only finitely many minimal elements. (Here by a tree is meant a connected graph which contains no cycles.) See [18], [19], [21], [22].

The above mentioned theorem of Kruskal is not provable in ATR_0. This nonprovability result is the first result of Friedman which we shall present in the present paper.

For our purposes it is appropriate to use a slightly modified formulation of Kruskal's Theorem. For us, a finite tree will be not a graph but rather a certain kind of finite partially ordered set (Definition 1.1). Nevertheless the content of our Theorem 1.3 and that of the above-mentioned graph theoretic theorem of Kruskal are essentially identical.

The second nonprovability result of Friedman concerns finite trees with certain additional structure. Labels from the finite set $\{0,1,\ldots,n-1\}$ will be assigned to the nodes of the trees. Between such labeled trees a certain embeddability relation will be defined. The embeddings will be required to preserve labels and also to satisfy a certain "gap condition" (Definition 4.1). It then turns out that the corresponding generalization of Kruskal's Theorem is true but not provable in $\Pi^1_1\text{-}CA_0$. These results of Friedman will be presented in §§4 and 5 of the present paper.

Unfortunately Kruskal's Theorem and Friedman's generalization of it do not belong unequivocally to the domain of pure finite combinatorics. The difficulty is that these theorems contain a quantifier ranging over infinite sets. But with the help of König's Lemma [17], one can finitely miniaturize Kruskal's Theorem and Friedman's generalization. The resulting purely finite combinatorial theorems are admittedly somewhat more complicated than their infinitistic prototypes. Namely, their statements involve a linear growth restriction on the sizes of the finite trees. Nevertheless, these finitely miniaturized combinatorial theorems possess a certain appeal. Friedman's main result (Theorems 3.2 and 5.16) is that these finitely miniaturized theorems are not provable in ATR_0, respectively $\Pi^1_1\text{-}CA_0$.

Friedman's method is based in part on the exploitation of a certain connection between finite trees and the ordinal notations which occur in proof theory. The existence of such a connection was noticed earlier, for example by Schmidt [27]. But Schmidt did not obtain any independence results. On the other hand, there is a notable resemblance between Friedman's labled trees (Definition 4.1) and the ordinal diagrams of Takeuti [38], [39]. (Compare also Definition 5.6.) The biggest difference between Friedman's labeled trees and Takeuti's ordinal diagrams is that each labeled tree has only finitely many predecessors (Definition 4.1). The other essential component of Friedman's method is his miniaturization procedure based on a linear growth restriction (Definitions 3.1 and 3.4, Theorem 3.5). This idea apparently has no precedent.

We thank Harvey Friedman for permission to publish the present paper. We also thank the members of the Mathematical Logic Seminar of the University of Munich, especially Drs. Buchholz, Jäger, Osswald and Pohlers, for their help with the preparation of this paper.

§1. Proof of Kruskal's Theorem.

1.1 Definition. A finite tree is a finite partially ordered set T such that:

(i) T has a smallest element (This element is called the root of T.);

(ii) for each $b \in T$ the set $\{a \in T : a \leq b\}$ is a totally ordered subset of T.

1.2 Definition. For finite trees T_1 and T_2, an embedding of T_1 into T_2 is a one-to-one mapping $f : T_1 \to T_2$ such that $f(a \wedge b)=f(a) \wedge f(b)$ for all $a,b \in T_1$. (Here $a \wedge b$ denotes the infimum of a and b.) We write $T_1 \leq T_2$ to mean that there exists an embedding $f : T_1 \to T_2$.

1.3 Kruskal's Theorem [18]. For every infinite sequence of finite trees $\langle T_k : k < \omega \rangle$, there exist indices i and j such that $i < j < \omega$ and

$T_i \leq T_j$. (In other words, there is no infinite set of pairwise nonembeddable finite trees.)

For the proof of Kruskal's Theorem, we need the following abstract formulation:

1.4 Definition. A quasiordering is a set $\mathcal{Q}$ together with a reflexive, transitive relation defined on $\mathcal{Q}$. A well quasiordering (abbreviated WQO) is a quasiordering with the property that for any infinite sequence $\langle Q_k : k < \omega \rangle$ of elements $Q_k \in \mathcal{Q}$, there exist indices i and j such that $i < j < \omega$ and $Q_i \leq Q_j$.

1.5 Example. Let $\mathcal{T}$ be the set of all finite trees with the embeddability relation $\leq$ of 1.2. Then obviously $\mathcal{T}$ is a quasiordering. Kruskal's Theorem 1.3 is the assertion WQO($\mathcal{T}$). For further developments in general well quasi-ordering theory, see [19], [20], [21], [31].

For an arbitrary quasiordering $\mathcal{Q}$, an infinite sequence $\langle Q_k : k < \omega \rangle$ of elements of $\mathcal{Q}$ is called bad if $Q_i \nleq Q_j$ for all i and j with $i < j < \omega$. Thus $\mathcal{Q}$ is wqo if and only if it contains no bad sequence.

1.6 Lemma (Higman [12]). If $\mathcal{Q}$ is an arbitrary wqo, then $\mathcal{Q}^{<\omega}$ is also a wqo. Here $\mathcal{Q}^{<\omega}$ denotes the set of all finite sequences of elements of $\mathcal{Q}$, quasiordered by putting

$$\langle Q_0, Q_1, \ldots, Q_{m-1} \rangle \leq \langle Q_0', Q_1', \ldots, Q_{n-1}' \rangle$$

if and only if there exist $i_0, i_1, \ldots, i_{m-1}$ such that $i_0 < i_1 < \ldots < i_{m-1} < n$ and $Q_0 \leq Q_{i_0}', Q_1 \leq Q_{i_1}', \ldots, Q_{m-1} \leq Q_{i_{m-1}}'$.

Proof. Assume that $\mathcal{Q}^{<\omega}$ is not wqo. Then there is at least one bad sequence from $\mathcal{Q}^{<\omega}$. We build a minimal bad sequence $\langle t_k : k < \omega \rangle$, $t_k \in \mathcal{Q}^{<\omega}$, as follows. Let $t_0 \in \mathcal{Q}^{<\omega}$ be of minimal length such that there exists a bad sequence $\langle t'_k : k < \omega \rangle$ with $t'_0 = t_0$; then let $t_1 \in \mathcal{Q}^{<\omega}$ be of minimal length such that there exists a bad sequence $\langle t'_k : k < \omega \rangle$ with $t'_0 = t_o$ and $t'_1 = t_1$; and so on.

It is clear that $t_k \neq \langle\,\rangle$ for each $k < \omega$. We set

$$t_k = \langle Q_k \rangle ^\frown s_k$$

where Q_k is the first element of t_k. Then $s_k \in \mathcal{Q}^{<\omega}$ and for the length we have $\ell h(t_k) = 1 + \ell h(s_k)$.

We claim that the sequence $\langle s_k : k < \omega \rangle$ possesses no bad subsequence. Otherwise there would be a bad sequence $\langle s_{k_i} : i < \omega \rangle$ with $k_i < k_j$ for all i and j with $i < j < \omega$. Then the sequence $\langle t_0, t_1, \ldots, t_{k_0 - 1}, s_{k_0}, s_{k_1}, \ldots \rangle$ would also be bad. But this would contradict the minimality of t_{k_0}. Our claim is proved.

By the above claim and Ramsey's Theorem [25], it is clear that the sequence $\langle s_k : k < \omega \rangle$ must contain a subsequence $\langle s_{k_i} : i < \omega \rangle$ such that $k_i < k_j$ and $s_{k_i} \leq s_{k_j}$ for all i and j with $i < j < \omega$. Since $\mathcal{Q}$ is wqo, there exist i and j such that $i < j < \omega$ and $Q_{k_i} \leq Q_{k_j}$. But then we have immediately

$$t_{k_i} = \langle Q_{k_i} \rangle ^\frown s_{k_i} \leq \langle Q_{k_j} \rangle ^\frown s_{k_j} = t_{k_j}$$

which contradicts the badness of $\langle t_k : k < \omega \rangle$.□

<u>Proof of Theorem</u> 1.3 (Nash-Williams [22]). Theorem 1.3 is the assertion that $\mathcal{T}$ is wqo. If not, let $\langle T_k : k < \omega \rangle$ be a minimal bad sequence from $\mathcal{T}$. For each $k < \omega$ let T_k^m, $1 \leq m \leq n_k$ be the components of $T_k \setminus \{\mathrm{root}(T_k)\}$. Set

$$S = \{T_k^m : k < \omega,\ 1 \le m \le n_k\}.$$

Thus S is a subset of T.

We claim that S is wqo. Otherwise there would be a bad sequence from S of the form $\langle T_{k_i}^{m_i} : i < \omega \rangle$ where $1 \le m_i \le n_{k_i}$ and $k_i < k_j$ for all i and j with $i < j < \omega$. Then the sequence $\langle T_0, T_1, \ldots, T_{k_0}^{m_0}, T_{k_1}^{m_1}, \ldots \rangle$ would also be bad, which would contradict the minimality of T_{k_0}. Our claim is thus proved.

By Lemma 1.6 we have also $\mathrm{WQO}(S^{<\omega})$. In particular the sequence $\langle t_k : k < \omega \rangle$ is not bad, where $t_k = \langle T_k^1, T_k^2, \ldots, T_k^{n_k} \rangle \in S^{<\omega}$. Let i and j be such that $i < j < \omega$ and $t_i \le t_j$. It follows immediately that $T_i \le T_j$ which contradicts the badness of $\langle T_k : k < \omega \rangle$.□

There is a still more general formulation which includes both Higman's Lemma 1.6 and Kruskal's Theorem 1.3:

1.7 Definition. If Q is an arbitrary quasiordering, we set

$$T(Q) = \{(T,\ell) : T \in T \text{ and } \ell : T \to Q\}.$$

Thus an element of $T(Q)$ is a finite tree with labels from Q. (The function ℓ is called a labeling function.) We define $(T_1,\ell_1) \le (T_2,\ell_2)$ if and only if there exists an embedding $f : T_1 \to T_2$ with $\ell_1(b) \le \ell_2(f(b))$ for all $b \in T_1$. $T(Q)$ is quasiordered by this relation.

1.8 Theorem (Kruskal [18]).

$$\forall Q\ (Q \text{ wqo} \to T(Q) \text{ wqo}).$$

Proof. The proof is a slight modification of the proof of Theorem 1.3. In the last part of the proof one must again apply Ramsey's Theorem and the WQOness of $\mathcal{Q}$, exactly as in the last part of the proof of Lemma 1.6.□

§2. Nonprovability of Kruskal's Theorem in ATR_0.

The purpose of this section is to prove that Kruskal's Theorem 1.3, namely WQO($\mathcal{T}$), is not provable in a certain logical system ATR_0. We first make some remarks on ATR_0.

ATR_0 is a subsystem of second order arithmetic. The principal axiom of ATR_0 is arithmetical transfinite recursion, i.e. the assertion that arithmetical comprehension can be iterated along any countable well ordering. In particular ATR_0 includes the formal system ACA_0 of arithmetical comprehension. Therefore ATR_0 permits a convenient development of a large part of ordinary mathematics, such as e.g. the theory of continuous functions of several real variables, the Riemann integral, the theory of countable fields, the topology of complete separable metric spaces, the structure theory of separable Banach spaces and in general all applicable mathematics. (The details of the development of ordinary mathematics within ACA_0 are presented in e.g. [11], [34], [36], [40].) In addition ATR_0 permits a good theory of countable well orderings. On this basis, the axioms of ATR_0 suffice to prove many mathematical theorems which are not provable in ACA_0. For example, in ATR_0 one obtains: (i) the fact that every uncountable analytic set contains a perfect subset; (ii) Lusin's Theorem on analytic sets; (iii) determinacy of open games in ω^ω; (iv) the Ramsey property for open subsets of $[\omega]^\omega$; (v) the Ulm structure theory for countable reduced abelian p-groups. Details on ATR_0 can be found in [10], [11], [14], [32], [33], [36].

In this section we shall use Friedman's theorem ([7], [10]) that the proof theoretic ordinal of ATR_0 is exactly Γ_0. It had long been known that Γ_0 is also the proof theoretic ordinal of Feferman's formal system IR of so-called predicative analysis [5]. (See also Schütte [28].) But from the standpoint of ordinary mathematical practice, IR is much weaker than ATR_0 [10], [11], [36]. For this reason we emphasize ATR_0 rather than IR.

We use the following primitive recursive notation system for the ordinals less than Γ_0. For each ordinal $\alpha \in On$ we define a function $\varphi_\alpha : On \to On$. Namely $\varphi_0(\beta) = \omega^\beta$ and, for $\alpha > 0$, $\varphi_\alpha(\beta)$ = the βth simultaneous fixed point of the functions $\varphi_{\alpha'}$, $\alpha' < \alpha$. Then Γ_0 is the smallest ordinal $\gamma > 0$ such that $\alpha+\beta < \gamma$ and $\varphi(\alpha,\beta) < \gamma$ for all $\alpha,\beta < \gamma$. In this way each $\alpha < \Gamma_0$ is denoted by a term in $0,+,\varphi$; e.g.

$$\varepsilon_2 = \varphi(1,2) = \varphi(\varphi(0,0),\varphi(0,0)+\varphi(0,0)).$$

The ordering relation on these terms is primitive recursive.

2.1 Lemma.

1. If $\alpha \le \beta$ then $\beta \le \alpha+\beta \le \beta+\alpha \le \varphi(\alpha,\beta) \le \varphi(\beta,\alpha)$.

2. If $\alpha_1 \le \alpha_2$ and $\beta_1 \le \beta_2$, then $\alpha_1+\beta_1 \le \alpha_2+\beta_2$ and $\varphi(\alpha_1,\beta_1) \le \varphi(\alpha_2,\beta_2)$.

Proof. Clear.□

We define a primitive recursive mapping $o : \mathcal{T} \to \Gamma_0$ where $\mathcal{T}$ is the set of all finite trees. The number of elements of T will be denoted $|T|$. By induction on $|T|$ we define $o(T) < \Gamma_0$.

I. If $|T| = 1$ we set $o(T)=0$.

II. Otherwise root(T) has finitely many immediate successors $b_1,b_2,\ldots,b_m, m \ge 1$. Let $T^i = \{c \in T : c \ge b_i\}$. Thus the subtrees $T^1,T^2,\ldots,T^m$ are just the components of $T\setminus\{\text{root}(T)\}$. Since $|T^i| < |T|$, we may assume as an induction hypothesis that $o(T^i)$ is already defined. We may further assume that $o(T^1) \ge o(T^2) \ge \ldots \ge o(T^m)$. We set $\beta = o(T^1)$, $\alpha = o(T^2)$, and

$$o(T) = \begin{cases} \beta & \text{if} \quad m = 1, \\ \alpha+\beta & \text{if} \quad m = 2, \\ \beta+\alpha & \text{if} \quad m = 3, \\ \varphi(\alpha,\beta) & \text{if} \quad m = 4, \\ \varphi(\beta,\alpha) & \text{if} \quad m \geq 5. \end{cases}$$

Thus $o(T)$ is defined for all $T \in \mathcal{T}$.

2.2 Lemma. For each $\alpha < \Gamma_0$ there exists a finite tree $T \in \mathcal{T}$ with $o(T)=\alpha$.

Proof. Clear.□

For each $c \in T$ we set $T^c = \{d \in T : d \geq c\}$.

2.3 Lemma. If $c \leq d$ then $o(T^c) \leq o(T^d)$.

Proof. Clear by 2.1.1 and 2.1.2.□

2.4 Lemma. If $f : T_1 \to T_2$ is an embedding, then $o(T_1^a) \leq o(T_2^{f(a)})$ for each $a \in T_1$.

Proof. By induction on $|T_1^a|$. For $|T_1^a| = 1$ we have $o(T_1^a) = 0$ and the lemma is trivial. Let m (respectively n) be the number of immediate successors of a in T_1 (respectively of $f(a)$ in T_2). For each immediate successor b_i of a, $1 \leq i \leq m$, let c_i be the unique immediate successor of $f(a)$ in T_2 such that $f(a) < c_i \leq f(b_i)$. By the induction hypothesis and Lemma 2.3, we have $o(T_1^{b_i}) \leq o(T_2^{f(b_i)}) \leq o(T_2^{c_i})$. Also, for $1 \leq i < j \leq m$ we have $a = b_i \wedge b_j$; hence $f(a) = f(b_i) \wedge f(b_j) = c_i \wedge c_j$ and hence $c_i \neq c_j$. Thus it is clear that $m \leq n$. By 2.1.1 and 2.1.2 it follows that $o(T_1^a) \leq o(T_2^{f(a)})$.□

2.5 Lemma. If $T \leq T'$ then $o(T) \leq o(T')$.

Proof. Immediate from Lemmas 2.3 and 2.4.□

We write $WO(\Gamma_0)$ as an abbreviation for the assertion that the primitive recursive notation system for the ordinals $< \Gamma_0$ is well ordered.

2.6 Lemma. The implication $WQO(\mathcal{T}) \to WO(\Gamma_0)$ is provable in ACA_0.

Proof. We reason within ACA_0. Let $\langle \alpha_k : k < \omega \rangle$ be an arbitrary sequence of notations for ordinals $< \Gamma_0$. By Lemma 2.2 there is a corresponding sequence of finite trees $\langle T_k : k < \omega \rangle$ with $o(T_k) = \alpha_k$ for each $k < \omega$. By the assumption $WQO(\mathcal{T})$, there exist i and j such that $i < j < \omega$ and $T_i \leq T_j$. By Lemma 2.5 it follows that $\alpha_i = o(T_i) \leq o(T_j) = \alpha_j$. Thus $\langle \alpha_k : k < \omega \rangle$ is not a descending sequence. This proves $WO(\Gamma_0)$.□

2.7 Lemma (Friedman). $WO(\Gamma_0)$ is not provable in ATR_0.

Proof. See [10], [14], [32].□

2.8 Theorem (Friedman [8]). Kruskal's Theorem $WQO(\mathcal{T})$ is not provable in ATR_0.

Proof. Since $ATR_0 \supseteq ACA_0$, it follows from Lemma 2.6 that the implication $WQO(\mathcal{T}) \to WO(\Gamma_0)$ is provable in ATR_0. If $WQO(\mathcal{T})$ were provable in ATR_0, then $WO(\Gamma_0)$ would be provable in ATR_0 contradicting Lemma 2.7.□

2.9 Remark. It is clear that we could replace Γ_0 by somewhat larger ordinals, for instance Γ_{ε_0}, Γ_{Γ_0}, etc. This is interesting because e.g. according to Friedman and Jäger (see [8], [14], [32]), Γ_{ε_0} is the proof theoretic ordinal of the formal system ATR. Thus it follows that WQO(T) is not provable in ATR. But there is a limit to this: Friedman [8] has shown that WQO(T) (actually $\forall Q(Q \text{ wqo} \to T(Q) \text{ wqo})$) is provable in the formal system Π^1_2-TI_0 (cf. [32]). The proof theoretic ordinal of Π^1_2-TI_0 is strictly smaller than the Howard ordinal $\Psi_0(\varepsilon_{\Omega_1+1})$. For a more exact computation of certain ordinals which are associated with Kruskal's Theorem, see also Schmidt [27]. But [27] contains no results like e.g. Theorem 2.8.

3. A Finite Miniaturization of Kruskal's Theorem.

The main problem with which we are concerned in the present paper is the following: to find mathematically meaningful, finite combinatorial theorems which are not provable in relatively strong logical systems.

If we consider Theorem 2.8 (the nonprovability of WQO(T) in ATR_0) from the standpoint of this problem, we notice a defect. Namely, the statement WQO(T) does not belong to the domain of pure finite combinatorics, because it contains a quantifier over infinite sequences. In other words, the statement WQO(T) is not finite combinatorial because its syntactic form is Π^1_1 rather than arithmetical.

In this section we shall see how Friedman overcomes this objection. The idea is to replace the Π^1_1 statement WQO(T) with its so-called Π^0_2 finite miniaturization. In general, a finite miniaturization should have two indeispensable properties: (i) It should be mathematically not much more complicated than its infinite prototype. (ii) It should nevertheless be unprovable in relatively strong logical systems.

3.1 Definition (Friedman [8]). $LWQO(\mathcal{T})$ is the following assertion: For any c there exists a constant k which is so large that, for any finite sequence $\langle T_0, T_1, \ldots, T_k \rangle$ of finite trees with $|T_i| \le c \cdot (i+1)$ for all $i \le k$, there exist indices i and j such that $i < j \le k$ and $T_i \le T_j$.

In the statement of $LWQO(\mathcal{T})$, $|T|$ denotes the number of elements (i.e. nodes) of the finite tree T. The condition $|T_i| \le c \cdot (i+1)$ is a linear growth restriction on the sequence $\langle T_0, T_1, \ldots, T_k \rangle$. The truth of $LWQO(\mathcal{T})$ follows easily from $WQO(\mathcal{T})$ and König's Lemma [17].

We shall see that $LWQO(\mathcal{T})$ is the desired finite miniaturization of $WQO(\mathcal{T})$. It is clear that Π^0_2 assertion $LWQO(\mathcal{T})$ is a mathematically meaningful theorem of finite combinatorics which is not much more complicated than $WQO(\mathcal{T})$. As to the second desired property, we have:

3.2 Theorem (Friedman [8]). The combinatorial principal $LWQO(\mathcal{T})$ is not provable in ATR_0.

It seems appropriate to present the method of proof of this theorem in a rather high degree of generality. Let B be a fixed "reasonable" primitive recursive system of ordinal notations. (All of the notation systems which are usual in proof theory [1], [3], [13], [15], [29] are "reasonable". For the proof of Theorem 3.2, the relevant example is of course $B=\Gamma_0$.)

3.3 Definition. $WO(B)$ is the assertion that B is well ordered. For each $\beta \in B$, $WO(\beta)$ is the assertion that B is well ordered up to β, i.e. B contains no infinite descending sequence beginning with β. $PRWO(B)$ is the assertion that B is primitive recursively well ordered, i.e. B contains no infinite primitive recursive descending sequence. In an analogous way we define $PRWO(\beta)$ for each $\beta \in B$. Note that the assertions $PRWO(B)$ and $PRWO(\beta)$ are Π^0_2.

3.4 <u>Definition</u>. For $\beta \in B$ we denote by $|\beta|$ the number of symbols in β. An infinite sequence $\langle \beta_i : i < \omega \rangle$ is called <u>slow</u> if $\exists c \; \forall i \; (|\beta_i| \leq c\cdot(i+1))$. LWO(B) is the assertion that B is <u>slowly well ordered</u>, i.e. B contains no slow infinite descending sequence. By König's Lemma, LWO(B) is equivalent to the following Π^0_2 assertion: For any c there exists a k so large that B contains no finite descending sequence $\beta_0 > \beta_1 > \ldots > \beta_k$ such that $|\beta_i| \leq c \cdot (i+1)$ for all $i \leq k$. This Π^0_2 assertion is also denoted LWO(B). In an analogous way we define LWO(β) for $\beta \in B$.

3.5 <u>Theorem</u> (Friedman [8]). In ACA_0 we can prove that the following assertions are pairwise equivalent:

1. Π^0_2 soundness of the formal system $ACA_0 + \{WO(\beta) : \beta \in B\}$.
2. PRWO(B).
3. LWO(B).

<u>Proof</u>. We reason within ACA_0. The implication $2 \to 1$ is a theorem of classical proof theory (see e.g. [39], [1]). Implications $1 \to 2$ and $1 \to 3$ are trivial since PRWO(β) and LWO(β) are Π^0_2 consequences of WO(β). It remains to prove the implication $3 \to 2$.

3.6 <u>Lemma</u>. Let $\mathbb{N}$ be the set of natural numbers. For each primitive recursive function $f : \mathbb{N} \to \mathbb{N}$ there exists a primitive recursive function $g : \mathbb{N}^2 \to \omega^\omega$ such that:

(i) $g(n,m) > g(n,m+1)$ for all $m < f(n)$;

(ii) $|g(n,m)| <$ constant $\cdot$ $(n+m+1)$.

<u>Proof</u>. We shall obtain $g : \mathbb{N}^2 \to \omega^k$ for some $k < \omega$ depending on f.

I. For $f(n)=n+1$ we may take $g(n,m) = n+2 \dot{-} m$. In this case we have $g : \mathbb{N}^2 \to \omega$.

II. Let us assume that $g : \mathbb{N}^2 \to \omega^k$ satisfies the desired condition for a particular f. We shall define a $g' : \mathbb{N}^2 \to \omega^{k+1}$ which satisfies the same condition for f' where $f^0(n)=n$, $f^{i+1}(n) = f(f^i(n))$, and $f'(n) = f^n(n)$. Namely

$$g'(n,m) = \omega^k \cdot (n-i) + g(f^i(n),j)$$

where $m=f(n)+f^2(n)+\ldots+f^i(n)+j$, $i < n$, $j < f^{i+1}(n)$. We compute:

$$\begin{aligned} |g'(n,m)| &\leq \text{constant} \cdot n + \text{constant} \cdot (f^i(n)+j+1) \\ &\leq \text{constant} \cdot (n+m+1). \end{aligned}$$

III. The Grzegorczyk hierarchy is defined by $f_0(n)=n+1$, $f_{k+1}(n)=f_k^n(n)$. From I and II we obtain for each f_k a corresponding g_k which satisfies conditions (i) and (ii) for f_k. But it is known that every primitive recursive function $f(n)$ is dominated by some $f_k(\text{constant} + n)$. Thus $g_k(\text{constant} + n,m)$ satisfies conditions (i) and (ii) for f.□

3.7 <u>Lemma</u>. Given a primitive recursive descending sequence $\langle \beta_n : n \in \mathbb{N} \rangle$ in B, we can produce a slow primitive recursive descending sequence $\langle \alpha_m : m \in \mathbb{N} \rangle$ in B.

<u>Proof</u>. By Lemma 3.6 let $g : \mathbb{N}^2 \to \omega^\omega$ be primitive recursive with $g(n,j) > g(n,j+1)$ for all $j < |\beta_{n+1}|$, and $|g(n,j)| \leq \text{constant} \cdot (n+j+1)$ for all n and j. We set

$$\alpha_m = \omega^\omega \cdot \beta_n + g(n,j),$$

where $m = |\beta_1| + |\beta_2| + \ldots + |\beta_n| + j$, $j < |\beta_{n+1}|$. We compute:

$$|\alpha_m| \leq \text{constant} \cdot |\beta_n| + \text{constant} \cdot (n+j+1)$$
$$\leq \text{constant} \cdot (m+1). \square$$

Lemma 3.7 gives the implication $3 \rightarrow 2$. This completes the proof of Theorem 2.5.□

We shall now apply Theorem 3.5 to the special case $B = \Gamma_0$.

3.8 Lemma. $LWO(\Gamma_0)$ is not provable in ATR_0.

Proof. In [10] it was shown within ACA_0 that $ACA_0 + \{WO(\beta) : \beta < \Gamma_0\}$ is an axiomatization of the Π^1_1 sentences which are provable in ATR_0. By Gödel's Second Incompleteness Theorem it follows that the consistency of $ACA_0 + \{WO(\beta) : \beta < \Gamma_0\}$ is not provable in ATR_0. By Theorem 3.5 it follows that $LWO(\Gamma_0)$ is also not provable in ATR_0. □

3.9 Lemma. For each $\alpha < \Gamma_0$ there is a finite tree $T \in \mathcal{T}$ such that $o(T)=\alpha$ and $|T| \leq 4 \cdot |\alpha|$. (Compare Lemma 2.2.)

Proof. Clear.□

3.10 Lemma. The implication $LWQO(\mathcal{T}) \rightarrow LWO(\Gamma_0)$ is provable in ACA_0.

Proof. We reason within ACA_0. Let the constant c be given. By the assumption $LWQO(\mathcal{T})$ there is a k so large that there exists no bad sequence $\langle T_0, T_1, \ldots, T_k \rangle$ such that $|T_i| \leq 4c \cdot (i+1)$ for all $i \leq k$. By Lemma 3.9 and 2.5 we see that there is also no descending sequence $\Gamma_0 > \alpha_0 > \alpha_1 > \ldots > \alpha_k$ such that $|\alpha_i| \leq c \cdot (i+1)$ for all $i \leq k$. This proves $LWO(\Gamma_0)$.□

Proof of Theorem 3.2. The Theorem is an immediate consequence of Lemmas 3.8 and 3.10 and the fact that $ATR_0 \supseteq ACA_0$. □

§4. An Extension of Kruskal's Theorem.

In this section we shall describe a certain extension of Kruskal's Theorem. The proof of this Extended Kruskal Theorem is a little more complicated than that of Kruskal's Theorem itself. In the next section we shall see that the Extended Kruskal Theorem is not provable in the formal system $\Pi_1^1\text{-CA}_0$. Both the Extended Kruskal Theorem and its nonprovability in $\Pi_1^1\text{-CA}_0$ are due to Friedman.

4.1 Definition (Friedman [9]). For $n < \omega$, $\mathcal{T}_n$ is the set of all finite trees with labels from n. I.e. $(T,\ell) \in \mathcal{T}_n$ if and only if $T \in \mathcal{T}$ and $\ell : T \to \{0,1,\ldots,n-1\}$. The set $\mathcal{T}_n$ is quasiordered by putting $(T_1,\ell_1) \leq (T_2,\ell_2)$ if and only if there exists an embedding $f : T_1 \to T_2$ with the following properties:

(i) for each $b \in T_1$ we have $\ell_1(b) = \ell_2(f(b))$;

(ii) if b is an immediate successor of $a \in T_1$, then for each $c \in T_2$ in the interval $f(a) < c < f(b)$ we have $\ell_2(c) \geq \ell_2(f(b))$.

The condition (ii) in the above definition is called a gap condition.

4.2 Theorem (Friedman [9]). For each $n < \omega$, $\mathcal{T}_n$ is a well quasiordering.

For the proof of this Theorem, we need the following generalized version. For $b \in T \in \mathcal{T}$ we write $T^b = \{a \in T : a \geq b\}$. If $T^b = T$, b is called the root of T. If $T^b = \{b\}$, b is called an end node of T. If $T^b \neq \{b\}$, b is called an interior node of T.

4.3 Definition. Given $n < \omega$ and an arbitrary quasiordering $\mathcal{Q}$, let $\mathcal{T}_n(\mathcal{Q})$ be the set of all finite labeled trees in which the end nodes are labeled from $\mathcal{Q}$ and the interior nodes are labeled from n. I.e. $(T,\ell) \in \mathcal{T}_n(\mathcal{Q})$ if and only if $T \in \mathcal{T}$, $\ell : T \to \mathcal{Q} \cup \{0,1,\ldots,n-1\}$, $\ell(b) \in \mathcal{Q}$ for all end nodes b, and $\ell(b) \in \{0,1,\ldots,n-1\}$ for all interior nodes b. The set $\mathcal{T}_n(\mathcal{Q})$ is quasiordered by putting $(T_1,\ell_1) \leq (T_2,\ell_2)$ if and only if there exists an embedding $f : T_1 \to T_2$ with the following properties:

(i) If b is an end node of T_1, then $f(b)$ is an end node of T_2 and $\ell_1(b) \leq \ell_2(f(b))$. (Compare Definition 1.7.)

(ii) If b is an interior node of T_1, then $f(b)$ is an interior node of T_2 and $\ell_1(b) = \ell_2(f(b))$.

(iii) If a and b are interior nodes of T_1 with b an immediate successor of a, then for each $c \in T_2$ in the interval $f(a) < c < f(b)$ we have $\ell_2(c) \geq \ell_2(f(b))$.

We also need the following variant:

4.4 <u>Definiton</u>. $\mathcal{T}_n^+(\mathcal{Q})$ is the same set of finite labeled trees as $\mathcal{T}_n(\mathcal{Q})$. The quasiordering of $\mathcal{T}_n^+(\mathcal{Q})$ is defined exactly as for $\mathcal{T}_n(\mathcal{Q})$ but with the following additional gap condition on the embeddings $f : T_1 \to T_2$:

(iv) If $b = \text{root}(T_1)$ is an interior node of T_1, then for each $c \in T_2$ in the interval $c < f(b)$ we have $\ell_2(c) \geq \ell_2(f(b))$.

We shall show by induction on $n < \omega$ that $\forall\mathcal{Q}(\mathcal{Q}\text{ wqo} \to \mathcal{T}_n^+(\mathcal{Q})\text{ wqo})$. In particular $\forall\mathcal{Q}(\mathcal{Q}\text{ wqo} \to \mathcal{T}_n(\mathcal{Q})\text{ wqo})$, and then the WQOness of $\mathcal{T}_n$ is essentially the special case where $\mathcal{Q}$ is the trivial 1-element well quasiordering.

For a given $(T,\ell) \in \mathcal{T}_n(\mathcal{Q})$ we shall often write simply T instead of (T,ℓ). Then for each $b \in T$ we shall write T^b instead of $(T^b, \ell|T^b)$, and $\text{label}(b)=\ell(b)$.

Given $T \in \mathcal{T}_{n+1}^+(\mathcal{Q})$ with $T \neq \{\text{root}(T)\}$, we set $m = \mu(T) = \min\{\text{label}(b):$ b is an interior node of $T\}$. We then define T^* to be the tree which results when we, first, convert each interior node b of T which is minimal with label m to an end node of T^* with label T^b, and second, subtract 1 from the label of each remaining interior node of T. Thus $T^* \in \mathcal{T}_n^+(\mathcal{T}_{n+1}(\mathcal{Q}))$.

4.5 Lemma. If $\mu(T_1) = \mu(T_2) = m$ and $T_1^* \leq T_2^*$ in $\mathcal{T}_n^+(\mathcal{T}_{n+1}(\mathcal{Q}))$, then $T_1 \leq T_2$ in $\mathcal{T}_{n+1}^+(\mathcal{Q})$.

Proof. Straightforward.□

4.6 Lemma. If $\forall \mathcal{Q}(\mathcal{Q}$ wqo $\to \mathcal{T}_n^+(\mathcal{Q})$ and $\mathcal{T}_{n+1}(\mathcal{Q})$ wqo), then $\forall \mathcal{Q}(\mathcal{Q}$ wqo $\to \mathcal{T}_{n+1}^+(\mathcal{Q})$ wqo).

Proof. Let $\langle T_k : k < \omega \rangle$ be an arbitrary sequence from $\mathcal{T}_{n+1}^+(\mathcal{Q})$. We want to show that $\exists i\, \exists j\ (i < j$ and $T_i \leq T_j)$. We may assume that $\forall k(T_k \neq \{\text{root}(T_k)\})$ and $\exists m\, \forall k(\mu(T_k) = m)$. We consider the sequence $\langle T_k^* : k < \omega \rangle$ from $\mathcal{T}_n^+(\mathcal{T}_{n+1}(\mathcal{Q}))$. By assumption $\mathcal{T}_n^+(\mathcal{T}_{n+1}(\mathcal{Q}))$ is wqo. Hence $\exists i \exists j(i < j$ and $T_i^* \leq T_j^*)$. But then by Lemma 4.5 it follows that $T_i \leq T_j$.□

4.7 Lemma. If $\forall \mathcal{Q}(\mathcal{Q}$ wqo $\to \mathcal{T}_n^+(\mathcal{Q})$ wqo), then $\forall \mathcal{Q}(\mathcal{Q}$ wqo $\to \mathcal{T}_{n+1}(\mathcal{Q})$ wqo).

Proof. If not, let $\langle T_k : k < \omega \rangle$ be a minimal bad sequence (compare §1) from $\mathcal{T}_{n+1}(\mathcal{Q})$. We set

$$S = \{T_k^b : k < \omega,\ b \in T_k \setminus \{\text{root}(T_k)\}\}.$$

We first claim that S is wqo as a subset of $\mathcal{T}_{n+1}(\mathcal{Q})$. Otherwise there would be a bad sequence of the form $\langle T_{k_i}^{b_i} : i < \omega \rangle$ where $k_i < k_j$ and $b_i \in T_{k_i} \setminus \{\text{root}(T_{k_i})\}$ for all i and j with $i < j < \omega$. But then the sequence $\langle T_0, T_1, \ldots, T_{k_0 - 1}, T_{k_0}^{b_0}, T_{k_1}^{b_1}, \ldots \rangle$ would also be bad, contradicting the minimality of T_{k_0}.

Next we claim that S is wqo as a subset of $\mathcal{T}_{n+1}^+(\mathcal{Q})$. Set $S_\infty = \{T \in S : T = \{\text{root}(T)\}\}$. Since $\mathcal{Q}$ is wqo, it is clear that S_∞ is wqo in $\mathcal{T}_{n+1}^+(\mathcal{Q})$. For each $m \leq n$ set $S_m = \{T \in S : \mu(T) = m\}$ and

$S_m^* = \{T^* : T \in S_m\}$. Thus $S_m^* \subseteq T_n^+(S)$ and by our assumption it follows that S_m^* is wqo in $T_n^+(T_{n+1}(Q))$. Hence by Lemma 4.5 S_m is wqo in $T_{n+1}^+(Q)$. It now follows that

$$S = S_\infty \cup \bigcup_{m=0}^{n} S_m$$

is wqo in $T_{n+1}^+(Q)$. Our claim is proved.

Now for all $k < \omega$ let $T_k^1, T_k^2, \ldots, T_k^{n_k}, n_k \geq 0$ be the components of $T_k \setminus \{\text{root}(T_k)\}$. The finite sequences $\langle T_k^1, T_k^2, \ldots, T_k^{n_k} \rangle$, $k < \omega$ belong to $S^{<\omega}$ and therefore, by Higman's Lemma 1.6, are well quasiordered in $(T_{n+1}^+(Q))^{<\omega}$. Thus there exist i and j such that $i < j < \omega$ and label(root(T_i)) = label(root(T_j)) and $\langle T_i^1, T_i^2, \ldots, T_i^{n_i} \rangle \leq \langle T_j^1, T_j^2, \ldots, T_j^{n_j} \rangle$ in $(T_{n+1}^+(Q))^{<\omega}$. It follows immediately that $T_i \leq T_j$ in $T_{n+1}(Q)$ (or even in $T_{n+1}^+(Q)$). But this contradicts the badness of $\langle T_k : k < \omega \rangle$ in $T_{n+1}(Q)$.□

4.8 Theorem (Friedman [9]). For each $n < \omega$, $\forall Q(Q \text{ wqo} \rightarrow T_{n+1}^+(Q) \text{ wqo})$.

Proof. From Lemmas 4.6 and 4.7 by induction on $n < \omega$. As the basis of the induction we take $\forall Q(Q \text{ wqo} \rightarrow T_0^+(Q) \text{ wqo})$, which is trivial since $T_0^+(Q) = Q$.□

Proof of Theorem 4.2. Theorem 4.2 is a special case of Theorem 4.8.□

5. Nonprovability of the Extended Kruskal Theorem in Π_1^1-CA_0.

In this section we shall that Friedman's Extended Kruskal Theorem 4.2, namely $\forall n\ \text{WQO}(T_n)$, is not provable in the logical system Π_1^1-CA_0. Here Π_1^1-CA_0 is the subsystem of second order arithmetic with the Π_1^1 comprehension axiom, $\exists X \forall n(n \in X \leftrightarrow \varphi(n))$ where φ is an arbitrary Π_1^1 formula, and the restricted induction axiom $(0 \in X \wedge \forall n(n \in X \rightarrow n+1 \in X)) \rightarrow \forall n(n \in X)$.

The system Π^1_1-CA_0 is very strong when viewed from the standpoint of ordinary mathematical practice. The axioms of Π^1_1-CA_0 are much stronger than those of ATR_0. Thus they lead to an improved theory of countable well orderings. On this basis Π^1_1-CA_0 suffices to prove many theorems of e.g. classical descriptive set theory which are not provable in ATR_0. For example one obtains in Π^1_1-CA_0: (i) the Cantor-Bendixson Theorem; (ii) Kondo's Uniformization Theorem; (iii) Silver's theorem on coanalytic equivalence relations; (iv) the Ulm structure theory for arbitrary countable abelian groups; (v) the Ramsey property for G_δ subsets of $[\omega]^\omega$. For details on Π^1_1-CA_0 see [7], [11], [30], [32], [36].

It is not an exaggeration to say that almost all theorems of ordinary mathematics which are expressible in the language of second order arithmetic are already provable in Π^1_1-CA_0. It is therefore especially interesting that the combinatorial theorem $\forall n\ WQO(T_n)$ is not provable in Π^1_1-CA_0.

From the standpoint of mathematical logic Π^1_1-CA_0 is also very interesting. It is known that Π^1_1-CA_0 is equivalent to a certain formal system of first order arithmetic with Ramsey quantifiers [26]. Π^1_1-CA_0 is also equivalent to the system $ID^{<\omega}$ of [2]. The proof theoretic ordinal of Π^1_1-CA_0 is $\Psi_0(\Omega_\omega)$ ($= \theta\,\Omega_\omega 0$ in the notation system of [29]). Here $\Omega_\omega = \aleph_\omega$ (the smallest singular uncountable initial ordinal) and $\Psi_0 : On \to \Omega_1$ is a collapsing function.

We use the following notation system for the provable ordinals of Π^1_1-CA_0.

5.1 <u>Definition</u> (Buchholz [1]). We have constants $\Omega_0 = 0$ and $\Omega_n = \aleph_n$, $1 \le n < \omega$. We have a 2-place function $\alpha + \beta$ and a 1-place function ω^α. We also have collapsing functions $\Psi_i(\alpha)$, $i < \omega$, which are defined by induction on α. First let $C_i(\alpha)$ be the smallest set of ordinals such that:

1. $\{\Omega_n : n < \omega\} \cup \Omega_i \subseteq C_i(\alpha)$;
2. if $\xi, \eta \in C_i(\alpha)$, then also $\xi+\eta,\ \omega^\xi \in C_i(\alpha)$;

3. if $\xi \in C_i(\alpha) \cap \alpha$, then also

$$\Psi_k(\xi) \in C_i(\alpha) \quad \text{for each } k \geq i, \; k < \omega.$$

Then $\Psi_i(\alpha)$ is defined to be the smallest β with $\beta \notin C_i(\alpha)$.

5.2 Facts. 1. $\Omega_i < \Psi_i(\alpha) < \Omega_{i+1}$ for all α.

2. If $\alpha \leq \beta$ then $\Psi_i(\alpha) \leq \Psi_i(\beta)$.

3. $\Psi_i(\alpha) = C_i(\alpha) \cap \Omega_{i+1}$.

4. If $\xi < \Psi_i(\alpha)$ then also $\omega^\xi < \Psi_i(\alpha)$. In other words, $\Psi_i(\alpha)$ is an ε-number.

5. If $\alpha \notin C_i(\alpha)$, then $C_i(\alpha+1) = C_i(\alpha)$ and $\Psi_i(\alpha+1) = \Psi_i(\alpha)$.

6. If $\alpha \in C_i(\alpha)$ then $\Psi_i(\alpha+1) = \Psi_i(\alpha)^+$ where β^+ denotes the smallest ε-number $> \beta$.

7. For each limit ordinal δ, $\Psi_i(\delta) = \sup \{\Psi_i(\alpha) : \alpha < \delta\}$.

5.3 Facts. 1. Let $\xi = \xi_1 + \ldots + \xi_k$ be given, where $\xi_1 \geq \ldots \geq \xi_k$ and $\xi_1, \ldots, \xi_k$ are additively indecomposable. If $\xi \in C_i(\alpha)$ then also $\xi_1, \ldots, \xi_k \in C_i(\alpha)$.

2. If $\omega^\xi \in C_i(\alpha)$ then also $\xi \in C_i(\alpha)$.

3. For each $\eta < \Omega_\omega$ and $k < \omega$, there is a unique ξ with $\Psi_k(\xi) = \Psi_k(\eta)$ and $\xi \in C_k(\xi)$.

4. If $\Psi_k(\xi) \in C_i(\alpha)$ where $\xi \in C_k(\xi)$ and $k \geq i$, then also $\xi \in C_i(\alpha) \cap \alpha$.

Remark. The proofs of Facts 5.2 and 5.3 except for 5.3.4 are relatively easy. The proof of 5.3.4 is somewhat more difficult. From 5.3.4 it follows that, in 5.1.3, we could have strengthened the assumption $\xi \in C_i(\alpha) \cap \alpha$ to say that $\xi \in C_i(\alpha) \cap \alpha \cap C_k(\xi)$. With this change in the definition, 5.3.4

and the other essential properties of the collapsing functions Ψ_i, $i < \omega$ could have been proved more easily. For details on the ordinal notation system 5.1, see [1], [3], [4], [13], [15].

5.4 Definition. By induction we define the set NF of terms in normal form.

1. For each $i < \omega$ we have $\Omega_i \in$ NF.

2. If $\alpha_1 \geq \ldots \geq \alpha_k > 0$, $k \geq 2$ with $\alpha_i \in$ NF and additively indecomposable, then also the sum $\alpha_1 + \ldots + \alpha_k \in$ NF.

3. If $\alpha \in$ NF with $\alpha < \omega^\alpha$, then also $\omega^\alpha \in$ NF.

4. If $\alpha \in$ NF and $\alpha \in C_i(\alpha)$, then also $\Psi_i(\alpha) \in$ NF.

In this way each $\alpha \in C_0(\Omega_\omega)$ is denoted by a unique NF term. Both the set of all NF terms and the corresponding ordering relation are primitive recursive [1], [4], [13], [15]. In what follows we shall identify $C_0(\Omega_\omega)$ with NF.

Following the pattern of §2, we shall prove in ACA_0 that $\forall n(WQO(T_n) \rightarrow WO(\Psi_0(\Omega_n)))$ holds. We need a certain primitive recursive mapping o. The domain of o is a certain primitive recursive subset of $\cup\{T_n : n < \omega\}$. The range of o is $C_0(\Omega_\omega)$.

Given $T \in \cup\{T_n : n < \omega\}$; we set $i = \text{label}(\text{root}(T))$. The definition of $o(T)$ is as follows:

I. If $T = \{\text{root}(T)\}$, then we set $o(T) = \Omega_i$. Otherwise root(T) has finitely many immediate successors $b_1,\ldots,b_m, m \geq 1$. We may assume that $\beta_1 \geq \ldots \geq \beta_m$ where $\beta_j = o(T^{b_j})$, $1 \leq j \leq m$.

II. If $m=2$, $\text{label}(b_1) = i$, $\beta_1 = \alpha_1+\ldots+\alpha_{k-1}$, $\beta_2 = \alpha_k$, $\alpha_1 \geq \ldots \geq \alpha_{k-1} \geq \alpha_k > 0$, $k \geq 2$ where each α_j is additively indecomposable, then we set $o(T) = \beta_1 + \beta_2$.

III. If $m=3$, $\text{label}(b_1) = i$, $\beta_1 < \omega^{\beta_1}$, $\beta_2 = \beta_3 = 0$, then we set $o(T) = \omega^{\beta_1}$.

IV. If $m=4$, $\beta_1 \in C_i(\beta_1)$, $\beta_2 = \beta_3 = \beta_4 = 0$, then we set $o(T) = \Psi_i(\beta_1)$.

V. Otherwise $o(T)$ is undefined.

5.5 Lemma. For each $\alpha \in C_0(\Omega_\omega)$ there is a unique tree $T \in \bigcup\{\mathcal{T}_n : n < \omega\}$ such that $\alpha = o(T)$. In this situation we have: (i) $\text{label}(\text{root}(T)) = i$ if and only if $\Omega_i \leq \alpha < \Omega_{i+1}$; (ii) $T \in \mathcal{T}_n$ if and only if $\alpha \in C_0(\Omega_n) \cap \Omega_n$.

Proof. Let $\alpha \in C_0(\Omega_\omega)$ be given. Then α is denoted by a unique NF term. From the definition of o it is clear that to each NF term α there corresponds a unique labeled tree T with $o(T) = \alpha$. Assertion (i) is clear by induction on $|T|$. Assertion (ii) is a consequence of the observation that $\alpha \in C_0(\Omega_n) \cap \Omega_n$ if and only if the corresponding NF term contains no occurrences of Ω_k or Ψ_k, $k \geq n$.□

5.6 Definition. For each $i < \omega$ and $\alpha \in NF$ we define the i-subterms of α by induction on $|\alpha|$, where $|\alpha|$ denotes the number of symbols in α.

1. Every $\alpha \in NF$ is itself an i-subterm of α.

2. For $\alpha = \alpha_1 + \ldots + \alpha_k \in NF$, every i-subterm of α_j, $1 \leq j \leq k$ is also an i-subterm of α.

3. For $\omega^\gamma \in NF$, every i-subterm of γ is also an i-subterm of ω^γ.

4. For $k \geq i$ and $\Psi_k(\gamma) \in NF$, every i-subterm of γ is also an i-subterm of $\Psi_k(\gamma)$.

5.7 Lemma. Let β be an i-subterm of $\alpha \in NF$.

1. If $\alpha \in C_i(\gamma)$ then $\beta \in C_i(\gamma)$.

2. If $\beta < \Omega_{i+1}$ then $\beta \leq \alpha$.

Proof. Part (1) is easily proved by induction on $|\alpha|$ with the help of Lemma 5.3. The proof of part (2) is also carried out by induction on $|\alpha|$.

First we note that the desired conclusion $\beta \le \alpha$ is an immediate consequence of the induction hypothesis, provided β is an i-subterm of α via 5.6.1 or 5.6.2 or 5.6.3. There remains the case where β is an i-subterm of α via 5.6.4, i.e. $\alpha = \Psi_k(\gamma)$ where $k \ge i$ and β is an i-subterm of γ. If $k > i$, we have $\beta < \Omega_{i+1} < \Psi_k(\gamma) = \alpha$ by 5.2.1. So we may assume that $k=i$. Since $\Psi_i(\gamma) = \alpha \in NF$, we have $\gamma \in C_i(\gamma)$. Hence by part (1) we get $\beta \in C_i(\gamma)$. Then by the assumption $\beta < \Omega_{i+1}$ and 5.2.3, it follows that $\beta < \Psi_i(\gamma) = \alpha$.□

5.8 Lemma. If $o(T)$ is defined and $b \in T$ with $\text{label}(c) \ge \text{label}(b)$ for all $c < b$, then $o(T^b) \le o(T)$.

Proof. Set $\beta = o(T^b)$ and $i = \text{label}(b)$. By 5.5 (i) we have $\beta < \Omega_{i+1}$. By our assumption, it is clear that β is an i-subterm of $o(T)$. By 5.7.2 we conclude $\beta \le o(T)$.□

5.9 Lemma. If $f : T_1 \to T_2$ is an embedding in $\mathcal{T}_n$, then $o(T_1^a) \le o(T_2^{f(a)})$ for all $a \in T_1$ (provided $o(T_1^a)$ and $o(T_2^{f(a)})$ are defined).

Proof. By induction on $|T_1^a|$. We set $i = \text{label}(a) = \text{label}(f(a))$.

I. If $T_1^a = \{a\}$, then we have $o(T_1^a) = \Omega_i \le o(T_2^{f(a)})$ by 5.5 (i).

Otherwise let $b_1,\dots,b_m$ (respectively $c_1,\dots,c_n$) be the immediate successors of a in T_1 (respectively of $f(a)$ in T_2). Thus $2 \le m \le n \le 4$. For each j with $1 \le j \le m$, we may assume that $f(a) < c_j \le f(b_j)$. We have $o(T_1^{b_j}) \le o(T_2^{f(b_j)})$ by the induction hypothesis, and $o(T_2^{f(b_j)}) \le o(T_2^{c_j})$ because of the gap condition 4.1 (ii) and Lemma 5.8. Thus $\beta_j \le \gamma_j$ where $\beta_j = o(T_1^{b_j})$ and $\gamma_j = o(T_2^{c_j})$. We may assume that $\beta_1 \ge \dots \ge \beta_m$. Set $\alpha = o(T_1^a)$ and $\delta = o(T_2^{f(a)})$. With this notation, our lemma asserts that $\alpha \le \delta$.

II. If $m=2$, we have $\alpha = \beta_1 + \beta_2$ where $\beta_1 \geq \beta_2 > 0$. Hence $\gamma_1 > 0$ and $\gamma_2 > 0$. Hence we must have $n=2$. Thus either $\alpha = \beta_1 + \beta_2 \leq \gamma_1 + \gamma_2 = \delta$ (if $\gamma_1 \geq \gamma_2$) or $\alpha = \beta_1 + \beta_2 \leq \gamma_2 + \gamma_1 = \delta$ (if $\gamma_2 \geq \gamma_1$).

III. If $m=3$, we have $\alpha = \omega^{\beta_1}$. If also $n=3$, then $\alpha = \omega^{\beta_1} \leq \omega^{\gamma_1} \leq \delta$. Now assume that $n=4$. We have $\text{label}(f(b_1)) = \text{label}(b_1) = i$. By the gap condition 4.1 (ii), $o(T_2^{f(b_1)})$ is a proper i-subterm of δ. By 5.7.2 and the uniqueness of NF terms, it follows that $o(T_2^{f(b_1)}) < \delta$. Thus $\beta_1 = o(T_1^{b_1}) \leq o(T_2^{f(b_1)}) < \delta$. By 5.2.4 it follows that $\alpha = \omega^{\beta_1} < \delta$.

IV. If $m=4$, we have $\alpha = \Psi_i(\beta_1) \leq \Psi_i(\gamma_1)$ and necessarily $n=4$, hence $\Psi_i(\gamma_1) \leq \delta$, so $\alpha \leq \delta$.□

5.10 Lemma. If $o(T_1)$ and $o(T_2)$ are defined and $\text{label}(\text{root}(T_1))=0$ and $T_1 \leq T_2$, then $o(T_1) \leq o(T_2)$.

Proof. Set $a = \text{root}(T_1)$. By 5.9 we have $o(T_1) = o(T_1^a) \leq o(T_2^{f(a)})$. Also $\text{label}(f(a)) = \text{label}(a) = 0$, hence $o(T_2^{f(a)}) \leq o(T_2)$ by 5.8.□

5.11 Lemma. $\forall n(\text{WQO}(\mathcal{T}_{n+1}) \to \text{WO}(\Psi_0(\Omega_{n+1})))$ is provable in ACA_0.

Proof. We reason within ACA_0. Let $\langle \alpha_k : k < \omega \rangle$ be an arbitrary sequence of NF terms $< \Psi_0(\Omega_{n+1})$. By Lemma 5.5 let $\langle T_k : k < \omega \rangle$ be a corresponding sequence of labeled trees $T_k \in \mathcal{T}_{n+1}$ such that $o(T_k) = \alpha_k$. Then $\text{label}(\text{root}(T_k)) = 0$ for all $k < \omega$. By the assumption $\text{WQO}(\mathcal{T}_{n+1})$, there exist indices i and j such that $i < j$ and $T_i \leq T_j$. Hence $\alpha_i \leq \alpha_j$ by 5.10. Thus $\langle \alpha_k : k < \omega \rangle$ is not a descending sequence. This proves $\text{WO}(\Psi_0(\Omega_{n+1}))$.□

5.12 Lemma. The implication $(\forall n \; \text{WQO}(\mathcal{T}_n)) \to \text{WO}(\Psi_0(\Omega_\omega))$ is provable in ACA_0.

Proof. Clear from 5.11 since $\Psi_0(\Omega_\omega) = \sup\{\Psi_0(\Omega_n) : n < \omega\}$.□

5.13 Lemma (Buchholz). $WO(\Psi_0(\Omega_\omega))$ is not provable in $\Pi^1_1\text{-}CA_0$.

Proof. See [1], [4], [15], [29].

5.14 Theorem (Friedman [9]). $\forall n\ WQO(\mathcal{T}_n)$ is not provable in $\Pi^1_1\text{-}CA_0$.

Proof. The theorem follows immediately from 5.13 and 5.12 since $ATR_0 \supseteq ACA_0$.□

Exactly as in §3, we can formulate a Π^0_2 finite miniaturization of the Π^1_1 combinatorial principle $\forall n\ WQO(\mathcal{T}_n)$:

5.15 Definition. $LWQO(\mathcal{T}_n)$ is the following assertion: For any c there exists a k so large that for every finite sequence $\langle T_0, T_1, \ldots, T_k \rangle$ such that $T_i \in \mathcal{T}_n$ and $|T_i| \leq c \cdot (i+1)$ for all $i \leq k$, there exist indices i and j with $i < j \leq k$ and $T_i \leq T_j$.

5.16 Theorem (Friedman [9]). The Π^0_2 combinatorial principle $\forall n\ LWQO(\mathcal{T}_n)$ is not provable in $\Pi^1_1\text{-}CA_0$.

Proof. Exactly as for Theorem 3.2. Along with 3.5 for $B = \Psi_0(\Omega_\omega)$, we also need the proof theoretic result that $PRWO(\Psi_0(\Omega_\omega))$ is not provable in $\Pi^1_1\text{-}CA_0$. See [1], [24], [39].□

References

[1] W. Buchholz, A new system of proof-theoretic ordinal functions, in prepartion. (See also [1a].)

[1a] W. Buchholz, An independence result for $(\Pi^1_1\text{-}CA)+BI$, Annals of Pure and Applied Logic, to appear.

[2] W. Buchholz, S. Feferman, W. Pohlers und W. Sieg, Iterated Inductive Definitions and Subsystems of Analysis: Recent Proof-Theoretical Studies, Lecture Notes in Mathematics 897, Springer-Verlag, 1981, 383 pages.

[3] W. Buchholz and K. Schütte, Ein Ordinalzahlensystem für die beweistheoretische Abgrenzung der Π^1_2-Separation und Bar-Induktion, Sitzungsberichte der Bayerischen Akademie der Wiss., Math.-Naturw. Klasse, 1983, pp. 99-132.

[4] W. Buchholz and K. Schütte, Proof Theoretic Ordinals of Impredicative Subsystems of Analysis, in preparation.

[5] S. Feferman, Systems of predicative analysis, Journal of Symbolic Logic 29 (1964), pp. 1-30; 33 (1968), pp. 193-220.

[6] H. Friedman, Some systems of second order arithmetic and their use, Procedings of the International Congress of Mathematicians (Vancouver, 1974), vol. 1, Canadian Mathematical Congress, 1975, pp. 235-242.

[7] H. Friedman, Systems of second order arithmetic with restricted induction (abstracts), Journal of Symbolic Logic 41 (1976), pp. 557-559.

[8] H. Friedman, Independence results in finite graph theory I-VII, unpublished manuscripts, Ohio State University, February-March 1981, 76 pages.

[9] H. Friedman, Beyond Kruskal's theorem I-III, unpublished manuscripts, Ohio State University, June-July 1982, 48 pages.

[10] H. Friedman, K. McAloon and S. G. Simpson, A finite combinatorial principle which is equivalent to the 1-consistency of predicative analysis, Patras Logic Symposion (edited by G. Metakides), North-Holland, 1982, pp. 197-230.

[11] H. Friedman, S. G. Simpson and R. L. Smith, Countable algebra and set existence axioms, Annals of Pure and Applied Logic 25 (1983), pp. 141-181.

[12] G. Higman, Ordering by divisibility in abstract algebras, Proc. Lond Math. Soc. 2 (1952), pp. 326-336.

[13] G. Jäger, ρ-inaccessible ordinals, collapsing functions, and a recursive notation system, Archiv f. math. Logik u. Grundlagenf. 24 (1984), pp. 49-62.

[14] G. Jäger, The strength of admissibility without foundation, Journal of Symbolic Logic 49 (1984), pp. 233-245.

[15] G. Jäger and W. Pohlers, Admissible Proof Theory, Springer-Verlag, in prepartion.

[16] L. Kirby and J. Paris, Accessible independence results for Peano arithmetic, Bull. London Math. Soc. 14 (1982), pp. 285-293.

[17] D. König, Über eine Schlussweise aus dem Endlichen ins Unendliche, Acta Litterarum ac Scientarum (Ser. Sci. Math.) Szeged 3 (1927), 121-130.

[18] J. Kruskal, Well-quasi-ordering, the tree theorem, and Vázsonyi's conjecture, Trans. Amer. Math. Soc. 95 (1960), pp. 210-225.

[19] R. Laver, Well-quasi-orderings and sets of finite sequences, Math. Proc. Cambridge Phil. Soc. 79 (1976), pp. 1-10.

[20] R. Laver, Better-quasi-orderings and a class of trees, Studies in Foundations and Combinatorics, Adv. in Math. Supplementary Studies 1 (1978), pp. 31-48.

[21] W. Mader, Wohlquasigeordnete Klassen endlicher Graphen, Journal of Combinatorial Theory B, 12 (1972), pp. 105-122.

[22] C.St.J.A. Nash-Williams, On well-quasi-ordering finite trees, Proc. Cambridge Phil. Soc. 59 (1963), pp. 833-835.

[23] J. Paris and L. Harrington, A mathematical incompleteness in Peano arithmetic, Handbook of Mathematical Logic (edited by J. Barwise), North-Holland, 1977, pp. 1133-1142.

[24] W. Pohlers, An upper bound for the provablility of transfinite induction in systems with n-times iterated inductive definitions, Proof Theory Symposion (Kiel 1974), Lecture Notes in Mathematics 500, Springer-Verlag, 1975, pp. 271-289.

[25] F. P. Ramsey, On a problem of formal logic, Proc. London Math. Soc. 30 (1930), pp. 264-286.

[25a] N. Robertson and P. D. Seymour, Generalizing Kuratowski's Theorem, preprint, 1984, 10 pages.

[26] J. H. Schmerl and S. G. Simpson, On the role of Ramsey quantifiers in first order arithmetic, Journal of Symbolic Logic 47 (1982), pp. 423-435.

[27] D. Schmidt, Well-partial orderings and their maximal order types, Habilitationsschrift, Heidelberg, 1979, 77 pages.

[28] K. Schütte, Predicative well-orderings, Formal Systems and Recursive Functions (edited by J. Crossley and M. Dummett), North-Holland, 1965, pp. 280-303.

[29] K. Schütte, Proof Theory, Springer-Verlag, 1977, 299 pages.

[30] S. Shelah, On logical sentences in PA, Spring 1983, 15 pages.

[31] S. G. Simpson, BQO theory and Fraisse's conjecture, Chapter 9 of: R. Mansfield and G. Weitkamp, Descriptive Set Theory, Oxford Logic Guides, 1985, 127 pages.

[32] S. G. Simpson, Σ_1^1 and Π_1^1 transfinite induction, Logic Colloquim '80 (edited by D. van Dalen, D. Lascar and J. Smiley), North-Holland, 1982, pp. 239-253.

[33] S. G. Simpson, Set theoretic aspects of ATR_0, Logic Colloquim '80 (edited by D. van Dalen, D. Lascar and J. Smiley), North-Holland, 1982 pp. 254-271.

[34] S. G. Simpson, Which set existence axioms are needed to prove the Cauchy/Peano theorem for ordinary differential equations?, Journal of Symbolic Logic 49 (1984), pp. 361-380.

[35] S. G. Simpson, Reverse Mathematics, Proceedings of the AMS Summer Institute in Recursion Theory (edited by A. Nerode and R. Shore), in press.

[36] S. G. Simpson, Subsystems of Second Order Arithmetic, in preparation.

[37] J. Steel, Determinateness and subsystems of analysis, Ph.D. Thesis, Berkeley, 1976, 107 pages.

[38] G. Takeuti, Ordinal diagrams, J. Math. Soc. Japan 9 (1957), pp. 386-394; 12 (1960), pp. 385-391.

[39] G. Takeuti, Proof Theory, North-Holland, 1975, 372 pages.

[40] G. Takeuti, Two Applications of Logic to Mathematics, Iwanami Shoten and Princeton University Press, 1978, 139 pages.

Correction. The inequality $\beta+\alpha \leq \varphi_\alpha(\beta)$ of Lemma 2.1.1 can fail if $\varphi_\alpha(\beta)=\beta$. Thus Lemma 2.4 does not hold unqualifiedly. However, it is straightforward to repair the proof of the key Lemma 2.6. This can be done by restricting attention to terms which contain no subterm of the form $\varphi_\alpha(\beta)$ where β is a fixed point of φ_α. (Compare the use of normal forms in §5.)

HARVEY FRIEDMAN'S RESEARCH ON THE FOUNDATIONS
OF MATHEMATICS, L.A. Harrington et al. (editors)

The Consistency Strengths of Some Finite Forms of the Higman and Kruskal Theorems

Rick L. Smith

Section 1: Introduction

If T is a theory and σ is a sentence in the language of T, we say that σ is independent from T if neither σ nor its negation is provable from T. The classical work of Gödel shows how to produce an independent sentence from any consistent, axiomatizable theory which contains a sufficiently strong fragment of arithmetic. This proof settled an important issue, but it left open the possibility that mathematicians, in the course of their customary enterprise, may never formulate a mathematical sentence which is independent of a reasonably strong theory. The combined work of Gödel and Cohen showed that mathematicians had formulated sentences which were independent of Zermelo-Fraenkel set theory, namely the Axiom of Choice and the Continuum Hypothesis. Later, work of Paris and Harrington [6] gave an example of a finite combinatorial statement, in fact a finite combinatorial variation of Ramsey's Theorem, which is independent from Peano Arithmetic (PA). In this paper we will give recent results of Harvey Friedman. Friedman shows how finite combinatorial variations of the theorems of Higman and Kruskal are independent from some subsystems of second order arithmetic. (In [2], he shows that the original Higman and Kruskal Theorems are independent from such systems).

Let Q be the set of all finite trees and write $T_1 \leq T_2$ if there is a homeomorphic (infimum preserving) embedding of T_1 into T_2. The basic Higman-Kruskal theorem says

If $T_1, T_2, \ldots$ is an infinite sequence of trees, then there exist $i < j$ such that $T_i \leq T_j$.

Higman proved this if a bound is placed on the splitting (valence) of the trees (different notation). Later, Kruskal proved the full theorem.

This is a Π^1_1 sentence. There are several ways to turn it into a closely related Π^0_2 sentence. Let $|T|$ = number of vertices in T. We consider two statements asserting the "combinatorial well-foundedness" of Q.

$$\mathrm{CWF}(Q) : \forall k \exists n \forall (T_1, \ldots, T_n)(|T_i| \leq k+i \rightarrow \exists i < j \; T_i \leq T_j)$$

A stronger form is

$$\mathrm{SCWF}(Q) : \forall k \exists n \forall (T_1, \ldots, T_n)(|T_i| \leq i \rightarrow \exists i_1 < \ldots < i_k (T_{i_1} \leq \ldots \leq T_{i_k}))$$

Yet another form is given in Simpson [8].

As given, both of these statements are independent from $\Pi^1_2 - BI_0$ but provable in $\Pi^1_2 - BI$. Some rather simple variations in the class of trees will produce corresponding variations in the subsystem of second order arithmetic. Let Q be the set of a binary trees. For this class CWF(Q) and SCWF(Q) are independent of ACA_0, but provable in ACA. Let Q be the set of all binary, structured trees on two labels. In this case we require the homeomorphisms to preserve structure and labels. In this case CWF(Q) and SCWF(Q) are independent from ATR_0, but provable in ATR. In Section 3 there are many more similar results.

Consider what happens when k is specialized to a natural number in either CWF(Q) or SCWF(Q). We obtain a Σ^0_1 statement which is provable. The proof, however, is quite long. In Section 4 we consider the set of all finite trees on six labels. For $k = 1$ the Σ^0_1 statement is provable in $\Pi^1_2 - BI_0$, but _any proof formalized in_ $\Pi^1_2 - BI_0$ _requires at least_ $2^{[1000]}$ _symbols_.

In section 2 we describe the develop and basic properties of the theory ATI($\prec$) for an ordinal notation system $\prec$. The results of Section 3 are applications of the work in Section 2.

This paper is not self contained. Simpson's paper [8] in this volume gives a self contained exposition of a result like those discussed in this paper. Simpson's paper [9] in this volume has a discussion of subsystems of second order arithmetic.

In unpublished work [3] Friedman has given a new variant of Ramsey's Theorem which is independent from PA.

Let $[A]^r$ deonte the set of r-element subsets of A and $[n] = \{1,\ldots,n\}$. For all $k,r,s \geq 1$ there is an $n \geq 1$ such that for all $F_1,\ldots,F_s:[n]^r \to [n]$ there exists an $A \in [2,n]^k$ such that each F_i takes at most one value $\leq \min(A)$ on $[A]^r$.

Section 2: The Theory ATI($\prec$).

We begin with the analysis of the two statements asserting the combinatorial well-foundedness of an ordinal notation system.

Definition 1: An elementary recursive ordinal notation system (ERONS) for a limit ordinal λ is structure $(A,\prec,+,\cdot,\omega^x,0,1)$ obeying

(1) $A \subseteq \mathbb{N}$ and $(A,\prec)$ has order type λ.

(2) $0 \prec 1$ and they are the first two elements in the $\prec$ ordering.

(3) $+,\cdot,\omega^x$ are elementary recursive functions.

(4) $\prec$ is elementary and hence $A = \{x \in \mathbb{N} : 0 \preceq x\}$ is elementary.

(5) If $\text{ord} : A \to \lambda$ is the unique order isomorphism, then
$\text{ord}(x+y) = \text{ord}(x) + \text{ord}(y)$, $\text{ord}(a \cdot b) = \text{ord}(a) \cdot \text{ord}(b)$, $\text{ord}(\omega^a) = \omega^{\text{ord}(a)}$.

Definition 2: A normed ERONS is an ERONS equipped with a norm $\| \ \| : A \to$ which is elementary and

(1) $\|0\|, \|1\| \leq 1$

(2) There is a uniform elementary bound on $\{x \in A : \|x\| \leq n\}$.

(3) There is a $c \in \mathbb{N}$ such that $\|a + b\|, \|a \cdot b\| \leq \|a\| + \|b\| + c$ for all $a,b \in A$.

We consider two Π_2^0 formulations of the combinatorial well-foundedness of a normed ERONS $(A, \prec)$.

$$\mathrm{CWF}(\prec) : \forall k \exists n \forall x_1,\dots,x_n (\|x_i\| \leq k+i \to \exists i < j\ (x_i \preceq x_j)) .$$

The stronger formulation

$$\mathrm{SCWF}(\prec) : \forall k \exists n \forall x_1,\dots,x_n (\|x_i\| \leq i \to \exists i_1 < \dots < i_k (x_{i_1} \preceq \dots \preceq x_{i_k})) .$$

<u>Lemma</u> 3: $\mathrm{SCWF}(\prec)$ holds and $\mathrm{SCWF}(\prec) \to \mathrm{CWF}(\prec)$ is provable over PRA for a normed ERONS $(A, \prec)$.

<u>Proof</u>: Let $T = \{(x_1,\dots,x_n) \in A^{<\omega} : \|x_i\| \leq i$ for all i and there is no sequence $i_1 < \dots < i_k$ such that $x_{i_1} \preceq \dots \preceq x_{i_k}\}$. Assuming $\mathrm{SCWF}(\prec)$ fails for k, we see that T is an infinite finite branching tree. By König's Lemma there is an infinite path $(x_1, x_2, \dots)$ through T. Now apply Ramsey's Theorem to this path to obtain either a descending chain through $(A, \prec)$ or a sequence $i_1 < \dots < i_k$ with $x_{i_1} \preceq \dots \preceq x_{i_k}$.

Now assume $\mathrm{SCWF}(\prec)$ and let k be given. Choose n' in $\mathrm{SCWF}(\prec)$ sufficiently large to handle $k + 2$. Let $x_1,\dots,x_{n'} \in A$ where $\|x_i\| \leq k + i$. By $\mathrm{SCWF}(\prec)$ there is a $\preceq$-increasing subsequence of length $k + 2$ of the sequence $y_1,\dots,y_k,x_1,\dots,x_{n'}$. Thus there exists $i < j$ with $x_i \preceq x_j$.

We define the Skolem functions ϕ and Ψ by

$$\phi(k) = \text{least } n[\forall x_1,\dots x_n \in A(\|x_i\| \leq k + i \to \exists i < j(x_i \preceq x_j))] .$$

Ψ is defined analogously for $\mathrm{SCWF}(\prec)$. Notice that the above proof gives

$\phi(k) \leq \Psi(k+2)$ for all k. ϕ_b and Ψ_b are these functions for the ordering restricted to $\{x \in A : x \prec b\}$.

Definition 4: Let g and h be elementary functions and b a natural number such that $h(n,m) \prec b$ for all (n,m). We write $f = D(g,h)$ if $f(m) = g(\mu n[h(n,m) \preceq h(n+1,m)])$, where μm indicates the least m in the ordering on the integers. We say f is descent recursive on $\prec$. A full discussion of this concept appears in [5].

Theorem 5: If f is descent recursive on $\prec$, then ϕ eventually dominates f, i.e. $f(n) \leq \phi(n)$ for all sufficiently large n.

Proof: We assume $\prec$ is a normed ERONS with constant $c \geq 1$ and $\|\bar{\omega}\| \geq 1$. Here we write $\bar{\omega}$ for ω^1 and for natural numbers n, $\bar{n}$, is the sum of 1 n times. We also use the notation $2^{[n]}(t)$ defined recursively by $2^{[0]}(t) = t$ and $2^{[n+1]}(t) = 2^{2^{[n]}(t)}$.

Lemma 6: For all p and sufficiently large t there is a chain $x_n \prec \ldots \prec x_1$ where $n = 2^{[p]}(t)$, $\|x_i\| \leq t + i$, and $x_1 \quad \bar{\omega} \cdot \overline{(4p+1)}$.

Proof of Lemma 6: By induction on p.

Let $p = 0$ and define b_i for $i \leq \log t + 1$ by $b_1 = 1$ and $b_{i+1} = \bar{2} \cdot b_i$. We see that $\|b_i\| \leq d \log t$ where d is independent of t. Let $x_1, \ldots, x_n$ be an enumeration of all sums of the form $z_{i_1} + \ldots + z_{i_k}$ where $z_{i_k} \prec \ldots \prec z_{i_1}$ is a subset of the b_i. Notice that $n = 2^{\lfloor \log t + 1 \rfloor} \geq t$ and $\|x_i\| \leq d \cdot \log t(\log t + 1) \leq t$ for t sufficiently large.

Notice that we can get $n - 2^{\lfloor t^{1/3} \rfloor}$ by replacing $\lfloor \log t \rfloor$ by $\lfloor t^{1/3} \rfloor$ and $t \geq (2 + 2c)^3$.

Assume the result holds for p and we have $x_n \prec \ldots \prec x_1$ where $n = 2^{[p]}(t)$, $\|x_i\| \leq t + i$ and $x_1 \prec \bar{\omega} \cdot \overline{(4p+1)}$. Let $k = \|\bar{\omega}\| + c$. Choose

k terms $z_k \prec \ldots \prec z_1$ with $x_1 \prec z_k$ and $\|z_i\| \le t + i$. This produces a new sequence

$$y_n \prec \ldots \prec y_1$$

where $n = 2^{[p]}(t)$, $\|y_i\| \le t + i$, $\bar{\omega} \prec y_n$, and $y_1 \prec \bar{\omega} \cdot \overline{(4p + 3)}$.

By the $p = 0$ case there are

$$y'_m \prec \ldots \prec y'_1$$

where $y'_1 \prec \bar{\omega}$, $\|y_i\| \le 2^{[p]}(t) + i$, and $m = 2^{\lfloor q^{1/3} \rfloor}$, $q = 2^{[p]}(t)$. Truncating m terms from the y_i's and the y'_i's gives a sequence

$$z_m \prec \ldots \prec z_1$$

where $\|z_i\| \le t + i$ and $z_1 \prec \bar{\omega} \cdot \overline{(4p + 3)}$.

Now use the z_i's as the starting point of the above process instead of the x_i's to get a sequence of length $2^{[p+1]}(t)$.

This proves Lemma 6.

Let f be descent recursive on $\prec$ and $f = D(g,h)$ where g,h are elementary and $h(n,m) \prec b$. There is a p where $g(n) \le 2^{[p]}(n)$ and $\|h(n,m)\| \le 2^{[p]}(\max(n,m))$, since $\| \ \|$ is elementary. We are going to define a sequence $\alpha_0, \alpha_1, \ldots$ such that each $\alpha_k = (\alpha_k(0), \ldots, \alpha_k(r-1))$ where $r = 2^{[p+1]}(n + k)$ for n sufficiently large.

For $k = 0$, let $x_r \prec \ldots \prec x_1$ where $r = 2^{[p+1]}(n)$, $\|x_i\| \le n + i$, $x_1 \prec \bar{\omega} \cdot \overline{(4p + 5)}$. Now add $\omega^2 \cdot \bar{b}$ to each x_i and (following Lemma 6) place $\|\omega^2 \cdot \bar{b}\| + c$ terms in front. Let α_0 be the first r terms of the resulting sequence.

Assume α_k has been defined and choose $x_r \prec \ldots \prec x_1$ where $r = 2^{[p+1]}(n + k + 1)$, $\|x_i\| \le n + k + i + 1$, $x_1 \prec \bar{\omega} \cdot \overline{(4p + 5)}$. Now set

$\alpha_{k+1}(i) = \omega^2 \cdot h(n,k+1) + x_{i+1}$.

Let $m = \mu m[h(n,m) \preceq h(n,m+1)]$ and let α be the concatention of $\alpha_0, \alpha_1, \ldots, \alpha_m$. Notice that the length of $\alpha \geq$ length of $\alpha_m \geq g(m) = f(n)$. By the construction α is strictly descending and $\|\alpha_i\| \leq n + i$. Thus $f(n) \leq \phi(n)$.

<u>Corollary</u> 7: Suppose $f = D(g,h)$ with $h(n,m) < b_0$ and let $b = \omega^2 \cdot b_0$. Then ϕ_b and Ψ_b eventually dominate f.

<u>Proof</u>: Immediate from the proof above.

The next theorem requires some proof theory. The proof theory used comes from Friedman and Pearce [5]. Significantly more information (and proofs!) about descent recursion and the formal systems ATI($\prec$) and DRA($\prec$) will be available there.

<u>Definition</u> 8: An ERONS is <u>accessible</u> if there are elementary functions H and I such that for each limit a , H(a,n) a is strictly increasing in n, and for b a, b H(a,I(a,b)). An ERONS is <u>provably accessible</u> if all properties of the elementary functions in the definition of an accessible ERONS $(+,\cdot,\omega^x,\| \ \|,H,I)$ are provable over Peano Arithemtic. (Here we also require that, e.g., for limits b, $\sup_{x \prec b} (a+x) = a + \sup_{x \prec b} x$, for the three operations, to be provable.)

In Friedman and Pearce [5] it is shown that the descent recursive functions on an accessible ERONS $\prec$ are closed under the scheme

$$f(n) = g(\mu m[h(n,m) \preceq h(n,m+1)]) .$$

The ordinal notation systems associated with proof-theoretic studies are probably accessible ERONS. <u>Hereafter, we assume</u> $(A, \prec)$ <u>is provably accessible</u>.

ATI($\prec$) is a first order theory extending PRA (primitive recursive arithmetic) with the following axioms:

(1) Induction (on $\prec$) for all formulas.

(2) $\prec$ is given by an elementary definition.

(3) For each $b \in A$ and formula $\phi(x,a_1,\ldots a_n)$,

$$\exists x \prec \bar{b}\phi(x,a_1,\ldots,a_n) \to (\exists x(\phi(x,a_1,\ldots,a_n) \wedge \forall y(\phi(y,a_1,\ldots,a_n) \to x \preceq y)).$$

Theorem (Friedman and Pearce [5])

(1) The provably recursive functions of $ATI(\prec)$ are the descent recursive functions on $\prec$.

(2) The following are equivalent over PA

(i) 1-consistency of $ATI(\prec)$.

(ii) $\prec$ has no elementary recursive descending sequences.

(iii) $\prec$ has no primitive recursive descending sequences.

The 1-consistency of a theory T is the assertion: if ϕ is a Σ_1^0 sentence provable from T, then ϕ is true.

$ETI(\prec)$ is an axiomatization of the Π_2^0 consequences of $ATI(\prec)$. $ETI(\prec)$ is a first order theory extending EFA (elementary function arithmetic) with the following axioms:

(1) Function induction on $\prec$, i.e. for each elementary function symbol f, $(f(0) = 0 \wedge \forall n(f(n) = 0 \to f(n+1) = 0)) \to \forall n(f(n) = 0)$, (parameters allowed).

(2) For each elementary function symbol f and $b \in A$, $\forall \vec{x} \exists y$ $(f(\vec{x},y) \preceq f(\vec{x},y+1) \vee \bar{b} \preceq f(\vec{x},y))$.

(3) Axioms asserting $\prec,+,\cdot,\omega^x,H,I$ form an accessible ERONS. (Note, this includes such axioms as for limits b, $\sup_{x \prec b}(a + x) = a + \sup_{x \prec b} x$ instead of (1), (5).)

Theorem (Friedman and Pearce [5]). $ATI(\prec)$ is conservative over $ETI(\prec)$ for Π_2^0 sentences. A provable Π_2^0 has a Skolem function which is descent recursive on $\prec$.

Theorem 9: Let $(A, \prec)$ be a provably accessible ERONS and $b \in A$. Then ϕ_b and Ψ_b are descent recursive on $\prec$.

Proof: By Lemma 3 the Π^0_2 statements CWF(b) and SCWF(b) are provable in ATI($\prec$) , and hence in ETI($\prec$) . Thus ϕ_b and Ψ_b are descent recursive.

Theorem 10: (PA) Let $(A,\prec)$ be a provably accessible ERONS. CWF($\prec$) is equivalent to "$\prec$ has no elementary descending sequences".

Proof: One direction is immediate by Theorem 5. Assume that $\prec$ has no elementary descending sequences. Thus ATI($\prec$) is 1-consistent. In the statement of CWF($\prec$) fix k to get a Σ^0_1 statement. We claim this Σ^0_1 statement is provable in ATI($\prec$) . Suppose not, then for each n there is a sequence $x_n \prec \ldots \prec x_1$ such that $\|x_i\| \leq k + i$. By Weak König's Lemma there is an infinite descending sequence. This can be taken to be an arithmetical descending sequence, but this contradicts the principle axiom of ATI($\prec$), i.e. transfinite induction up to $b = \sup\{x : \|x\| \leq k + 1\}$.

Given an ERONS $\prec$ there is an ERONS $\prec^+$ which represents the next ε-number after the ε-number represented by $\prec$. The proof of SCWF($\prec$) given in Lemma 3 can be formalized in PA + 1-consistency of ATI($\prec^+$) .

Theorem 11: Let $(A,\prec)$ be a provably accessible ERONS using H,I . The following theories have the same Π^0_2 consequences in the language of EFA:

(1) ATI($\prec$)

(2) ETI($\prec$)

(3) EFA + CWF(b) for all $b \in A$ + " $\prec$ is an accessible ERONS using H,I".

(4) EFA + SCWF(b) for all $b \in A$ + " $\prec$ is an accessible ERONS using H,I".

Proof: For (1) and (2), this result was cited earlier. Also CWF(b) and SCWF(b) are Π^0_2 consequences of ETI($\prec$) , as in Theorem 9. Finally, for any Π^0_2 consequence of ETI($\prec$) , the Skolem function is descent recursive on $\prec$. Thus by the proof of Theorem 5, the Skolem function is provably dominated by both ϕ and Ψ , and hence the Π^0_2 statement is provable in both (3) and (4).

This completes the analysis of the well-orderings.

<u>Definition</u> 12: $(Q,\leq,|\ |)$ is a <u>normed elementary recursive well quasi-ordering</u> if

(1) $\leq$ is transitive, reflexive, and for all sequences $(q_1,q_2,\ldots)$ there exists $i < j$ such that $q_i \leq q_j$.

(2) $|\ | : Q \to \mathbb{N}$ such that $\{x \in Q : |x| \leq n\}$ has a uniform elementary recursive bound.

(3) $\leq$ and $|\ |$ are elementary recursive.

Call this a normed ERWQO.

For $(Q,\leq,|\ |)$ it is possible to define the statements $CWF(\leq)$ and $SCWF(\leq)$ as before. For $q \in Q$, $CWF(q)$ is the restriction to $\{x \in Q : q \leq x\}$. Similarly for $SCWF(q)$. ϕ_Q denotes the Skolem function for $CWF(\leq)$. Similarly for Ψ_Q, ϕ_q, and Ψ_q.

<u>Theorem</u> 13: Let $(A,\preceq,\|\ \|)$ be a normed accessible ERONS and let $(Q,\leq,|\ |)$ be a normed ERWQO. Suppose $f : Q \to A$ is an elementary order-preserving onto map with an elementary right inverse, where $|q| \leq \|f(q)\|$ for all $q \in Q$. Then

(1) ϕ_Q dominates ϕ_A.

(2) Ψ_Q dominates Ψ_A.

(3) For each $a \in A$ there is a $q \in Q$ such that ϕ_q dominates ϕ_a.

(4) For each $a \in A$ there is a $q \in Q$ such that Ψ_q dominates Ψ_a.

<u>Proof</u>: (1) Let $k \in \mathbb{N}$ and consider $a_1,\ldots,a_n \in A$ where $\|a_i\| \leq k + i$. There are $q_1,\ldots,q_n \in Q$ such that $f(q_i) = a_i$ and $|q_i| \leq \|a_i\| \leq k + i$. Now if $n = \phi_Q(k)$, then there exist $i < j$ with $q_i \leq q_j$ and hence $a_i \preceq a_j$. Thus $\phi_A(k) \leq \phi_Q(k)$. (2) is analogous.

(3) Let $a \in A$ and choose $q \in Q$ such that $f(q) = a$. Let ϕ' be the Skolem function for $CWF(\leq)$ restricted to $\{x \in Q : f(x) \prec a\}$, then $\phi'(k) \leq \phi_q(k)$ for all k. By (1), $\phi_a(k) \leq \phi'(k)$. (4) is analogous.

Corollary 14: If we assume that the hypothesis of Theorem 13 is provable over PA, then each of the following is provable over PA :

(1) $CWF(\leq) \rightarrow CWF(\prec)$

(2) $SCWF(\leq) \rightarrow SCWF(\prec)$

(3) $\forall a \in A \exists q \in Q(CWF(q) \rightarrow CWF(a))$

(4) $\forall a \in A \exists q \in Q(SCWF(q) \rightarrow SCWF(a))$.

Proof: Immediate.

Section 3: Applications

The tools developed in Section 2 will be applied here. With a few exceptions, which will be noted, these applications depend on Friedman [2]. The well quasi-orderings under consideration here are finite trees with various additional attributes. A tree is labelled if there is a function which assigns each vertex to a label from some set of labels. A tree is structured if the sons of each vertex are linearly ordered.

I. PA and ACA_0

Let $(A, \prec)$ be a standard provably accessible ERONS for ε_0 and $\|x\|$ = the number of symbols in x . The following are equivalent:

(1) $CWF(\prec)$

(2) 1-consistency of PA

(3) 1-consistency of ACA_0

(4) 1-consistency of $ATI(\prec)$

We are going to list five well quasi-orderings $(Q,\leq)$. In each case $CWF(Q)$ is equivalent to $CWF(\varepsilon_0)$ and hence both $CWF(Q)$ and $SCWF(Q)$ are independent from ACA_0 and PA . These results are obtained by applying Corollary 14. Unless otherwise stated, the hypothesis to Corollary 14 is proven in Friedman [2], as well as the fact that $CWF(Q)$ and $SCWF(Q)$ are provable from the 1-consistency of PA , and from ACA .

(i) Let Q be the set of all binary trees with $T_1 \leq T_2$ iff there is a homemorphic (inf-preserving) embedding of T_1 into T_2.

(ii) Let Q be the set of all binary structured trees with $T_1 \leq T_2$ iff there is a structure preserving homeomorphic embedding.

(iii) Let Q be the set of all exactly binary trees (i.e. each vertex has either no sons or two sons) with $T_1 \leq T_2$ iff there is a homeomorphic embedding.

(iv) For each $n \in \mathbb{N}$ let Q_n be the set of trees of height n under homeomorphic embedding. In this case $CWF(\varepsilon_0)$ is equivalent to, for all $n \in \mathbb{N}$, $CWF(Q_n)$ and for all $n \in \mathbb{N}$, $SCWF(Q_n)$.

(v) For each $n \in \mathbb{N}$, let Q_n be the set of finite sequences on $\{0,\dots,n-1\}$. For $\sigma \in Q_n$, let $|\sigma|$ be the length of σ and define $\sigma \leq \tau$ iff there is an $h : \{0,\dots,n-1\} \to \{0,\dots,n-1\}$ such that $\sigma(i) = \tau(h(i))$ and for $f(i) < j < f(i+1)$, $\tau(j) \geq \tau(f(i+1))$. As above, $CWF(\varepsilon_0)$ iff for all $n \in \mathbb{N}$ $CWF(Q_n)$, and the 1-consistency of PA is equivalent to, for all $n \in \mathbb{N}$, $CWF(Q_n)$. See Schütte and Simpson [7].

II. ATR_0

Let $(A, \prec)$ be a standard provably accessible ERONS for Γ_0, and $\|x\|$ be as above. Let Q be the set of all binary structured trees on two distinct labels under structure preserving, label-preserving homeomorphisms. The following are equivalent (see Friedman, McAloon, Simpson in the Patras Logic Symposion, North-Holland, 1982).

(1) $CWF(\prec)$

(2) 1-consistency of ATR_0

(3) 1-consistency of $ATI(\prec)$

(4) $CWF(Q)$

Thus, both CWF(Q) and SCWF(Q) are independent from ATR_0 . At the same time CWF(Q) and SCWF(Q) are provable from ATR . This case is also considered in Simpson [8] . See Friedman [2] for the hypothesis to Corollary 14, and proofs of CWF(Q), SCWF(Q) from the 1-consistency of ATR_0 , and from ATR .

III. $\Pi^1_2 - BI_0$

Let $(A, \prec)$ be a standard provably accessible ERONS for the ordinal $\theta_{\Omega^\omega}(0)$, and $\|x\|$ be as above.

Let Q be the set of all finite trees under homeomorphic embeddings (with or without structure preservation). The following are equivalent (see [4]):

(1) $CWF(\prec)$

(2) 1-consistency of $\Pi^1_2 - BI_0$

(3) 1-consistency of $ATI(\prec)$

(4) CWF(Q)

Thus, both CWF(Q) and CSWF(Q) are independent from $\Pi^1_2 - BI_0$. The equivalence of (2) and (3) is from Friedman and Pearce [4]. See Friedman [2] for the hypothesis to Corollary 14, and proofs of CWF(Q), SCWF(Q) from the 1-consistency of $\Pi^1_2 - BI_0$, and from $\Pi^1_2 - BI$.

Now let Q_n be the set of finite trees such that each vertex has at most n sons under homeomorphic embeddings (with or without structure preservation). As above, the 1-consistency of $\Pi^1_2 - BI_0$ is equivalent to, for all $n \in \mathbb{N}$, $CWF(Q_n)$, and also to, for all $n \in \mathbb{N}$, $SCWF(Q_n)$.

IV. $\Pi^1_1 - CA_0$

We refer to the book of Buchholz, Feferman, Pohlers and Sieg [1] for a discussion of the systems ID_n for $n \in \mathbb{N}$ and the fact that $\Pi^1_1 - CA_0$ has the same first order consequences as $\bigcup_{n \in \mathbb{N}} ID_n$.

Now let Q_n be the set of all finite structured trees with n labels under homeomorphic, label-preserving, and structure-preserving embedding with the gap condition of Application 1 (v). $\Pi^1_1 - CA_0$ proves $WF(Q_n)$ for each $n \in \mathbb{N}$ and $\Pi^1_1 - CA$ proves $\forall n CWF(Q_n)$. Furthermore, the 1-consistency of ID_{n+1} proves $CWF(Q_n)$ and $SCWF(Q_n)$, each of which proves the 1-consistency of ID_n , over PA . Thus, the 1-consistency of $\Pi^1_1 - CA_0$ is equivalent to, for all $n \in \mathbb{N}$, $CWF(Q_n)$ and also to, for all $n \in \mathbb{N}$, $SCWF(Q_n)$. See Friedman [2] and Simpson [8] for the order-preserving surjections.

All of the above results in the applications can be proven over PA .

Section 4: Practical Matters

In the statement SCWF(Q) , when k is specialized we obtain a Σ^0_1 statement. Of course, this Σ^0_1 statement can be proven in the system under consideration, but an actual formalization of the proof in the system may be quite lengthy. Friedman has given a lower bound for such a statement, and it turns out to be impressively large.

Let Q be the well quasi-ordering of all finite trees with 6 labels under label preserving homeomorphisms. Both CWF(Q) and SCWF(Q) are independent of $\Pi^1_2 - BI_0$, but specializations of both are of course provable in $\Pi^1_2 - BI_0$.

Theorem 15: Any proof of the existence of $\Psi_Q(1)$ in $\Pi^1_2 - BI_0$ requires at least $2^{[1000]}$ symbols.

The remainder of the paper is devoted to the proof of this result. (We actually prove a stronger form of this result.) We begin with a closer analysis of descending sequences in an ERONS.

Lemma 16: Let $t \geq 2^{25}(p + 2)(\|\bar{\omega}\| + c + 1)^3$. Then there is a sequence $x_n \prec \ldots \prec x_1$ where $n = 2^{[p]}(t)$, $\|x_i\| \leq t + 1$, and $x_1 \prec \bar{\omega} \cdot \overline{(4p + 1)}$.

Proof: As in Lemma 6.

Lemma 17: Let $f = D(g,h)$ and let p be so that $g(n) \leq 2^{[p]}(n)$ and $h(n,m) \leq 2^{[p]}(\max(n,m))$. Then

(i) If $n \geq 2^{25}(p + 3)(\|\bar{\omega}\| + \|b\| + 3c + 1)^3$, then $f(n) \leq \Psi_Q(n)$.

(ii) If $n \geq 2^{25}(p + 5)(\|\bar{\omega}\| + \|b\| + 3c + 1)^3$, then
$f(n + 2) \leq \Psi_Q(n) \leq \phi_Q(n + 2)$.

(iii) If $n \geq 2^{25}(p + 7)(\|\bar{\omega}\| + \|b\| + 3c + 1)^3$, then $f(n) \leq \phi_Q(n)$.

Proof: As in Theorem 5.

We say $(A, \prec, \| \ \|)$ is a strongly normed ERONS if it is a normed ERONS satisfying $\|\omega^a\| \leq \|a\| + c$. Given a provably accessible strongly normed ERONS, consider a number $d>c$ such that all the elementary functions in the ERONS have Turing tables bounded by d, the elementary functions are bounded by $2^{[d]}$, and the proofs of the properties of an accessible, strongly normed ERONS in PA have at most d symbols. The least such d is the critical constant of the ERONS.

Let $\pi_b(n)$ be the least integer m such that if $\exists x\phi(x)$ is any Σ^0_1 sentence provable in $ATI(\prec)$ using $\leq n$ symbols and transfinite induction $\preceq b$, then $\exists\, x \leq m\phi(x)$.

Lemma 18: Let d be the critical constant of the provably accessible strongly normed ERONS $(A, \prec, \| \ \|)$. Then $\pi_b = D(g,h)$ where $h(n,m) \prec \omega^{[d+1000]}(b + 1)$, g,h have Turing tables with at most $2^{[d+1000]}$ symbols, and g,h are bounded by $2^{[d+1000]}$.

The proof of this lemma will appear in [5].

Lemma 19: If $n \geq 2^{25}(d + 1007)(\|\bar{\omega}\| + \|b\| + (c + 1003)^2)^3$, then $\phi_Q(n), \Psi_Q(n) \geq \pi_b(n)$.

Proof: By Lemmas 17 and 18.

The standard notation system for $\varepsilon_{\lambda+1}$ where $\lambda = \Theta_{\Omega^\omega}(0)$ is a provably accessible strongly normed ERONS with a critical constant $d < 2^{[100]}$. Let Q' be the well quasi-ordering of all finite trees with two labels. As in application III we get $\phi_A \leq \phi_{Q'}$ and $\Psi_A \leq \Psi_{Q'}$, see [2].

Lemma 20: Let $\lambda = \Theta_{\Omega^\omega}(0)$. Then

$$\phi_{Q'}(2^{[1001]}) , \Psi_{Q'}(2^{[1001]}) \geq \pi_\lambda(2^{[1001]}) > \pi_\lambda(2^{[1000]}) .$$

Proof: By Lemma 19.

Lemma 21: $\Psi_Q(1) \geq 2^{[1001]}$.

Proof: Let T_1 be the tree which is only the root, labeled 6. Let T_2 be two joined vertices, both labeled 5. Let $T_3,\ldots,T_{29}$ be all the trees with exactly three linearly ordered vertices all labeled from $\{1,2,3,4\}$, and in which exactly one vertex is labeled 4. Let $T_{30},\ldots,T_{1032}$ be nonisomorphic trees consisting of a root labeled 2 with 29 immediate successors labeled from $\{2,3,4\}$ not all labeled 2. We now describe the trees $T_{1033},\ldots,T_{2^{[1001]}}$. Each tree has a root, labeled 3 and two sons, sons, the son of the left is labeled 3, the son on the right labeled 2. Off of each son there is a long linear order of vertoces. The left side has length i where $2 \leq i \leq 1027$ and all vertices are labeled 3. The right side has length $2+2^{[1027-i]}$ and the vertices are chosen from $\{2,3\}$. Since the right side has $2^{[1027-i+1]}$ different labellings from $\{2,3\}$, it is possible to enumerate these trees so that $|T_k| \leq k$. It is easy to check that $T_i \leq T_j$ for $i < j$.

Notice that in the sequence just constructed the label 1 is never used and no tree uses the label 2 exclusively. Thus if T is labelled from $\{1,2\}$, then $T_i \leq T$ for $1 \leq i \leq 2^{[1001]}$.

Lemma 22: $\Psi_Q(1) \geq \pi_\lambda(2^{[1000]})$ where $\lambda = \Theta_{\Omega^\omega}(0)$.

Proof: From Lemma 20 we know that $\Psi_{Q'}(2^{[1001]}) > \pi_\lambda(2^{[1001]})$. For $n + 1 = \Psi_{Q'}(2^{[1001]})$ there are trees $T'_1 , \ldots, T'_n$ where $|T'_k| \leq 2^{[1001]} + k$, and for $i < j$, $T'_i \leq T'_j$, and each tree uses labels from $\{1,2\}$. Let

$T_1,\ldots T_{2^{[1001]}}$ be the sequence constructed in Lemma 21. Let $S_1,\ldots S_{2^{[1001]}+n}$ be the sequence obtained by placing the T_i's in front of the T'_j's . Thus $|S_i| \leq i$ for all i and for $i < j$, $S_i \leq S_j$.

Now in Friedman and Pearce [4] the 1-consistency of $\Pi^1_2 - BI_0$ is proven by transfinite induction up through $\Theta_{\Omega^\omega}(0)$ by a proof shorter than $2^{[100]}$. Thus $\pi(2^{[1000]}) < \pi_\lambda(2^{[1000]}) \leq \Psi_Q(1)$, where $\lambda = \Theta_{\Omega^\omega}(0)$ and the π on the left hand side refers to $\Pi^1_2 - BI_0$. The theorem now follows. Actually Lemma 22 is a stronger result.

Friedman claims that similar results are obtainable for stronger systems using homeomorphic embedding subject to a gap condition as in application I(v).

REFERENCES

[1] Buchholz, W., S. Feferman, W. Pohlers, W. Sieg, Iterated Inductive Definitions and Subsystems of Analysis : Recent Proof-Theoretical Studies, Springer Lecture Notes 897, Springer Verlag, 1981.

[2] Friedman, H., The metamathematics of the Higman and Kruskal theorems, in preparation.

[3] Friedman, H., Unbounded colorings and Peano arithmetic I, unpublished manuscript, March 1983, The Ohio State University.

[4] Friedman, H. and J. Pearce, Fragments of admissible set theory and bar induction, to appear.

[5] Friedman, H. and J. Pearce, Transfinite elementary descent recursions and provably recursive functions, in preparation.

[6] Paris, J. and L. Harrington, A mathematical incompleteness in Peano arithmetic, Handbook of Mathematical Logic, ed. J. Barwise, North-Holland (1977), pp. 1133-1142, reprinted in this volume.

[7] Schütte, K. and S. G. Simpson, Ein in der reinen Zahlentheorie unbeweisbarer Staz über endliche Folgen von natürlichen Zahlen, to appear in Archiv f. Math. Logik und Grundl. der Math.

[8] Simpson, S. G., Unprovability of certain combinatorial properties of finite trees, to appear in Archiv f. Math. Logik und Grundl. der Math., reprinted in this volume.

[9] Simpson, S. G., Friedman's research on subsystems of second order arithmetic, this volume.

HARVEY FRIEDMAN'S RESEARCH ON THE FOUNDATIONS
OF MATHEMATICS, L.A. Harrington et al. (editors)

FRIEDMAN'S RESEARCH ON SUBSYSTEMS OF SECOND ORDER ARITHMETIC*

Stephen G. Simpson

Department of Mathematics
Pennsylvania State University
University Park, PA 16802

§0. Introduction.

Subsystems of second order arithmetic can be examined from two related but essentially disparate points of view. On the one hand, such systems have many interesting metamathematical properties which can be investigated using proof-theoretic and model-theoretic tools. On the other hand, subsystems of second order arithmetic are a natural vehicle for the formal axiomatic study of ordinary mathematics. (By ordinary mathematics we mean such branches of mathematics as geometry, number theory, differential equations, algebra, functional analysis etc.)

Harvey Friedman has made important contributions to both of the above-mentioned areas of research. But it is in the second area, that of the relationship between subsystems of second order arithmetic and ordinary mathematical practice, that Friedman's insights have been particularly influential and indeed decisive for later developments.

We use Z_2 to denote the formal system of second order arithmetic. All of the subsystems of Z_2 which we consider in this paper employ classical logic. For general background on subsystems of Z_2 see [25], [48], [50], [53], [54].

§1. Formal Hyperarithmetic Theory.

Friedman's earliest research on subsystems of Z_2 appears in Chapter II of his 1967 Ph.D. thesis [9]. That chapter is concerned mainly with formal systems

*Preparation of this paper was partially supported by NSF grant MCS-8317874.

for the hyperarithmetical sets. Before discussing Friedman's results, we review some background material concerning the class

$$\mathrm{HYP} = \left\{ X \subseteq \omega : X \text{ is hyperarithmetical} \right\}.$$

Let $\mathcal{O}$ be the Church-Kleene system of recursive ordinal notations [6]. For each $e \in \mathcal{O}$ we define H_e to be the subset of ω obtained by iterating the Turing jump operator along e, beginning with the empty set. (According to a theorem of Spector [56], the Turing degree of H_e depends only on $|e|$, the recursive ordinal denoted by e.) A subset of ω is said to be hyperarithmetical if it is recursive in H_e for some $e \in \mathcal{O}$.

A fundamental theorem of Kleene [34] asserts that $\mathcal{O}$ is a complete Π^1_1 subset of ω, and that the hyperarithmetical subsets of ω are just the Δ^1_1 sets, i.e. those which are both Σ^1_1 and Π^1_1 definable. These results relativize to an arbitrary set $X \subseteq \omega$. Thus $\mathcal{O}^X$ is the set of notations for ordinals which are recursive in X, and for all $e \in \mathcal{O}^X$ we have the relativized H-set H^X_e obtained by iterating the Turing jump operator along the well ordering e^X beginning with X. An excellent source of information on classical hyper-arithmetic theory is Harrison [28].

In a very important paper, Kleene [35] showed that the class HYP can be characterized as the smallest ω-model of the formal theory Δ^1_1-CA. The principal axiom of Δ^1_1-CA is the Δ^1_1 comprehension axiom

$$\forall n(\varphi(n) \leftrightarrow \psi(n)) \rightarrow \exists X \forall n(n \in X \leftrightarrow \varphi(n))$$

where the formulas $\varphi(n)$ and $\psi(n)$ are Σ^1_1 and Π^1_1 respectively. This axiom is expressed as a formula in the language of Z_2. A key step in Kleene's proof was his observation that, for each $e \in \mathcal{O}$, H_e is Δ^1_1 definable over the ω-model M_α, $\alpha = |e|$, consisting of all subsets of ω which are recursive in some $H_{e'}$, $|e'| < |e|$. Furthermore, the same Δ^1_1 definition of H_e works over all larger

ω-models $M \supseteq M_\alpha$. Thus the hyperarithmetical subsets of ω are built up by a kind of absolute, autonomous process of transfinitely iterated Δ^1_1 definability.

On the basis of Kleene's results, Kreisel [37] suggested that the class HYP and the formal theory Δ^1_1-CA should in some sense correspond to a certain a priori notion of "predicativity" in the foundations of mathematics. Under Kreisel's notion of predicativity, the infinite set ω is taken as a completed totality, but successive universes of subsets of ω are built up by a process of transfinitely iterated definability. In this process, the definition of each particular set X is not allowed to contain quantifiers ranging over a universe which already contains X.

At about the same time, Kreisel [38] (see also Harrison [28]) introduced the Σ^1_1 axiom of choice, Σ^1_1-AC, written as

$$\forall n \exists X\, \varphi(n,X) \rightarrow \exists Y \forall n\, \varphi(n,(Y)_n),$$

and the closely related axiom of Σ^1_1 dependent choices, Σ^1_1-DC, written as

$$\forall n \forall X \exists Y\, \varphi(n,X,Y) \rightarrow \forall X \exists Z ((Z)_0 = X \wedge \forall n\, \varphi(n,(Z)_n,(Z)_{n+1})),$$

where in both axioms, the formula φ is Σ^1_1. Kreisel extended Kleene's result of [35] by showing that HYP is also the minimum ω-model of Σ^1_1-AC and of Σ^1_1-DC. He then pointed out that

$$\Sigma^1_1\text{-DC} \Rightarrow \Sigma^1_1\text{-AC} \Rightarrow \Delta^1_1\text{-CA}$$

and asked whether the converse implications hold.

Throughout the 1960's these ideas were pursued and elaborated. See for instance Kreisel's notion of the "hard core" [27] and Feferman's work on predicative provability and autonomous progressions [7].

Chapter II of Friedman's thesis [9] consists mainly of a study of the formal systems Δ_1^1-CA, Σ_1^1-AC, and Σ_1^1-DC. These are subsystems of Z_2 with full induction on the natural numbers. Friedman obtains three main results:

(1.1.) Σ_1^1-DC, and hence Σ_1^1-AC, is conservative over Δ_1^1-CA with respect to Π_2^1 sentences. Hence all three systems have the same logical strength. Friedman accomplishes this by means of an inner model construction. Namely, he defines the class HYP in such a way that the relativization of Σ_1^1-DC to HYP can be proved within Δ_1^1-CA. There are considerable technical obstacles here since Δ_1^1-CA is not strong enough to prove some very basic and elementary facts about recursive ordinals. For instance, Δ_1^1-CA does not prove that any two recursive well orderings are comparable. It is therefore necessary to consider pseudohierarchies, i.e. H-sets on recursive linear orderings which are not well orderings. Despite the difficulties, Friedman manages to push the construction through.

(1.2.) Friedman [9] computes the proof-theoretic ordinal of all three systems. Namely, he shows that

$$|\Delta_1^1\text{-CA}| = |\Sigma_1^1\text{-AC}| = |\Sigma_1^1\text{-DC}| = \varphi_{\varepsilon_0}(0).$$

This ordinal is much smaller than what had been expected on the basis of Feferman's proof-theoretic analysis of predicativity [7]. Once again Friedman employs a model-theoretic argument involving pseudohierarchies.

(1.3.) Friedman [9] shows that the implication from Σ_1^1-DC to Σ_1^1-AC cannot be reversed. His method here is to consider the true Π_2^1 sentence

$$(*) \qquad \forall X \forall e\ (e \in \mathcal{O}^X \to \exists\ H_e^X).$$

Because (*) is Π_2^1, it is fairly easy to see that Σ_1^1-DC plus (*) implies the existence of a countable ω-model of (*). Hence, by Gödel's Second Incompleteness Theorem, there exists an ω-model of (*) in which Σ_1^1-DC fails. On the other hand, Friedman shows (by means of a mysterious application of pseudohierarchies) that (*) implies Σ_1^1-AC.

Perhaps the most remarkable aspect of this work is its exploitation of model-theoretic and formal recursion-theoretic methods. Prior to Friedman's thesis, the traditional tool for establishing logical properties of subsystems of Z_2 had been the cut-elimination method of Gentzen-style proof theory. See for instance Takeuti [60], Schütte [43], and Howard [31]. This method suffers from the usual disadvantages of syntax as against semantics. Namely, it tends to reduce all problems to a common formalistic pattern. Friedman's model-theoretic and formal recursion-theoretic constructions are more flexible, easier to visualize, and more effective in bringing out the differences between various systems.

Friedman left open the problem of whether Δ_1^1-CA implies Σ_1^1-AC. This was solved negatively in subsequent research of Steel [57], [58] and Harrington (unpublished). Namely, Steel used a novel notion of forcing to construct an ω-model of Δ_1^1-CA in which Σ_1^1-AC fails.

§2. Other Relative Consistency Results.

In several papers subsequent to his Ph.D. thesis [9], Friedman continued to apply his model-theoretic and formal recursion-theoretic techniques to obtain relative consistency results for various subsystems of Z_2. We shall now briefly describe this work.

In [10] Friedman analyzes the relative logical strength of two subsystems of Z_2 known as Π_1^1-CA and BI. The principal axiom of Π_1^1-CA is Π_1^1 comprehension, i.e.

$$\exists X \forall n (n \in X \leftrightarrow \varphi(n))$$

where $\varphi(n)$ is Π_1^1. The system BI is axiomatized by the scheme

$$\forall X \forall e (e \in \mathcal{O}^X \rightarrow TI(e,X,\varphi))$$

which is known as bar induction. Here φ is an arbitrary formula in the language of Z_2, and $TI(e,X,\varphi)$ stands for transfinite induction with respect to φ along the given well ordering e^X. Friedman [10] used a formal version of the Gandy basis theorem to show that Π^1_1-CA proves the existence of an ω-model of BI. (This construction was refined in subsequent unpublished work of Harrington, who showed that BI proves the same arithmetical-in-Π^1_1 sentences as Feferman's [5] system ID_1. See also Howard [31].)

Later Friedman [12] obtained the following interesting result: BI is equivalent to the scheme of ω-model reflection, i.e.

$$\varphi(X_1,\ldots,X_n) \to \exists \text{ countable } \omega\text{-model of } \varphi(X_1,\ldots,X_n)$$

where φ is any formula in the language of Z_2. From this it follows by Gödel's Second Incompleteness Theorem that BI is not finitely axiomatizable. In fact, for any finite set F of true sentences in the language of Z_2, there exists an ω-model of F in which BI fails.

In [11] Friedman published the result of 1.2 above and some related results. In particular, Friedman [11] shows that Σ^1_{k+1}-AC is a conservative extension of $(\Pi^1_k\text{-CA})^{<\varepsilon_0}$ for Π^1_4 sentences, if $k > 1$; similarly for Π^1_3 sentences if $k=1$, and for Π^1_2 sentences if $k=0$.

Friedman's methods in [11] were of course model-theoretic. Subsequently the same results were obtained proof-theoretically by Sieg [45]. Somewhat later, Sieg [46] used his Gentzen-style methods to obtain analogous results for systems with restricted induction. These results were in turn rederived by Schmerl [41] using a model-theoretic method related to that of Friedman.

Over the years, other workers inspired by Friedman have adapted his model-theoretic methods to obtain a variety of relative consistency results for subsystems of Z_2. See for instance Barwise-Schlipf [2], Feferman [8], Schmerl-Simpson [42], Simpson [51], [52], and Harrington (unpublished). Thus it can be said that Friedman was the initiator of a fairly profound methodological shift in the study of subsystems of Z_2.

In [24] Friedman computed the proof-theoretic ordinal of a subsystem of Z_2 known as ATR_0. This is a system with restricted induction whose principal axiom is the Π^1_2 sentence (*) considered in 1.3 above. Friedman's result here is that $|ATR_0| = \Gamma_0$. The proof uses Gödel's Second Incompleteness Theorem plus pseudohierarchies. Subsequently, for the system ATR with full induction, Friedman and Jäger (see [51]) obtained $|ATR| = \Gamma_{\varepsilon_0}$ by a related method. (The letters ATR stand for arithmetical transfinite recursion.) Later Jäger [32] showed how to obtain these results by purely Gentzen-style methods.

Right up to the present day, a number of researchers who are interested in subsystems of Z_2 continue to shun semantical methods in favor of syntactical ones. Perhaps this tendency emanates from a Hilbertian distrust of "transcendental" notions. But, as Friedman has many times remarked in conversation, the relative consistency results obtained by his semantical methods are no less finitistic than those of Gentzen-style proof theory. (For a general result which helps to explain this phenomenon, see Friedman [21].)

§3. Reverse Mathematics.

In his 1974 address to the International Congress of Mathematicians, Friedman [12] undertook to study the following questions. "What are the proper axioms to use in carrying out proofs of particular theorems, or bodies of thoerems, in mathematics? What are those formal systems which isolate the essential principles needed to prove them?" He went on to explain that subsystems of Z_2 are of fundamental importance in this context.

Friedman was not the first person to study the formal development of ordinary mathematics within subsystems of Z_2. Hilbert and Bernays [30] had pointed out that most theorems of ordinary mathematics can be stated and proved in the full system Z_2. Weyl [61] had shown that a substantial part of ordinary mathematics can be developed within a certain "predicative" subsystem of Z_2, allowing ω-iterated arithmetical definability. (See also Zahn [62].) Kreisel [39] had given an example to show that the Cantor-Bendixson theorem is not true in the ω-model HYP. From this it follows that the Cantor-Bendixson theorem is not predicatively provable.

But apparently Friedman [12] was the first to initiate a systematic study of precisely which theorems of ordinary mathematics are provable in precisely which subsystems of Z_2.

Along with the systems Δ^1_1-CA, Σ^1_1-AC, Σ^1_1-DC, ATR, BI, and Π^1_1-CA which have already been mentioned, Friedman in [12] considered three weaker subsystems of Z_2. We now describe these briefly.

RCA. This is the weakest system considered by Friedman in [12]. It is based on the Δ^0_1 (i.e. recursive) comprehension axiom

$$\forall n(\varphi(n) \leftrightarrow \psi(n)) \rightarrow \exists X \forall n(n \in X \leftrightarrow \varphi(n))$$

where $\varphi(n)$ and $\psi(n)$ are Σ^0_1 and Π^0_1 respectively. Thus the minimum ω-model of RCA is just the class of recursive subsets of ω. The formal development of ordinary mathematics within RCA is similar to what is known as recursive analysis and recursive algebra; compare e.g. Aberth [1] and Fröhlich-Shepherdson [26].

ACA. This system is based on the arithmetical comprehension axiom

$$\exists X \ \forall n(n \in X \leftrightarrow \varphi(n))$$

where the formula $\varphi(n)$ is arithmetical, i.e. contains no set quantifiers. (By relativization one can restrict $\varphi(n)$ to be Σ^0_1. Thus ACA is equivalent to Σ^0_1-CA.) The minimum ω-model of ACA is the class of arithmetical subsets of ω.

WKL. This system is intermediate between RCA and ACA. Its principal axiom is weak König's lemma, i.e. the statement that every infinite tree of sequences of 0's and 1's has a path. The ω-models of WKL are just the Scott systems [44], [36] which have been useful in the study of complete extensions and models of first-order Peano arithmetic.

All of the above systems are formulated with full induction on the natural numbers.

The main theme of Friedman [12] is that, when a theorem of ordinary mathematics is provable in one of the above systems, then surprisingly often, the theorem is in fact provably equivalent to the principal axiom of the system needed to prove it. This theme is nowadays known as Reverse Mathematics (cf. [4], [24], [25], [50], [51], [53], [54], [55]).

In [12] Friedman presents (without proof) the following examples of Reverse Mathematics.

(3.1.) RCA proves the equivalence of WKL to each of:

(a) the compactness theorem for propositional calculus;

(b) the completeness theorem for the predicate calculus;

(c) Lindenbaum's lemma;

(d) the sequential Heine-Borel theorem, i.e. the statement that every covering of the closed unit interval by a sequence of open intervals has a finite subcovering.

(3.2.) RCA proves the equivalence of ACA to each of:

(a) König's lemma, i.e. the statement that every infinite, finitely branching tree has an infinite path;

(b) the statement that every bounded (or bounded increasing) sequence of real numbers has a least upper bound;

(c) the sequential Bolzano-Weierstrass theorem, i.e. the statement that every bounded sequence of real numbers has a convergent subsequence.

(3.3.) RCA proves the equivalence of Σ_1^1-AC to the statement that whenever any neighborhood of a real number x contains at least two distinct reals in some arithmetically definable collection of reals, then x is the limit of a sequence of reals in that collection.

(3.4.) RCA proves the equivalence of ATR to each of:

(a) the Perfect Set Theorem, i.e. the statement that every tree with uncountably many paths has a perfect subtree;

(b) the statement that any two well orderings of the natural numbers are comparable;

(c) the statement that, if any two distinct real numbers in an arithmetically definable collection of reals are at least one unit apart, then there is a sequence which includes all of the reals in that collection.

(3.5.) RCA proves the equivalence of Π^1_1-CA to each of:

(a) the Perfect Kernel Theorem, i.e. the statement that to every tree T there is a perfect subtree P (possibly empty) and a sequence of paths through T such that every path through T is either a path through P or a member of the sequence.

(b) the statement that every bounded, arithmetically definable collection of reals has a least upper bound.

We should mention that earlier, in 1973, Steel had investigated the axiom of determinacy in the context of subsystem of Z_2. At that time Steel proved the following:

(3.6.) (i) RCA proves the equivalence of ATR to the determinacy of open, or clopen, subsets of subsets of ω^ω;

(ii) RCA proves the equivalence of Π^1_1-CA to the determinacy of Boolean combinations of open subsets of ω^ω;

(iii) RCA proves the equivalence of Δ^1_2-CA to the determinacy of $F_\sigma \cap G_\delta$ subsets of ω^ω.

Thus Steel's results foreshadowed the theme of Reverse Mathematics as inaugurated in Friedman [12]. (Steel's results were not widely circulated until the appearance of his 1976 Ph.D. thesis [57].)

The program of Reverse Mathematics continues to be very active. For a fairly complete account of the recent activity, see Simpson [54]. Among the contributions to Reverse Mathematics subsequent to Friedman [12] we mention the following:

(3.7.) RCA proves the equivalence of WKL to each of:

(a) the statement that every countable field has a unique algebraic closure (Friedman-Simpson-Smith [25]);

(b) the statement that every formally real field has a real closure (Friedman-Simpson-Smith [25]);

(c) the statement that every countable commutative ring has a prime ideal (Friedman-Simpson-Smith [25]);

(d) the statement that every continuous function on the closed unit interval is bounded, or uniformly continuous, or Riemann integrable, or attains a maximum value (Simpson [54]);

(e) the local existence theorem for solutions of ordinary differential equations (Simpson [53]);

(f) the Hahn-Banach theorem for separable Banach spaces (Brown-Simpson [4]).

(3.8.) RCA proves the equivalence of ACA to each of:

(a) the statement that every countable vector space has a basis (Friedman-Simpson-Smith [25]);

(b) the statement that every countable field has a transcendence base (Friedman-Simpson-Smith [25]);

(c) the statement that every countable abelian group has a unique divisible closure (Friedman-Simpson-Smith [25]);

(d) the statement that every countable commutative ring has a maximal ideal (Friedman-Simpson-Smith [25]);

(e) the Ascoli lemma (Simpson [53]);

(f) Ramsey's theorem (Simpson [54]).

(3.9.) RCA proves the equivalence of ATR to each of:

(a) the Ramsey property for open, or clopen, subsets of $[\omega]^{\omega}$ (Simpson [49], [24]);

(b) the statement that every uncountable closed, or analytic, set of real numbers contains a perfect set (Simpson [54]);

(c) the statement that any two disjoint analytic sets can be separated by a Borel set (Simpson [54]);

(d) the statement that the domain of any single-valued Borel set in the plane is Borel (Simpson [54]);

(e) the Ulm structure theorem for countable reduced abelian p-groups (Friedman-Simpson-Smith [25]).

(3.10.) RCA proves the equivalence of Π^1_1-CA to each of:

(a) Kondo's Theorem, i.e. the statement that every coanalytic set in the plane can be uniformized by a single-valued coanalytic set (Simpson [54]);

(b) Silver's Theorem, i.e. the statement that for every coanalytic (or F_σ) equivalence relation with uncountably many equivalence classes, there exists a perfect set of inequivalent elements (Simpson [54]);

(c) the statement that every countable abelian group is the direct sum of a divisible group and a reduced group (Friedman-Simpson-Smith [25]).

It is noteworthy that the systems Δ^1_1-CA, Σ^1_1-AC, Σ^1_1-DC, and BI have played a relatively minor or nonexistent role in Reverse Mathematics. While these systems have many very interesting metamathematical properties, they appear to be of little or no interest from the viewpoint of the formal axiomatic study of ordinary mathematics.

The moral of Reverse Mathematics is as follows. For many key theorems of ordinary mathematics, there is a weakest natural subsystem of Z_2 in which the given theorem is provable. Furthermore, this weakest natural system often turns out to be one of the systems RCA, WKL, ACA, ATR, and Π^1_1-CA. These conclusions have obvious significance for any philosophical investigation of the role of set existence axioms in ordinary mathematics. (See also the discussion of reductionist programs in §5 below.)

§4. Restricted Induction.

All of the subsystems of Z_2 mentioned above are formulated with the scheme of <u>full induction</u>

$$(\varphi(0) \wedge \forall n(\varphi(n) \rightarrow \varphi(n+1))) \rightarrow \forall n\varphi(n)$$

where $\varphi(n)$ is any formula in the language of Z_2. In his abstracts [13] Friedman initiated the study of subsystems of Z_2 with <u>restricted induction</u>. This means that the full induction scheme is replaced by the induction axiom

$$(0 \in X \wedge \forall n(n \in X \to n+1 \in X)) \to \forall n(n \in X).$$

If S is any of the systems ACA, Δ^1_1-CA, Σ^1_1-AC, Σ^1_1-DC, ATR, BI, Π^1_1-CA, Δ^1_2-CA, ... discussed above, we denote by S_0 the corresponding system with restricted induction. Thus S_0 is in general weaker than S since it contains induction only for those sets which can be proved to exist within S_0. For instance, the axioms of ACA_0 include arthmetical comprehension and the induction axiom as stated above. Combining these, one can prove all arithmetical instances of the induction scheme, but not for instance Σ^1_1 or Π^1_1 induction.

For technical reasons we must exercise a little care in defining RCA_0 and WKL_0. We define them to consist of RCA and WKL with Σ^0_1 induction, i.e. we take as an axiom $(\varphi(0) \wedge \forall n(\varphi(n) \to \varphi(n+1))) \to \forall n\varphi(n)$ for all Σ^0_1 formulas $\varphi(n)$.

(These versions of RCA_0 and WKL_0 are superficially different from, but essentially equivalent to, Friedman's formulation in [13]. The superficial discrepancy arises because Friedman [13] uses a language with function variables rather than set variables. Thus Friedman is able to formulate RCA_0 and WKL_0 in terms of a quantifier-free induction axiom. Then, with the help of some other axioms for the existence of functions (including primitive recursion and the μ-operator), he is able within his version of RCA_0 to prove Σ^0_1 induction.)

At the end of this section we shall comment further on the role of Σ^0_1 induction.

In [13] Friedman does not explicitly discuss his reasons for concentrating on systems with restricted induction. However, with hindsight, one can say that the decision to do so was a sound one.

In the first place, the passage from S to S_0 does not appear to affect materially the formalization of ordinary mathematics. Admittedly, the lack of full induction in RCA_0 and WKL_0 causes occasional difficulties. Several of the proofs in §§2,3 of [25] would become easier in the presence of full induction. However, by working harder, one can usually get by with restricted induction. All of the results of Reverse Mathematics (3.1 through 3.10 above) go through with RCA, WKL, ACA, ... replaced by RCA_0, WKL_0, ACA_0, ... Moreover these results with restricted induction are more definitive than their full induction counterparts.

In the second place, the use of restricted induction is almost mandatory from the viewpoint of Reverse Mathematics. This is because full induction is itself equivalent to a set existence principle, namely the bounded comprehension scheme

$$\forall n \exists X \forall m (m \in X \leftrightarrow (m < n \wedge \varphi(m)))$$

where $\varphi(m)$ is any formula in the language of Z_2. The whole point of Reverse Mathemaics is to prove ordinary mathematical theorems using only the weakest possible set existence principles. Since full induction is a set existence principle, the reverse mathematician is constrained to use full induction as sparingly as possible.

In the third place, from the viewpoint of traditional mathematical logic, the systems with restricted induction appear to be much more natural than those with full induction. The proof-theoretic ordinals of the systems with restricted induction are often easier to calculate and have better closure properties. For instance, a simple model-theoretic argument shows that ACA_0 is a conservative extension of first order Peano arithmetic. Hence $|ACA_0| = \varepsilon_0$. In contrast $|ACA| = \varepsilon_{\varepsilon_0}$ (see Feferman [8] p. 959) and there is no simple characterization of the first-order part of ACA. (See however Ratajczyk [40].) In the case of arithmetical transfinite recursion, the situation is similar. Namely $|ATR_0| = \Gamma_0$

and ATR_0 proves the same Π^1_1 sentences as Feferman's system IR of predicative analysis [13], [24], [32]. In contrast $|ATR| = \Gamma_{\varepsilon_0}$ (see [51]) and there is no nice characterization of the arithmetical sentences which are provable in ATR. Finally, for Π^1_1 comprehension, we have that $|\Pi^1_1\text{-CA}_0| = \psi_0(\Omega_\omega)$ and that $\Pi^1_1\text{-CA}_0$ proves the same arithmetical sentences as first order arithmetic with Ramsey quantifiers [43], [42], but $|\Pi^1_1\text{-CA}| = \psi_0(\Omega_\omega \cdot \varepsilon_0)$ (see [5] p. 14) and there is no nice way to characterize the arithmetical sentences which are provable in Π^1_1-CA.

For a discussion of further benefits which flow from restricted induction, see §5 below.

We end this section with a remark on Σ^0_1 induction. Once one begins to examine the role of induction in ordinary mathematics, it is natural to wonder whether even Σ^0_1 induction is really needed. This question has been discussed by Friedman in an unpublished abstract [23]. Friedman defines a system $ERCA_0$ which is weaker than RCA_0 and whose minimum ω-model consists of the elementary recursive functions from ω into ω. He then states the following results [23]:

(4.1.) Over $ERCA_0$, Σ^0_1 induction is equivalent to the statement that any finite sequence of vectors in $\mathbb{R}^n$ has a linearly independent subsequence with the same span.

(4.2.) Over $ERCA_0$, Σ^0_2 induction is equivalent to the statement that every finite sequence of real numbers has a subsequence which is linearly independent over the rational numbers and has the same span.

Subsequently Simpson and Smith [55] considered another system RCA_0^* which is also weaker than RCA_0. It appears that in most instances, RCA_0^* is strong enough to replace RCA_0 as the base theory for Reverse Mathematics. However, Simpson and Smith [55] proved:

(4.3.) Over RCA_0^*, Σ^0_1 induction is equivalent to each of the following statements:

(a) For any countable field K, any polynomial from $K[x_1,\ldots,x_n]$ can be factored into polynomials over K which are irreducible over K.

(b) For any countable field K, any polynomial of the form $x^{2^n}+1$ has at least one irreducible factor over K.

These results show that sometimes Reverse Mathematics applies even to theorems which are provable in relatively weak systems such as RCA_0.

§5. Reductionist Programs.

In the foundations of mathematics, a reductionist is anyone who proposes to reduce a large part of mathematics to some restricted set of "acceptable" principles. For instance Hilbert [29] proposed to reduce all of mathematics to finitistic reasoning. One might also consider a program of predicative reductionism, whose purpose would be to reduce a large part of mathematical practice to reasoning which is predicative in the sense of Kreisel [37] and Feferman [7].

Results arising from the study of subsystems of Z_2 have gone a long way toward answering questions concerning the practicability of various reductionist proposals.

Consider for example Hilbert's program. Tait [59] has argued convincingly that Hilbert's concept of finitism is well explicated by the formal system of primitive recursive arithmetic, PRA. Using a model-theoretic construction due to Kirby and Paris [33], Friedman (unpublished, but see [22] and [54]) has shown that WKL_0 is a conservative extension of PRA with respect to Π^0_2 sentences. This means that any Π^0_2 sentence which is provable in WKL_0 is already provable in PRA and hence possesses a primitive recursive Skolem function. Subsequently Sieg [47] presented a Gentzen-style finitist relative consistency proof for this result.

Thus any theorem of ordinary mathematics which can be proved in WKL_0 is finitistically reducible in the sense of Hilbert's program. In particular, any Π^0_2 consequence of such a theorem is primitive recursively true.

Of course all of this would be pointless if WKL_0 were as weak as PRA with respect to ordinary mathematics. But such is not the case. The ongoing efforts of Friedman, Simpson, and others [3], [4], [13], [25], [53], [54] have shown that WKL_0 is rather strong from the standpoint of ordinary mathematics. Namely:

(5.1.) WKL_0 proves:

(a) the Heine-Borel covering theorem (Friedman [13], see also Simpson [53], [54]);

(b) basic properties of continuous functions of several real variables; for example, any continuous real-valued function on a closed bounded rectangle in $\mathbb{R}^n$ is uniformly continuous and Riemann integrable and attains a maximum value (Simpson [53], [54]);

(c) the local existence theorem for solutions of systems of ordinary differential equations (Simpson [53]);

(d) Alaoglu's theorem and the Hahn-Banach theorem for separable Banach spaces (Brown-Simpson [4], Brown [3]);

(e) the existence of prime ideals in countable commutative rings (Friedman-Simpson-Smith [25]);

(f) the existence and uniqueness of the algebraic closure of a countable field (Friedman-Simpson-Smith [25]);

(g) the Artin-Schreier theory for countable formally real fields (Friedman-Simpson-Smith [25]).

These examples seem to indicate that WKL_0 is strong enough to develop a very large part of ordinary mathematics, including several theorems which appear to have been of particular interest to Hilbert in connection with his finitist-reductionist program. Admittedly, the proofs of these theorems within WKL_0 are sometimes much more difficult than the most natural or obvious proofs; see for instance Simpson [53], Friedman-Simpson-Smith [25], and Brown-Simpson [4]. Nevertheless, combining all of these results, we seem to have in hand a rather far-reaching partial realization of Hilbert's program.

We now turn to the question of predicative reducibility.

It is important to distinguish carefully between predicative provability and pedicative reducibility. Feferman [7] has argued successfully that his formal system IR and others like it constitute a precise explication of predicative provability. Since HYP is a model of IR, it follows that any predicatively provable sentence in the language of Z_2 must be true in the ω-model HYP. Unfortunately, it appears that hardly any theorems of ordinary mathematics are true in HYP, beyond those that are already provable in ACA. For example, the theorem that any two countable well orderings are comparable is known to fail in HYP. The same goes for all of the ordinary mathematical theorems listed under 3.4, 3.5, 3.6, 3.9 and 3.10 above. Hence none of these thoerems is predicatively provable.

In contrast, when we turn from predicative provability to the weaker notion of predicative reducibility, the situation is much better. Friedman [24] (see also Jäger [32]) has given a finitistic reduction of ATR_0 to IR. His proof shows that ATR_0 and IR prove the same Π^1_1 sentences. Thus any ordinary mathematical theorem which is provable in ATR_0 is in a certain sense predicatively reducible. In particular, any Π^1_1 consequence of such a theorem is predicatively true. This applies to all of the theorems listed under 3.4, 3.6(i), and 3.9 above. For instance, ATR_0 is just strong enough to develop a good theory of countable ordinal numbers and Borel and analytic sets of real numbers. Hence this entire theory, although false in the ω-model HYP and therefore not predicatively provable, is predicatively reducible.

Thus the systems WKL_0 and ATR_0 emerge as being of special significance with respect to reductionist programs. The discovery of these systems is one of the major successes of Reverse Mathematics.

In this context, we would like to mention some further results which serve to highlight a certain parallelism between WKL_0 and ATR_0.

As is well known, the recursive sets form the minimum ω-model of Δ^0_1-CA_0 (i.e. RCA_0). The existence of a non-recursive set is provable in WKL_0. The weakest natural system in which to carry out the obvious diagonal construction of a non-recursive set is Σ_1^0-CA (i.e. ACA_0). Thus WKL_0 appears to lie midway between Δ_1^0-CA and Σ_1^0-CA. This impression is confirmed by the following result:

(5.9.) Over RCA_0, WKL_0 is equivalent to the Σ_1^0 separation principle, i.e. the axiom

$$\sim\exists n(\varphi_0(n) \wedge \varphi_1(n)) \rightarrow \exists X\forall n\ (\varphi_0(n) \rightarrow n \in X.\wedge.\varphi_1(n) \rightarrow n \notin X)$$

where $\varphi_0(n)$ and $\varphi_1(n)$ are Σ_1^0 formulas (Simpson [53]).

The situation for ATR_0 is analogous. The hyperarithmetical sets form the minimum ω-model of Δ_1^1-CA_0. The existence of a non-hyperarithmetical set is provable in ATR_0. The weakest natural system in which to carry out the obvious diagonal construction of a non-hyperarithmetical set is Σ_1^1-CA_0 (i.e. Π_1^1-CA_0). Thus ATR_0 appears to lie midway between Δ_1^1-CA_0 and Σ_1^1-CA_0. This is confirmed by:

(5.10.) Over RCA_0, ATR_0 is equivalent to the Σ_1^1 separation principle, i.e. just as above with $\varphi_0(n)$ and $\varphi_1(n)$ being Σ_1^1 formulas (Simpson [54]).

In addition, the following results have been announced by Friedman [12]:

(5.11.) There exists an ω-model of WKL_0 in which every definable set is recursive.

(5.12.) There exists an ω-model (in fact β-model) of ATR_0 in which every definable set is hyperarithmetical.

These results are presented by Friedman [12] as illustrations of a general theme: "Much more is needed to explicitly define a hard-to-define set of integers than merely to prove its existence."

We shall now finish by mentioning two other projects of Friedman which, strictly speaking, fall outside the scope of the present survey.

First, there are Friedman's papers [14], [15] on the foundations of Bishop-style constructive analysis. In them, Friedman presents some extremely strong formal systems which are more than adequate for constructive analysis, yet conservative over first-order Heyting arithmetic. The languages employed go far beyond that of second-order arithmetic.

Lastly, we mention Friedman's unpublished series of papers [16], [17], [18], [19], [20] which are concerned with calibrating the logical strength of "raw" mathematical texts. These papers represent a vast extension of the program of Reverse Mathematics. In them, formal languages and systems are tailored to the needs of ordinary mathematics, rather than vice versa. The potential applications to problems such as automatic theorem proving have remained largely unexplored.

Bibliography

[1] O. Aberth, Computable Analysis, McGraw Hill, New York, 1980, 187 pages.

[2] K. J. Barwise and J. A. Schlipf, On recursively saturated models of arithmetic, in: Model Theory and Algebra: a Memorial Tribute to Abraham Robinson, Springer-Verlag, Lecture Notes in Mathematics 498 (1975), pp. 42-55.

[3] D. K. Brown, Ph.D. Thesis, Pennsylvania State University, in preparation.

[4] D. K. Brown and S. G. Simpson, Which set existence theorems are needed to prove the Hahn-Banach theorem for separable Banach spaces?, Annals of Pure and Applied Logic, to appear.

[5] W. Buchholz, S. Feferman, W. Pohlers, and W. Sieg, Iterated Inductive Definitions and Subsystems of Analysis: Recent Proof-Theoretical Studies, Springer-Verlag, Lecture Notes in Mathematics 897 (1981), 383 pages.

[6] A. Church and S. C. Kleene, Formal definitons in the theory of ordinal numbers, Fund. Math. 28 (1937), pp. 11-21.

[7] S. Feferman, Systems of predicative analysis, J. Symbolic Logic 29 (1964), pp. 1-30; 33 (1968), pp. 193-220.

[8] S. Feferman, Theories of finite type related to mathematical practice, in: Handbook of Mathematical Logic, edited by J. Barwise, North-Holland, 1977, pp. 913-971.

[9] H. Friedman, Subsystems of set theory and analysis, Ph.D. thesis, M.I.T. (1967), 83 pages.

[10] H. Friedman, Bar induction and Π^1_1-CA, J. Symbolic Logic 34 (1969), pp. 353-362.

[11] H. Friedman, Iterated inductive definitions and Σ^1_2-AC, in: Intuitionism and Proof Theory, edited by J. Myhill, A. Kino and R. E. Vesley, North-Holland, 1970, pp. 435-442.

[12] H. Friedman, Some systems of second order arithmetic and their use, in: Proceedings of the International Congress of Mathematicians, Vancouver 1974, Vol. 1, Canadian Mathematical Congress, 1975, pp. 235-242.

[13] H. Friedman, Systems of second order arithmetic with restricted induction I, II (abstracts), J. Symbolic Logic 41 (1976), pp. 557-559.

[14] H. Friedman, Set theoretic foundations for constructive analysis, Annals of Mathematics 105 (1977), pp. 1-28.

[15] H. Friedman, A strong conservative extension of Peano arithmetic, in: Proceedings of the Kleene Symposium, Madison 1978, edited by K. J. Barwise, H. J. Keisler, and K. Kunen, North-Holland 1980, pp. 113-122.

[16] H. Friedman, The analysis of mathematical texts and their calibration in terms of intrinsic strength, I-IV, informally distributed reports, State University of New York at Buffalo, April-August 1975, 70 pages.

[17] H. Friedman, Analytic definitions of notions of modern logic, preprint, December 1975, 5 pages.

[18] H. Friedman, The arithmetic theory of sets and functions I, preliminary report, August 1976, 7 pages.

[19] H. Friedman, The logical strength of mathematical statements I, preliminary report, August 1976, 21 pages.

[20] H. Friedman, The syntax and semantics of mathematical text, preliminary report, April 1977 version, 52 pages.

[21] H. Friedman, Translatability and relative consistency, preprint, November 1976, 10 pages.

[22] H. Friedman, On fragments of Peano arithmetic, preprint, May 1979, 7 pages.

[23] H. Friedman, Provable equivalents of induction I, unpublished abstract, handwritten, July 1982, 6 pages.

[24] H. Friedman, K. McAloon, and S. G. Simpson, A finite combinatorial principle which is equivalent to the 1-consistency of predicative analysis, in: Patras Logic Symposion, edited by G. Metakides, North-Holland, 1982, pp. 197-230.

[25] H. Friedman, S. G. Simpson, and R. L. Smith, Countable algebra and set existence axioms, Annals of Pure and Applied Logic 25 (1983), pp. 141-181.

[26] A. Fröhlich and J. C. Shepherdson, Effective procedures in field theory, Trans. Royal Soc. London 248 (1956), pp. 407-432.

[27] R. O. Gandy, G. Kreisel, and W. W. Tait, Set existence, Bull. Acad. Polonaise des Sciences (Ser. des Sci. Math., Astr. et Phys.), vol. 8 (1960), pp. 577-582.

[28] J. Harrison, Recursive pseudo well orderings, Trans Amer. Math. Soc. 131 (1968), pp. 526-543.

[29] D. Hilbert, Über das Unendliche, Math. Annalen 95 (1926), pp. 161-190; translated in: From Frege to Gödel (edited by J. van Heijenoort), Harvard Univ. Press, 1967, pp. 369-392.

[30] D. Hilbert and P. Bernays, Grundlagen der Mathematik, vols. I, II, Springer-Verlag, 1934, 1939; 2nd edition 1968, 1970; 473 + 561 pages.

[31] W. A. Howard, Ordinal analysis of bar recursion of type zero, Comp. Math. 42 (1981), pp. 105-119.

[32] G. Jäger, The strength of admissibility without foundation, J. Symbolic Logic 49 (1984), pp. 867-879.

[33] L. Kirby and J. B. Paris, Initial segments of models of Peano's axioms, in: Set Theory and Hierarchy Theory V, Lecture Notes in Mathematics 619, Springer-Verlag, 1977, pp. 211-226.

[34] S. C. Kleene, On the forms of the predicates in the theory of constructive ordinals (second paper), American J. Math. 77 (1955), pp. 405-428.

[35] S. C. Kleene, Quantification of number-theoretic functions, Comp. Math. 14 (1959), pp. 23-40.

[36] J. Knight and M. Nadel, Expansions of models and Turing degrees, J. Symbolic Logic 47 (1982), pp. 587-604.

[37] G. Kreisel, La predicativité, Bull. Soc. Math. France 88 (1960), pp. 371-391.

[38] G. Kreisel, The axiom of choice and the class of hyperarithemtic functions, Indag. Math. 24 (1962), pp. 307-319.

[39] G. Kreisel, Analysis of the Cantor-Bendixson theorem by means of the analytic hierarchy, Bull. Acad. polon. des Scie. 7 (1959), pp. 621-626.

[40] S. Ratajczyk, Satisfaction classes and combinatorial sentences independent from PA, Zeitschr. f. math. Logik und Grundlagen d. Math. 28 (1982), pp. 149-165.

[41] J. H. Schmerl, Peano arithemtic and hyper-Ramsey logic, Trans. Amer. Math. Soc., to appear.

[42] J. H. Schmerl and S. G. Simpson, On the role of Ramsey quantifiers in first order arithmetic, J. Symbolic Logic 47 (1982), pp. 423-235.

[43] K. Schütte, Proof Theory, Springer-Verlag, 1977, 299 pages.

[44] D. Scott, Algebras of sets binumerable in complete extensions of arithmetic, in: Recursive Function Theory, Proc. Symp. Pure Math. 5, Amer. Math. Soc., 1962, pp. 117-121.

[45] W. Sieg, Proof theoretical equivalences between classical and constructive theories for analysis, in [5], pp. 78-142.

[46] W. Sieg, Conservation theorems for subsystems of analysis with restricted induction (abstract), J. Symbolic Logic 46 (1981), pp. 194-195.

[47] W. Sieg, Fragments of arithmetic, preprint, May 1983, 59 pages.

[48] S. G. Simpson, Notes on subsystems of analysis, unpublished lecture notes, Berkeley, 1973, 38 pages.

[49] S. G. Simpson, Sets which do not have subsets of every higher degree, J. Symbolic Logic 43 (1978), pp. 135-138.

[50] S. G. Simpson, Reverse Mathematics, in: Proceedings of the Recursion Theory Summer School, edited by A. Nerode and R. Shore, Proc. Symp. Pure Math., Amer. Math. Soc., to appear.

[51] S. G. Simpson, Σ_1^1 and Π_1^1 transfinite induction, in: Logic Colloquim '80, edited by D. van Dalen, D. Lascar, and J. Smiley, North-Holland, 1982, pp. 239-253.

[52] S. G. Simpson, Set theoretic aspects of ATR_0, in: Logic Colloquim '80, edited by D. van Dalen, D. Lascar, and J. Smiley, North-Holland, 1982, pp. 255-271.

[53] S. G. Simpson, Which set existence axioms are needed to prove the Cauchy/Peano theorem for ordinary differential equations?, J. Symbolic Logic 49 (1984), pp. 783-802.

[54] S. G. Simpson, Subsystems of Second Order Arithmetic, in preparation.

[55] S. G. Simpson and R. L. Smith, Factorization of polynomials and Σ_1^0 induction, preprint, 1984, 29 pages.

[56] C. Spector, Recursive well orderings, J. Symbolic Logic 20 (1953), pp. 151-163.

[57] J. R. Steel, Determinateness and subsystems of analysis, Ph.D. Thesis, Berkeley, 1976, 107 pages.

[58] J. R. Steel, Forcing with tagged trees, Annals of Math. Logic 15 (1978), pp. 55-74.

[59] W. W. Tait, Finitism, Journal of Philosophy, 1981, pp. 524-546.

[60] G. Takeuti, Proof Theory, North-Holland, 1975, 372 pages.

[61] H. Weyl, Das Kontinuum: Kritische Untersuchungen über die Grundlagen der Analysis, Veit & Co., Leipzig, 1918, iv + 84 pages; reprinted in: H. Weyl, E. Landau, and B. Riemann, Das Kontinuum, und andere Monographien, Chelsea, 1960, 1973.

[62] P. Zahn, Ein Konstruktiver Weg zur Masstheorie und Funktionalanalysis, Wissenschaftliche Buchgesellschaft, 1978, 350 pages.

HARVEY FRIEDMAN'S RESEARCH ON THE FOUNDATIONS
OF MATHEMATICS, L.A. Harrington et al. (editors)

Borel Structures for First-order and Extended Logics

Charles Steinhorn

§1. The aspects of model theory to be discussed in this chapter bear upon and blend together two general problems in model theory. The first, that of finding and examining "natural" logics that are more expressive than first-order logic, has been an active and stimulating area of research for at least the last twenty-five years. Work in this area typically has operated between two constraints: that the logic should express concepts beyond first-order logic that are mathematically significant, and also that which is imposed by Lindstrom's theorem, i.e., in order to say more, some of the properties of first-order logic that render it so manageable must be sacrificed. The second broad problem concerns avoiding pathology that might creep in when one studies all models of a particular set of axioms. For example, model theorists routinely build structures by utilizing uncountable transfinite inductions, or well-orderings of, e.g., the real numbers. Such constructions often can enable one to produce examples which run counter to the original spirit that prompted study of a set of axioms, and can wrap one up in set-theoretic dilemmas.

Friedman addresses these issues innovatively first by proposing that one restrict oneself to those structures whose domain and some subset of whose definable functions and relations are Borel subsets of $\mathbb{R}^n$. Not only are such "Borel structures" natural insofar as they are a class of models that one encounters frequently in analysis and topology, but the restriction to such structures provides one with a collection of uncountable models which is, to some extent, immune from the set-theoretic pathologies alluded to above. This remark can be made more precise by the observation that "Borel structures" have definitions that are absolute between models of set theory, and so retain many of their properties in the pass from one model of set theory to another.

Secondly, Friedman introduces the study of logics for Borel structures which include quantifiers that are intended to deal with measure, category and uncountability - those gauges of size that are most often encountered in analysis and topology. Such logics undoubtedly meet the requirement of naturalness that ought to be applied to extended logics. Moreover, the restriction to Borel structures permits one to sidestep the semantic hypothesis needed for Lindstrom's theorem. This makes it much more likely that these logics will share some of the good properties of first-order logic, and, as will be seen, these hopes are, to at least some extent, realized.

In the remainder of this introductory section, the necessary preliminaries and definitions are given. The second section deals with first-order Borel model theory. It divides between traditional first-order model theoretic results and theorems about the class of Borel models of particular first-order theories. The third, and last, section of this paper concerns logics for Borel structures with quantifiers for measure, category and uncountability. Most of the results in this paper are due to Friedman and were announced in Friedman [1978], [1979a] and [1979b]. Many fundamental problems remain to be solved. A selection of those which this author believes to be most central are scattered throughout the text.

All theories to be discussed in this chapter will be constructed from a countable vocabulary τ . We consider first-order logic and finitary extensions of it obtained by the addition in different combinations of the quantifiers Q, Q_c and Q_m . The quantifier Q is to be understood as expressing "there exist uncountably many," Q_c as "there exist not first-category many," and Q_m as "there exist non-(Lebesgue) measure 0 many." So, for example we might form the language $L(Q_c)$ in which only Q_c is added to first-order logic, or $L(Q,Q_m)$, in which first-order logic is extended by the adjunction of both Q and Q_m .

Any model discussed here will have some subset of $\mathbb{R}$, the real numbers, as its domain. With this understood, the new clauses in the definition of satisfaction for the additional quantifiers may be given easily. The clause for Q is as always. That is, if M is a τ-structure, $\mathrm{dom}(M) = M \subseteq \mathbb{R}$, and $a_1,\ldots,a_n \in M$, then

$$M \models Qx\psi(x,a_1,\ldots,a_n)$$

iff $$|\{x \in M : M \models \psi(x,a_1,\ldots,a_n)\}| \geq \aleph_1 .$$

Similarly, with M and $a_1,\ldots,a_n$ as above, we define

$$M \models Q_c x\psi(x,a_1,\ldots,a_n)$$

iff $\{x \in M : M \models \psi(x,a_1,\ldots,a_n)\}$ is not of first category

and

$$M \models Q_m x\psi(x,a_1,\ldots,a_n)$$

iff $\{x \in M : M \models \psi(x,a_1,\ldots,a_n)\}$ does not have measure 0 .

We observe that in the generality in which we have defined it, the Q_m clause above must be understood to be a partial definition of satisfaction at best, as it implicitly entails that $\{x \in M : M \models \psi(x,a_1,\ldots,a_n)\}$ be Lebesgue measurable. With this in mind, we make the following definitions.

A structure whose domain is a non-empty Borel subset of $\mathbb{R}$ and whose relations and functions are Borel in the usual sense of analysis and topology will be called a <u>Borel structure</u>. A model for any of the logics $\mathcal{L}$ described above is <u>totally Borel</u> if any relation defined using an $\mathcal{L}$-formula with parameters is Borel. In particular, for such a model, the partial truth definition for Q_m will be total.

Suppose now that ϕ is a formula of one of the logics $\mathcal{L}$, possibly including some or all of the quantifiers Q, Q_c and Q_m . We say that an $\mathcal{L}$-structure is <u>Borel for</u> ϕ if each relation definable in M from a subformula of ϕ is Borel. Similarly, if T is an $\mathcal{L}$-theory, then the $\mathcal{L}$-structure M is <u>Borel for</u> T if for each $\phi \in T$, M is Borel for ϕ .

As far as expressiveness is concerned, these logics allow us to state definable versions of many facts from analysis and topology. We present a few examples. First, the $L(Q,Q_m)$-sentence

$$\neg Q_y \exists x\, \phi(x,y) \wedge \forall y \neg Q_m x\phi(x,y) \rightarrow \neg Q_m x \exists y\phi(x,y)$$

yields, in any totally Borel structure, a definable instance of the fact that a countable union of measure 0 sets has measure 0 . Also, one may "dualize" the sentence by replacing each occurrence of "Q_m" by "Q_c" to obtain a definable instance of the proposition that the countable union of first-category sets is still first-category. This sentence will be true in any structure M with $\mathrm{dom}(M) \subseteq \mathbb{R}$. (cf., Oxtoby [1971]).

The sentence below expresses a definable instance of the statement, due to Baire originally, that the complement of a first-category subset of $\mathbb{R}$ must be dense:

$$Q_c x\psi(x) \rightarrow \forall y\, \forall z\, \forall w\, [z < y < w \rightarrow \exists x(\psi(x) \wedge z < x < w)] .$$

Again, this proposition may be dualized to obtain the corresponding measure-theoretic statement that is valid on all totally Borel models.

Lastly,

$$Q_m x Q_m y\psi(x,y) \leftrightarrow Q_m y Q_m x\psi(x,y)$$

represents a definable instance of Fubini's Theorem which, of course, is valid on all totally Borel models. Once more, by dualizing one obtains a definable instance of Kuratowski-Ulam theorem (cf., Oxtoby [1971]). This seemingly ubiquitous duality phenomenon will be further explored and explained in §3.

§2. We shall now discuss some first order Borel model theory. The most basic theorem here is:

2.1. Borel Completeness Theorem (Friedman [1978]). Let T be a first-order theory. Then T has an uncountable totally Borel model iff T has an infinite model.

We first remark that the real content of the theorem lies in the existence of an uncountable totally Borel model for T , as all countable subsets of $\mathbb{R}^n$, for any n , are Borel. Secondly, since it is known that any uncountable Borel subset of $\mathbb{R}$ is Borel isomorphic to $\mathbb{R}$, it follows that the uncountable Borel model whose existence is asserted actually may be assumed to have $\mathbb{R}$ as its domain.

Friedman's original proof of 2.1 involves carefully stretching a sequence of indiscernibles. The reader may consult Steinhorn [1984] for the proof. Because of the limitations of this method in general, it does not seem as though it will be well-suited for applications. We now will sketch an alternative proof which might be more promising. It might be viewed as an attempt to extract the first order content from Friedman's proof of the completeness of $L(Q_m)$ which is outlined in §4 of Steinhorn [1984]. Full details will appear in Steinhorn [1984a].

Sketch of the Proof of Theorem 2.1. First, we define an auxiliary language L^*. We let $L^* = \bigcup_{n<\omega} L^*_n$, where $L^*_n = L$, and L^*_{n+1} is the language built from L^*_n and a collection of new Skolem function symbols, $F_{\exists x\psi}$, for all formulas of the form $\exists x\psi$ in L^*_n. Then, for any $L' \subseteq L^*$, let $\mathcal{S}(L') \subseteq L^*$ be all formulas of the form $\forall x_1 \ldots \forall x_p [\exists x\, \psi(x,x_1,\ldots,x_p) \to \psi(F_{\exists x\psi}(x_1,\ldots,x_p), x_1,\ldots,x_p)]$, where $\exists x\, \psi(x,x_1,\ldots,x_p) \in L'$. In particular, observe that $\mathcal{S}(L')$ is finite if L' is. We shall also add a new set of constants $C = \{c_s : s \in \{0,1\}^{<\omega}\}$ to the vocabulary of L and shall consider that L^*-formulas which we enumerate in a fixed order as $\{\phi_i : i < \omega\}$.

By recursion on $n < \omega$ we define finite sets F_n of $L(C)^*$ formulas, and infinite sets X_s, for $s \in \{0,1\}^n$, all contained in some fixed $M_n \models T$. Let $F_0 = \phi$ and $X_\phi = \mathrm{dom}(M)$ for some infinite $M \models T$. Supposing that F_n and $\{X_s : s \in \{0,1\}^n\}$ have been defined, we take F_{n+1} and $\{X_s : s \in \{0,1\}^{n+1}\}$ so that

(a) $c_s \neq c_t \in F_{n+1}$ for all $c_s, c_t \in C$, $s \neq t$, and $s,t \in \{0,1\}^n$;

(b) if $\phi(c_{s_1},\ldots,c_{s_k}) \in F_n$, then for all $i_1,\ldots,i_k \in \{0,1\}$,

$$\phi(c_{s_1 \wedge <i_1>},\ldots,c_{s_k \wedge <i_k>}) \in F_{n+1};$$

(c) for every $i < n$, if the free variables of ϕ_i are $v_0,\ldots,v_{k-1}$, $k < n$, then for all distinct $c_{s_i},\ldots,c_{s_k}$ in C with each $s_j \in \{0,1\}^n$, either

$\phi_i(c_{s_i},\ldots,c_{s_k}) \in F_{n+1}$ or $\phi_i(s_1,\ldots,s_k) \in F_{n+1}$;

(d) $\mathcal{S}(F_n) \subseteq F_{n+1}$;

and (e) F_{n+1} is consistent in the following strong sense:

for all $< x_s : s \in \{0,1\}^{n+1} \wedge x_s \in X_s >$,

$$M_{n+1} \models \bigwedge_{\phi \in F_{n+1}} \phi[<c_s/x_s : s \in \{0,1\}^{n+1}>] ,$$

where the expression in brackets represents replacing each c_s by x_s in ϕ .

The only really difficult step is to satisfy (e). This is done by applying the polarized partition relations for finite sets as given in, e.g., Erdös-Rado [1956], §8, to the sequence of sets $<X_s : s \in \{0,1\}^n>$ that is given by induction hypothesis.

This done, one completes the construction of the desired Borel model for T exactly as in the proof of 4.1 in Steinhorn [1984].

One interesting possible avenue of research would be the development of Borel model theory along the lines of classical model theory. We next present two examples of theorems in this spirit. A proof of the first, which was observed by David Marker, can be found in Steinhorn [1984].

<u>2.2</u>. <u>Theorem</u>. Every countably infinite L-structure $M = <M,R_1,\ldots,R_n>$ has an uncountable totally Borel elementary extension that is recursively saturated.

As this theorem is not a particularly strong result from the point of classical model theory, one certainly must ask of how much saturation does a Borel model admit. Although there as yet is no definitive answer to this question, Theorem 2.4, below, suggests that it very well might be subtle and interesting.

The second example we present improves 2.1.6 of Steinhorn [1984]. Its proof, a sketch of which follows, draws out the "Borel content" of Shelah's theorem that a first-order theory with a two-cardinal model of type $(\aleph_\omega, \aleph_0)$ has one of type $(2^{\aleph_0}, \aleph_0)$ (cf., Shelah [1977]) (for complete details, cf., Steinhorn [1984a]).

<u>2.3</u>. <u>Theorem</u>. A first order theory T with distinguished unary predicate P that has a model M of power $\aleph_\omega$ in which $|P(M)| = \aleph_0$ has an uncountable totally Borel model N with $|P(N)| = \aleph_0$.

We notice that this is best possible, as the Continuum Hypothesis holds for Borel subsets of $\mathbb{R}$, i.e., every Borel subset of $\mathbb{R}$ has power $\aleph_0$ or $2^{\aleph_0}$.

<u>Sketch of proof of 2.3</u>. First, we recall two definitions from Shelah [1975]. Suppose that $s_1,\ldots,s_n, t_1,\ldots,t_n \in {}^{\omega>}2$. Then $\langle s_1,\ldots,s_n\rangle$ and $\langle t_1,\ldots,t_n\rangle$ are said to be <u>similar over</u> <u>m</u> if $\ell h(s_i), \ell h(t_i) \geq m$ and $s_i \quad m = t_i \quad m$ for all $i = 1,\ldots,n$ and for $i \neq j$, $s_i \neq s_j$. A collection $\{i_\eta : \eta \in {}^{\omega}2\}$ of an L-structure M is a set of <u>tree indiscernibles</u> iff for any sequences $\langle \eta_1,\ldots,\eta_n\rangle$ and $\langle \nu_1,\ldots,\nu_n\rangle$ of elements of ${}^{\omega}2$ that are similar over m, for some $m < \omega$, it is the case that for all L-formulas $\phi(v_1,\ldots,v_n)$,

$$M \models \phi(i_{\eta_1},\ldots,i_{\eta_n}) \leftrightarrow \phi(i_{\nu_i},\ldots,i_{\nu_n}) .$$

Similarly, if we adjoin constants to L for the elements of any $A \subseteq M$, we can speak of tree indiscernibles over A - i.e., tree indiscernibles for all $L(A)$-formulas.

In Shelah [1977] it is proved that if a theory T has a two-cardinal model of type $(\aleph_\omega, \aleph_0)$, then it has a model N containing a set of tree indiscernibles, $\{i_\eta : \eta \in {}^{\omega}2\}$, over $P(N)$, the set of elements satisfying the distinguished predicate $P(\cdot)$ in N. By taking the Skolem hull of the tree indiscernibles and $P(N)$ (without loss, we may suppose that L has built-in Skolem functions) one obtains the desired $(2^{\aleph_0}, \aleph_0)$ model. The remainder of the sketch of the proof of the theorem is given to showing that the procedure just described can be used to produce a Borel model. The analysis involved, not surprisingly, bears a certain resemblence to that required to carry out Friedman's original argument for Theorem 2.1, above (cf., Steinhorn [1984], §2).

Suppose that $\tau(v_1,\ldots,v_m)$ is a term and $\bar{s} = \langle s_1,\ldots,s_m\rangle$ is a lexicographically increasing sequence of elements in ${}^{<\omega}2$. With N and $\{i_\eta : \eta \in {}^{\omega}2\} \subseteq N$ as above, we define the <u>support for</u> τ <u>at</u> $\bar{s}$, $S_\tau(\bar{s})$, to be the least $S \subseteq \{1,\ldots,m\}$ so that for all $\langle\eta_1 \ldots,\eta_m\rangle$, $\langle\xi_1,\ldots,\xi_m\rangle \in ({}^{\omega}2)^m$ extending $\bar{s}$,

$$\langle\eta_1,\ldots,\eta_m\rangle \quad S = \langle\xi_1,\ldots,\xi_m\rangle \quad S = \tau(i_{\eta_1},\ldots,i_{\eta_m}) = \tau(i_{\xi_1},\ldots,i_{\xi_m}) \;\ldots\ldots\ldots(*)\;.$$

We leave the proof that $S_\tau(\bar{s})$ can be defined to the reader. Moreover, by the minimality of $S_\tau(\bar{s})$, if $\langle\eta_1,\ldots,\eta_m\rangle, \langle\xi_1,\ldots,\xi_m\rangle \in (\omega_2)^m$, each extends $\bar{s}$, and for some $i \in \{1,\ldots,m\}$,

$$\eta_i(\ell h(\bar{s}(i))) \neq \xi_i(\ell h(\bar{s}(i)))\text{ , then } \tau(\eta_1,\ldots,\eta_m) \neq \eta(\xi_1,\ldots,\xi_m)\;.$$

This done, for each term $\tau(v_1,\ldots,v_m)$, we can define a set $\mathcal{S}_\tau \subseteq ({}^{<\omega}2)^m$ of <u>support nodes for</u> τ that will satisfy

(i) for any $\bar{s} \in \mathcal{S}_\tau$, there is no $\bar{t} \supsetneq \bar{s}$ that also is in $\mathcal{S}_\tau$;

(ii) For any (lexicographically) increasing $\bar{t} \in ({}^{<\omega}2)^m$, there is some $\bar{s} \in \mathcal{S}_\tau$ so that either $\bar{t} \subseteq \bar{s}$ or $\bar{s} \subseteq \bar{t}$;

and

(iii) for any $\bar{s} \in \mathcal{S}_\tau$ and $\langle\eta_1,\ldots,\eta_m\rangle, \langle\xi_1,\ldots,\xi_m\rangle \in ({}^{\omega}2)^m$ that extend $\bar{s}$,

$$\text{if } \langle\eta_1,\ldots,\eta_m\rangle \quad S_\tau(\bar{s}) \neq \langle\xi_1,\ldots,\xi_m\rangle \quad S_\tau(\bar{s})$$
$$\text{then } \tau(\eta_1,\ldots,\eta_m) \neq \tau(\xi_1,\ldots,\xi_m)\;.$$

Although we omit the verification of details, $\mathcal{S}_\tau$ may be chosen to be the set of all $\bar{s} \in ({}^{<\omega}2)^m$ for which there is no $\bar{t} \supseteq \bar{s}$ with $S_\tau(\bar{t}) \subsetneq S_\tau(\bar{s})$ and which are minimal, with respect to inclusion, for this property.

Having obtained $\mathcal{S}_\tau$, we now define a Borel subset $V_\tau \subseteq ({}^{\omega}2)^m$, which we dub the set of <u>values</u> of τ. For $\bar{s} \in \mathcal{S}_\tau$ and $i \in \{1,\ldots,m\} \setminus S_\tau(\bar{s})$, fix $\eta_i^0 \in {}^{\omega}2$ extending $\bar{s}(i)$. Then let $V_\tau(\bar{s}) = \{<\eta_1,\ldots,\eta_m> : \eta_i \in {}^{\omega}2$ extending $\bar{s}(i)$ for $i \in S_\tau(\bar{s}) \wedge \eta_i = \eta_i^0$ for $i \in \{1,\ldots,m\} \setminus S_\tau(\bar{s})\}$, and

$V_\tau = \bigcup_{s \in \mathcal{S}_\tau} V_\tau(\bar{s})$.

The last ingredient that goes into the construction of the Borel two-cardinal models concerns the relationship between different terms. That is, suppose that $\tau(v_1,\ldots,v_m)$ and $\tau'(u_1,\ldots,u_n)$ are terms, and that for some $\bar{s} \in \mathcal{S}_\tau$ and $\bar{t} \in \mathcal{S}_{\tau'}$, there are $<\eta_1,\ldots,\eta_m> \in ({}^{\omega}2)^m$ and $<\xi_1,\ldots,\xi_n> \in ({}^{\omega}2)^n$ extending $\bar{s}$ and $\bar{t}$, respectively, so that $\tau(\eta_1,\ldots,\eta_m) = \tau'(\xi_1,\ldots,\xi_n)$. Then using tree-indiscernibility, and the properties of $\mathcal{S}_\tau$ and $\mathcal{S}_{\tau'}$, one can show that $|S_\tau(\bar{s})| = |S_{\tau'}(\bar{t})|$ and $\bar{s} \quad S_\tau(\bar{s}) = t \quad S_{\tau'}(\bar{t})$.

Finally, we sketch the construction of the Borel two-cardinal model, which will be the Skolem hull of the set $\{i_\eta : \eta \in {}^{\omega}2\}$ of tree-indiscernibles. Once and for all, we fix homeomorphism of $({}^{\omega}2)^m$, $m < \omega$, and the Cantor set. We shall identify the two where convenient in what follows. Also, enumerate the terms in the language as $\{\tau_i : i < \omega\}$. First, we identify $\{i_\eta : \eta \in {}^{\omega}2\}$ with the Cantor set on $[0,1]$. Next, identify V_{τ_0} with its canonical image (a Borel subset of the Cantor set) on the interval $[2,3]$. Now, suppose we have dealt with $\tau_0,\ldots,\tau_{n-1}$, and we wish to identify V_{τ_n} with its image in $\mathbb{R}$. Let us enumerate $\mathcal{S}_{\tau_n}$ as $\{\bar{s}_i : i < \omega\}$. Suppose that we have appropriately identified $V_{\tau_m}(\bar{s}_j)$ for $j < i$, and now we wish to do the same for $V_{\tau_n}(\bar{s}_i)$. If there is no $k < n$ and $\bar{t} \in \mathcal{S}_{\tau_k}$ so that τ_n, τ_n, $\bar{s}_i$ and $\bar{t}$ satisfy the conditions in the preceding paragraph, just map $V_{\tau_n}(\bar{s}_i)$ canonically into $[2(n+1), 2n+3]$. If not, let k_0 be the least such k as above, and identify $V_{\tau_n}(\bar{s}_i)$ with $V_{\tau_{k_0}}(\bar{t})$ (there is only one such $\bar{t}$ by appeal to the properties of $\mathfrak{S}_\tau$, for any term τ) . As $V_{\tau_n} = \bigcup_{S \in \mathfrak{S}_{\tau_n}} V_{\tau_n}(\bar{s})$, the construction is complete.

As in the proof of 2.1.1 in Steinhorn [1984], one now may verify that the model constructed here is totally Borel.

In addition to the consideration of the Borel analogues of problems in classical model theory, the study of Borel models offers an interesting avenue of research not available to the classical model theorist, viz., the study of the structural properties of classes of Borel structures. As an easy and obvious example, the class of all Borel structures obeys the continuum hypothesis. A much more subtle result arises in the case of the "Borel version" of the Suslin problem.

A Suslin line is a linear order which is not separable and has no uncountable collection of pairwise disjoint open intervals. The statement which asserts that there are no Suslin lines, i.e., that up to isomorphism, $(\mathbb{R}, <)$ is the only unbounded dense complete linear order that does not have an uncountable collection of pairwise disjoint open intervals, is known as Suslin's Hypothesis. It is, of course, well-known that Suslin's Hypothesis is independent of ZF set theory. It is of interest, then, that we have the following result of Shelah (for a proof, cf., Steinhorn [1984 §2).

2.4. Theorem (Shelah). A Borel linear order either is separable or has uncountably many pairwise disjoint open intervals. Moreover, this theorem has been sharpened by Friedman to:

2.5. Theorem (Friedman). A Borel linear order either is separable or contains a perfect totally isolated set, i.e., a perfect set A so that for any

$a \in A$ there is an open I with $a \in I$ and $I \cap A = \phi$.

These theorems suggest a very general and potentially very interesting program. They demonstrate that certain set-theoretic propositions that are undecidable may be better-behaved if attention is restricted to appropriate "Borel versions" of the problem.

The last theorem we mention in this section reveals that the full saturation possible in classical model theory may not be possible always if one specializes to classes of Borel structures.

§3. Here we describe progress that has been made in the study of the logics obtained by adjoining various combinations of Q, Q_c and Q_m to first-order logic. At this stage in the development of the area, the techniques of proof seem to be just as important as the results themselves. We therefore choose to survey results (e.g., 3.1) which are subsumed by other theorems (in the case of 3.1, 3.10) but whose proofs differ substantially from those of the stronger propositions. We begin with some axiomatizability, or "abstract completeness" theorems.

3.1. Theorem (Friedman [1978]). The set of sentences of $L(Q, Q_c)$ that are valid in all totally Borel $L(Q, Q_c)$-structures is recursively enumerable.

This theorem follows from the lemma below, which really is the heart of the matter.

3.2. Lemma (Friedman). Let ϕ be an $L(Q, Q_c)$ sentence. Then there is a totally Borel model of ϕ iff there is a model of second-order arithmetic, Z_2 , which also satisfies, "there is a totally Borel model of ϕ ."

The proof of 3.2 (cf., Steinhorn [1984]) cleverly combines forcing with Cohen generic reals and absoluteness. It strongly turns on the fact that a Cohen generic real avoids all meager Borel sets with codes in the model over which the Cohen real is generic. It does not surprise, then, to learn that if one carries out the same argument, but instead forcing with random reals - i.e., those which miss all measure 0 Borel sets coded in the ground model - rather than with Cohen reals, then we have

3.3. Theorem (Friedman [1978]). The set of sentences of $L(Q, Q_m)$ that are valid in all totally Borel $L(Q, Q_m)$-structures is recursively enumerable.

The apparently close relationship between 3.1 and 3.3 can be drawn even more sharply by modifying the proofs of 3.1 and 3.3. (cf., Steinhorn [1984]). The result explains the duality noted in §1, above.

3.4. Theorem (Friedman [1978]). For a sentence ϕ of $L(Q, Q_c)$, let ϕ^* be the sentence in $L(Q, Q_m)$ obtained from ϕ by replacing each "Q_c" by "Q_m". Then, ϕ has a totally Borel $L(Q, Q_c)$-model iff ϕ^* has a totally Borel $L(Q, Q_m)$-model.

More strongly, yet one may ask about axiomatizability, and full duality - that is, the interchangeability of "Q_c" and "Q_m" in an $L(Q, Q_c, Q_m)$ sentence - for $L(Q, Q_c, Q_m)$. These problems, however, are open. Finally, further modification of the arguments for 3.1 and 3.3, as indicated in Steinhorn [1984], yields:

3.5. Theorem (Friedman [1978]). $L(Q, Q_c)$ is countably compact.

3.6. Theorem (Friedman [1978]). $L(Q, Q_m)$ is countably compact.

The axiomatizability results, above, provide important information and their proofs illustrate stimulating techniques. Ultimately, though, they cannot be as satisfying as completeness theorems in which explicit complete sets of axioms are given, and in whose proofs genuine model building tools are developed. Friedman also has found such theorems, however. We shall survey these results below, beginning with $L(Q_c)$. The axioms for $L(Q_c)$ are:

(A) All the usual axiom schemas for first-order logic (e.g., as in Chang and Keisler [1973])

(C0) $\neg(Q_c x)x = y$

(C1) $(Q_c x)\psi(x,\ldots) \leftrightarrow (Q_c y)\psi(y,\ldots)$, where $\psi(x,\ldots)$ is an $L(Q_c)$ formula in which y does not occur, and $\psi(y,\ldots)$ results by replacing each free occurrence of x by y .

(C2) $(Q_c x)(\phi \vee \psi) \rightarrow (Q_c x)\phi \vee (Q_c x)\psi$

(C3) $[(Q_c x)\phi \wedge (\forall x)\phi \rightarrow \psi] \rightarrow (Q_c x)\psi$

(C4) $(Q_c x)(Q_c y)\phi \rightarrow (Q_c y)(Q_c x)\phi$.

Observe that axiom (C4) expresses a definable version of the Kuratowski-Ulam theorem referred to in §1. Let K_c be the proof system given by the axioms (A), (C0) - (C4) , and the rules of modus ponens and generalization. Then,

3.7. Theorem (Friedman [1979a]). A set of sentences T in $L(Q_c)$ has a totally Borel model iff T is consistent in K_c .

Next, let the axioms (M0) - (M4) in $L(Q_m)$ be the result of everywhere replacing "Q_c" by "Q_m" in (C0) - (C4) , and K_m be the resulting proof system for $L(Q_m)$.

3.8. Theorem (Friedman [1979a]). A set of sentences T in $L(Q_c)$ has a totally Borel model iff T is consistent in K_m .

The proof of 3.8, and so, mutatis mutandis, 3.7 is sketched in Steinhorn [1984].

We now consider axioms for $L(Q)$. These are given in a vocabulary expanded to include a unary predicate $N(\cdot)$, whose intended interpretation is N , and a binary function symbol $F(\cdot,\cdot)$, which, as the first coordinate varies, is intended to represent one-to-one functions from the universe of the given totally Borel structure to all perfect subsets of the structure. The axioms then are:

(Q0) The usual axioms for $L(Q)$ (cf., Keisler [1970])

(Q1) $(\forall x)(\forall y)(\forall z)[F(x,y) = F(x,z) \rightarrow y = z]$

(Q2) $\neg(Qx)N(x)$

(Q3) $(Qy)\phi \rightarrow (\exists x)(\forall y)(\forall z)[y = F(x,z) \rightarrow \phi(y)]$

(Q4) $(Qy)\phi \rightarrow (\exists x)(\forall y)[\phi \rightarrow (\exists z)(N(z) \wedge F(x,z) = y])$

(Q5) $(Qx)(x = x) \rightarrow (\exists x)(Qy)(\forall z)[F(x,z) \neq y]$

(Q6) $(Qx)(x = x) \rightarrow [(\forall y_1) \ldots (\forall y_n)(\forall z)(N(z) \wedge \phi(y_1,\ldots,y_n,z))$

$\rightarrow (\exists z)(\exists x_1) \ldots (\exists x_n)(\forall y_1) \ldots (\forall y_n)$

$(N(z) \wedge \phi(F(x_1, y_1),\ldots,F(x_n, y_n),z))]$,

where $x_1,\ldots,x_n$ are not free in ϕ .

Axiom (Q5) expresses that any perfect set contains two disjoint perfect subsets. Also, (Q6) says, in a definable sense, of course, that a Borel partition of a product of n perfect sets, $\prod_{i=1}^{n} P_i$, into countably many pieces has a homogeneous subset of the form $\prod_{i=1}^{n} P'_i$, in which each $P'_i \subseteq P_i$ and is perfect. Let K_u be the proof system based on axioms (A) and (U0) - (U6) .

<u>3.9</u>. <u>Theorem</u> (Friedman [1979a]). A set of sentences T in $L(Q)$ <u>in the original vocabulary</u> has a totally Borel model iff T is consistent in K_u .

The proof in the direction from right-to-left resembles that of 3.7 and 3.8. Moreover, the other direction - soundness - no longer is a trivial matter since K_u includes symbols not in the formulas in T . That is, there is no reason to believe that every totally Borel model for the original vocabulary can be expanded to a totally Borel model for the expanded vocabulary, and so it is not immediately apparent how to prove that any sentence in the original vocabulary which is a consequence of K_u is valid. Friedman sidesteps this obstacle by a clever absoluteness argument (cf., Steinhorn [1984], §4).

Lastly, we come to the logics $L(Q, Q_c)$ and $L(Q, Q_m)$. Let

(CU) $(Q_c x)(\exists y)(\phi(x,y) \wedge N(y)) \rightarrow (\exists y)(N(y) \wedge (Q_c x)\phi(x,y))$

be the axiom which asserts that the union of countably many sets of first category is still of first category. Also, let (MU) be the result of replacing "Q_c" everywhere in (CU) by "Q_m". If $K_{c,u}$ is the system for $L(Q, Q_c)$ based on axioms (A), (C0) - (C4), (Q0) - (Q6) and (CU), and $K_{m,u}$ is the comparable system for $L(Q, Q_m)$, then we have

3.10. Theorem (Friedman [1979a]). A set of sentences T in $L(Q, Q_c)$ (resp., $L(Q, Q_m)$) *in the original vocabulary* has a totally Borel model iff T is consistent in $K_{c,u}$ (resp., $K_{m,u}$).

It would be of considerable interest to develop the model theory for these logics. Also, as remarked upon above, resolution of the status of $L(Q, Q_c, Q_m)$ would be of significance.

We conclude with a theorem which offers yet another angle on the logics treated here.

3.11. Theorem (Friedman [1979a]). The logic $L(Q_m)$ without first-order quantifiers is undecidable.

Sketch of proof. Let $R(\cdot,\cdot)$ be a binary relation symbol. For any first-order sentence ϕ whose only non-logical symbol is R, we shall define an $L(Q_m)$-sentence ϕ^* without first-order quantifiers in a larger, but fixed, finite vocabulary so that

(*) ϕ is satisfiable iff ϕ^* is totally Borel satisfiable.

This suffices to prove the theorem by well-known results (cf., Lewis [1979], part II B).

For the remainder of the proof, we let "$\check{Q}_m$" abbreviate "$\neg Q_m$". We begin with the consideration of totally Borel models for $L(Q_m)$ without first-order quantifiers in which the only non-logical symbols are the two-place $R(\cdot,\cdot)$ and $\cdot \approx_1 \cdot$. For such a model M, we introduce the defined binary relation

$$M \models x \approx_2 y \leftrightarrow \neg(Q_m z)[(x \approx_1 z \wedge \ y \approx_1 z) \vee (y \approx_1 z \wedge \ \neg x \approx_1 z)] .$$

It is not difficult to see that $\approx_2$ is an equivalence relation in M .

Let α be the sentence

$$\alpha \equiv (\check{Q}_m x)(Q_m y)x \approx_2 y ,$$

asserting that for almost all x , the $\approx_2$-equivalence class of x has positive measure. Weclaim that if $M \models \alpha$, then $L = \{x \in M :$ the measure of $[x]_{\approx_2}$ is greater than $0\}$ includes all but a subset of measure 0 of M . For if $M \setminus L$ had positive measure, then since $M \models \alpha$, some point in $M \setminus L$ would have to be in L , which is ludicrous. Next, let β be the sentence

$$\beta \equiv (\check{Q}_m x)(\check{Q}_m y)[\ \ (x,y) \wedge (Q_m u)x \approx_2 u \wedge (Q_m u)y \approx_2 u$$

$$\to (\check{Q}_m u)(\check{Q}_m v)(x \approx_2 u \wedge y \approx_2 v \to R(u,v))] .$$

In other words, β as closely as possible asserts that R is a congruence on points whose $\approx_2$-class has positive measure.

We now are prepared to define ϕ^* as required for the verification of (*) , above. For a formula ϕ in first-order logic whose only non-logical symbol is $R(\cdot,\cdot)$, let ϕ^* be $\phi' \wedge \alpha \wedge \beta$, in which ϕ' is defined recursively by:

$$R(x,y)' \equiv (Q_m u)x \approx_2 u \wedge (Q_m u)y \approx_2 u$$

$$\wedge \neg(Q_m u) \ \ \neg(Q_m v)[x \approx_2 u \wedge y \approx_2 v \wedge \ \neg R(u,v)] ;$$

$$(\phi \wedge \psi)' \equiv \phi' \wedge \psi' ;$$

$$(\neg\phi)' \equiv \neg\phi' ;$$

and $((\forall x)\phi)' \equiv \check{Q}_n x\phi'$.

We now can establish (*) .

For the direction from left-to-right in (*), begin with a countable model M of the first-order sentence ϕ. Enumerate M as $\{a_i : i \in \mathbb{Z}\}$. We define a totally Borel model whose universe is $\mathbb{R}$ which satisfies ϕ^*. First, for $r_1, r_2 \in \mathbb{R}$, let

$$r_1 \approx_1 r_2 \quad \text{iff} \quad (\exists j \in \mathbb{Z})\ r_1, r_2 \in [j, j+1) .$$

Then, for $r_1, r_2 \in \mathbb{R}$, define

$$R(r_1, r_2) \quad \text{iff} \quad M \models R(a_{i_1}, a_{i_2}) \quad \text{and} \quad r_1 \in [i_1, i_1 + 1),\ r_2 \in [i_2, i_2 + 1] .$$

It is apparent that $\langle \mathbb{R}, \approx_1, R \rangle = \alpha \wedge \beta$. Moreover, by induction on complexity, one can verify that $\langle \mathbb{R}, \approx_1, R \rangle$ is totally Borel and

$$M \models \psi(a_{i_1}, \ldots, a_{i_m})$$

$$\text{iff} \quad (\forall r_{i_1} \in [i_1, i_1 + 1)) \ldots (\forall r_{i_m} \in [i_m, i_m + 1)) \langle \mathbb{R}, \approx_1, R \rangle \models \psi'(r_{i_1}, \ldots, r_{i_m}) ,$$

which certainly shows that ϕ^* is totally Borel satisfiable.

For the remaining direction of (*), suppose that $\langle \mathbb{R}, \approx_1, R \rangle$ is a totally Borel model of ϕ^* (we may assume without loss of generality that the totally Borel model has domain $\mathbb{R}$). We must construct a model M of ϕ. For the universe of $\mathcal{M}$, we take those $\approx_2$-equivalence classes of $\langle \mathbb{R}, \approx_1, R \rangle$ which have measure greater than 0. Then, for any such classes $[r], [s]$, let

$$M \models R([r], [s]) \quad \text{iff for almost all} \quad r' \in [r],\ s' \in [s]$$
$$\langle \mathbb{R}, \approx_1, R \rangle \models R(r', s') .$$

Since $\langle \mathbb{R}, \approx_1, R \rangle \models \alpha \wedge \beta$, this definition makes sense. Then, by induction on complexity, one proves

$$M \models \psi([r_1], \ldots, [r_m]) \quad \text{iff for almost all} \quad r'_1 \in [r_1], \ldots,$$
$$\text{for almost all} \quad r'_m \in [r_m],\ \langle \mathbb{R}, \approx_1, R \rangle \models \psi'(r'_1, \ldots, r'_m) .$$

Although we omit the details - which are somewhat involved - we see that $M \models \phi$, which proves the right-to-left direction of (*) .

In the closing, we remark that this last result invites one to study the decidability/undecidability strength of various prefix classes in $L(Q_m)$ without first-order quantifiers analogous to what has been done in first-order logic.

REFERENCES

[1] Chang, C. C., and Keisler, H. J., Model Theory. Amsterdam: North Holland, (1973).

[2] Erdös, P., and Rado, R., A partition calculus in set theory. Bulletin of the American Mathematical Society 62, 427-489, (1956).

[3] Friedman, H., On the logic of measure and category I, ms., (1978).

[4] Friedman, H., Addendum to "On the logic of measure and category I," ms., 1979a.

[5] Friedman, H., Borel structures in mathematics I, ms., (1979b).

[6] Harrington, L., and Shelah, S., Counting equivalence classes for co-κ-Suslin equivalence relations, ms., (1981).

[7] Keisler, H. J., Logic with the quantifier "there exist uncountably many." Annals of Mathematical Logic 1, 1-93, (1970).

[8] Lewis, H. R., Unsolvable Classes of Quantificational Formulas. Reading, Massachusetts: Addison-Wesley, (1979).

[9] Oxtoby, J., Measure and Cateogory. New York: Springer-Verlag, 1971.

[10] Shelah, S., A two-cardinal theorem. Proceedings of the American Mathematical Society 48, 207-213, (1975).

[11] Shelah, S., A two-cardinal theorem and a combinatorial theorem. Proceedings of the American Mathematical Society 62, 134-136, (1977).

[12] Steinhorn, C., Borel structures and measure and category logics, Model-Theoretic Logics, Chapter 16 (J. Barwise and S. Feferman, ed.). Amsterdam: North-Holland (in press), (1984).

[13] Steinhorn, C., Some first order Borel model theory, in preparation, (1984a).

HARVEY FRIEDMAN'S RESEARCH ON THE FOUNDATIONS
OF MATHEMATICS, L.A. Harrington et al. (editors)

Nonstandard Models and Related Developments

C. Smoryński

To the average mathematician, the adjective "nonstandard" attaches itself to the noun "analysis". However, there are also nonstandard models of first-order arithmetic, second-order arithmetic, and even set theory. While the study of nonstandard models of set theory postdates that of nonstandard models of the reals, that of second-order arithmetic stems from about the same time as nonstandard analysis, and the existence, if not the serious study thereof, of nonstandard models of arithmetic predates nonstandard analysis by several decades. From the beginning, nonstandard models of arithmetic were studied for their implications and not for themselves: Dedekind's categorical characterisation of the set of natural numbers was a normative goal; axiomatisations were supposed to characterise up to isomorphism the structures the theories of which were being axiomatised. Completeness--the ability of the theory to prove all sentences true in whatever structure a given theory is trying to capture--follows immediately from such a categoricity result. Of course, Gödel's Incompleteness Theorem implies that no decent axiomatisation of arithmetic can be complete, whence it cannot be categorical. This decency--an effectiveness criterion--is necessary for incompleteness, but not for non-categoricity: In the 1930s, Thoralf Skolem proved the existence of structures which were not isomorphic to the set of natural numbers and which, nonetheless, satisfied exactly the same first-order sentences satisfied by the set of natural numbers. Although Skolem's proof was specifically arithmetical, it was soon recognised that the phenomenon was quite general: A first-order theory could not hope to characterise any structure up to isomorphism unless the structure were finite. This is a fundamental weakness of first-order logic. Skolem considered it a criticism: in view of such things as nonstandard analysis, we presently consider it a tool.

There are three aspects to the study of nonstandard models: i. the study of the models themselves, ii. the use of the models to study formal systems, and iii. the application of such outside logic to ordinary mathematics. To a small extent, one can illustrate all three of these by an early use made of nonstandard models of arithmetic by Czeslaw Ryll-Nardzewski, who, in the 1950s, showed that no finitely axiomatised theory could yield all instances of induction in its own language (provided, of course, the language include that of arithmetic). Before outlining the contents of this chapter, I shall briefly sketch the proof of Ryll-Nardzewski's result.

The language of arithmetic contains $+$, $\cdot$, names $\bar{0}$, $\bar{1}$ for 0 , 1 respectively, whence names $\bar{n} = \bar{1} + \bar{1} + \ldots + \bar{1}$ for each number $n = 1 + 1 + \ldots + 1$. With $+$, one can express the ordering of the natural numbers:

$$x < y : \exists z(x + z + \bar{1} = y) .$$

One can derive all the usual properties of the ordering relation from the axioums of Peano Arithmetic, PA, which axioms include logical axioms, equality axioms, recursion equations for $+, \cdot$, axioms for the successor function $x \to x + 1$, and the schema of induction. Of particular interest is the fact, proven by meta-mathematical induction (i.e., induction outside the theory PA), that, for each $n > 0$, PA proves

$$\forall x(x < \bar{n} \to x = \bar{0} \vee \ldots \vee x = \overline{n - 1}) .$$

Interpreting this in any nonstandard model $\underset{\sim}{M} = (M,+,\cdot,0,1)$ of PA, we see that the _standard_ natural numbers 0 , 1 , $2,\ldots$ form an initial segment of M under the given ordering, i.e. the new _nonstandard_ elements of M come after (or: are larger than) all the standard natural numbers. We can emphasise this fact by saying that the standard integers are _finite_ and the nonstandard ones are infinite.

Ryll-Nardzewski's result is obtained by combining an infinite integer with the following property of PA: Let F be a function definable in the language of arithmetic for which PA proves the function total, i.e. assume some formula $\phi(x,y)$ defines the graph of a function F in that

i. $PA \vdash \forall x\, \exists y\, \phi(x,y)$

ii. $PA \vdash \forall xyz(\phi(x,y) \wedge \phi(y,z) \rightarrow y = z)$.

Then: PA similarly proves the existence of the function $G(x,y) = F^{(x)}(y)$ = the result of iterating application of F to y x times. Intuitively, this makes sense: G satisfies the recursion

$$G(0,y) = y$$
$$G(x+1,y) = F(G(x,y)) ,$$

and one can use the induction available in PA to prove the totality and single-valuedness of G . The one subtle point is the necessity of expressing G in the arithmetical language before applying the induction. This necessity can be met by coding tricks different from but entirely analogous to those used in axiomatic set theory to define functions by transfinite recursion.

Now, suppose we have a finite axiomatisation $\phi_1,\ldots,\phi_n$ of PA (or some consistent extension in an expanded language). Letting $\phi = \phi_1 \wedge \ldots \wedge \phi_n$, we can assume $n = 1$. Every sentence, including ϕ , can be put into <u>prenex normal form</u>, say

$$\forall x_1 \exists y_1 \ldots \forall x_m \exists y_m\, \psi(x_1,\ldots,x_m, y_1,\ldots,y_m) \qquad (*)$$

where ψ has no quantifiers. For, equivalences like

$$\exists x \psi x \vee \chi \leftrightarrow \exists x(\psi x \vee \chi) , \quad x \text{ not in } \chi$$
$$\forall x \psi x \vee \chi \leftrightarrow \forall x(\psi x \vee \chi) , \quad x \text{ not in } \chi ,$$

allow one to "pull the quantifiers outside". [This done, the form might not be exactly that of (*). One might have several like quantifiers in a row instead of an alternation. One can either loosen the description of (*) or introduce dummy variables, e.g. by replacing

$$\forall x_1 \forall x_2 \exists y_z\, \psi(x_1,x_2,y_2)$$

by its equivalent

$$\forall x_1 \exists y_1 \forall x_2 \exists y_2\, (\psi \wedge y_1 = y_1) .$$

For notational convenience, I shall assume the exact form (*) .] As I said, ϕ

can be assumed written in the form (*) . Let $\underset{\sim}{M}$ be some nonstandard model of ϕ , whence of PA. the truth of (*) means the existence of functions $F_1,\ldots,F_n$ defined by:

$$F_1(x_1) = \text{least } y_1 \ \forall x_2 \ldots \exists y_m \ \psi(x_1,\ldots,x_m, y_1,\ldots,y_m)$$
$$F_2(x_1,x_2) = \text{least } y_2 \ \forall x_3 \ldots \exists y_m \ \psi(x_1,\ldots,x_m, F_1(x_1),y_2,\ldots,y_m)$$
$$\vdots$$
$$F_m(x_1,\ldots x_m) = \text{least } y_m \ \psi(x_1,\ldots,x_m, F_1(x_1),\ldots,F_{m-1}(x_1,\ldots,x_{m-1}),y_m) .$$

Moreover, the closure of any subset X of M containing 0 , 1 under $F_1,\ldots,F_m$, + , $\cdot$ guarantees the structure $(X,+,\cdot,0,1)$ to model ϕ , whence to be a model of PA.

We can now get a contradiction to our assumption that ϕ yielded all axioms of PA fairly quickly. First, let

$$F(x) = x + 1 + \sup\{F_1(x_1),\ldots,F_m(x_1,\ldots,x_m) , x_1 + x_2 , x_1x_2 : x_1,\ldots,x_m \leqslant x\} .$$

Second, let $a \in M$ be an infinite integer. Third, define

$$X = \{x \in M : \text{for some } n \in \omega , x \leqslant F^{(n)}(a)\} .$$

Evidently, $\underset{\sim}{X} = (X,+,\cdot,0,1)$ is a model of ϕ , whence of PA. Unfortunately, it is <u>not</u> a model of PA: Consider the function $G(x,y) = F^{(x)}(y)$ defined in PA. If $\underset{\sim}{X}$ were indeed a model of PA, X would be closed under G . But $G(a,a)$ is too big to fit into X : Since

$$x < F(x) < F(x + 1)$$

holds for all x , a simple induction shows

$$x_1 < x_2 \rightarrow G(x_1,y) < G(x_2,y) ,$$

whence, for each finite n ,

$$F^{(n)}(a) = G(n,a) < G(a,a) ,$$

i.e. $G(a,a) \notin X$. With this contradiction, we see that ϕ does not imply all axioms of PA, in particular, it does not imply the instance of induction needed prove the totality of G .

How has this example illustrated the three aspects of the study of nonstandard models? i. The information it provides about nonstandard models themselves is minimal: The nonstandard integers are infinite. ii. The information provided on formal systems is readily stated: Induction is inherently infinite and cannot be finitely axiomatised. ii. We haven't really concluded anything about arithmetic itself, but we have in this argument a glimpse of things that were eventually to come: An independence result,

$$PA \nvdash \forall x \exists y\, \psi(x,y) ,$$

where $\forall x \exists y\, \psi(x,y)$ is true, could be proven by starting with an infinite integer a in a nonstandard model $\underset{\sim}{M}$ and obtaining a new model $\underset{\sim}{I}$ on an initial segment $I \subseteq M$ containing a and closed under all the functions needed to guarantee the truth in $\underset{\sim}{I}$ of the axioms of PA and yet in which $\exists y \psi(\bar{a},y)$ nonetheless fails. If this can be done with any degree of generality, it follows that the growth of the function

$$F(x) = \text{least } y\ \psi(x,y) .$$

is phenomenal: F grows faster than any function probably total in PA. In fact, this can be done for subsystems of PA and Ryll-Nardzewski's proof is of this form for the subsystem axiomatised by ϕ .

So with this example we have seen something of what nonstandard models are good for and how they are used. Friedman's work also exhibits these three aspects of study, though at a deeper level. The object of this chapter is to discuss some of Friedman's work with nonstandard models. I have delayed this so far to give the above example and shall delay it further to discuss, first, a few generalities about nonstandard models. This will be done in the immediately following section 1. In section 2, I shall say just a few words about the extent of Friedman's work with nonstandard models and follow this in sections 3, 4, and 5 with more detailed discussions of specific results. In section 3, I will discuss Friedman's "Countable models of set theory", the paper that brought about a resurgence of interest in nonstandard models of arithmetic in the 1970s. This work displays our first cited aspect of the study of nonstandard models, namely, the study of the models themselves. In section 4, we discuss a key result of Friedman obtained in joint work with Robert Flagg on Epistemic Arithmetic, a curious subject the nature of which will be explained in that section. This work illustrates the second aspect of the study of nonstandard models, namely their use in establishing metamathematical results. The reader's natural expectation that section 5 will cover an instance of the third aspect of the study of nonstandard models will not be met: In section 5, I present something entirely different.

I can explain the bearing of section 5 on nonstandard models by analogy to a simpler result. A problem Haim Gaifman raised at the first ASL meeting in Jerusalem in December of 1975 called for a sequence of sentences of decreasing proof theoretical strength, i.e. he wanted a sequence $\phi_0, \phi_1, \ldots$ of sentences consistent with PA such that, for all n,

$$PA + \phi_n \vdash Con(PA + \phi_{n+1}),$$

where $Con(T)$ is the assertion of the consistency of T. Robert Solovay and the author independently proved that this could not uniformly be done, i.e. one could not find ϕ_n such that

$$PA \vdash \forall x Pr_{PA+\phi_x}(\ulcorner Con(PA + \phi_{x+1})\urcorner).$$

If we drop the uniformity requirement, however, it can be done. Nonstandard models offer a good heuristic: Define $Con^n(PA)$ by

$$Con^1(PA) = Con(PA)$$
$$Con^{n+1}(PA) = Con(PA + Con^n(PA)).$$

Let a be a nonstandard integer and let

$$\phi_n = Con^{a-n}(PA).$$

This can be done because there is a single formula $C(x)$ such that

$$\mathrm{Con}^n(\mathrm{PA}) \to C(\bar{n}) .$$

Thus, $C(\bar{a})$, i.e. $\mathrm{Con}^a(\mathrm{PA})$, is assertable--modulo the name $\bar{a}$ and the integer a . If we demand $\phi_0, \phi_1, \ldots,$ all to lie in the language of PA, then we see that the sentences produced, including the constant $\bar{a}$, do not offer a legitimate solution to the problem.

These are two ways of getting a solution to the problem from the nonstandard attempt. These are due to Solomon Feferman and Friedman, respectively. Feferman's solution is rather interesting and worth citing for its nonstandard content: One can iterate the construction $\mathrm{Con}^n(\mathrm{PA})$ into the transfinite and define $\mathrm{Con}^\alpha(\mathrm{PA})$ for _ordinal notations_ α-numerical codes for ordinals. In the arithmetic language, one cannot completely define the set of codes that denote ordinal numbers; one can merely define a good set of _candidates_, the actual ordinal notations being those candidates satisfying the additional inexpressible well-ordering condition. Now, the set of candidates, for which the iteration works is definable, but the set of ordinal notations is not. Hence there must be sentences $\mathrm{Con}^\alpha(\mathrm{PA})$ for _nonstandard_ ordinals α . A descending sequence $\alpha_0 > \alpha_1 > \ldots$ of such nonstandard ordinals yields the sequence

$$\phi_n = \mathrm{Con}^{\alpha_n}(\mathrm{PA})$$

desired. (A small point: Although the ordinals α_n are nonstandard, their numerical codes are standard integers, whence the sentences ϕ_n are all in the language of arithmetic.)

I have been very brief in my description of Feferman's solution. Some aspects of it should be more intelligible after reading the next section, but some of it won't be: The construction of ordinal notations and the details of the iteration into (and beyond, as it were) the transfinite will have to be left unexplained. The point is that the nonstandard heuristic is transformed into a nonstandard solution--but now with nonstandard ordinals.

Friedman's solution, though still based on the nonstandard heuristic, eliminates all direct reference to the nonstandard and, in so doing, manages to avoid the sophisticated machinery invoked by Feferman: One begins with the formula $C(x)$ with free variable x and satisfying

$$Con^n(PA) \leftrightarrow C(\bar{n}) ,$$

for each n. In place of $\phi_n = C(\bar{a} - \bar{n})$ for some nonstandard a, Friedman chooses

$$\psi_n : \forall y[Prov_{PA}(y, \ulcorner \exists x \neg C(x) \urcorner) \wedge \forall z < y \neg Prov_{PA}(z, \ulcorner \exists x \neg C(x) \urcorner)$$
$$\rightarrow C(y - \bar{n})] ,$$

i.e. ψ_n asserts: If y is the least (number coding a) proof that $\forall xC(x)$ is false, then $Con^{y-\bar{n}}(PA)$ holds. The sequence $\psi_0, \psi_1, \ldots$ is easily shown to have the properties desired. (This is an easy exercise: Consider separately the cases where y exists and y does not exist.)

The material in section 5 has the same nonstandard character as Friedman's solution to Gaifman's probem. The nonstandard modellist will immediately be interested in the results, even though others may find it hard to see the non-standardness of it. As for what it is, let me say here only that it concerns the equivalence (in a sense) of translatability with provable relative consistency.

1. Nonstandard Models

In the case of arithmetic, the standard model is the structure $\underset{\sim}{N} = (\omega,+,\cdot,0,1)$; any other model either is isomorphic to $\underset{\sim}{N}$ and can be identified with $\underset{\sim}{N}$ or is not isomorphic to $\underset{\sim}{N}$ and can be labelled nonstandard. Whether we are looking at all models of true arithmetic (i.e. $TH(\underset{\sim}{N})$) or all models of Peano Arithmetic, there is, up to isomorphism, only one standard model, namely $\underset{\sim}{N}$. The one big structural fact we know about nonstandard models of arithmetic is that it consists of a copy of the natural numbers followed by a lot of infinite integers. We will learn a bit more shortly, but first we shall meet some other nonstandard models.

"Second-order arithmentic" is an ambiguous term. Its standard model contains, in addition to the natural numbers, the collection of all subsets of such: $\underset{\sim}{N}_2 = (\omega,P(\omega) ; \in,+,\cdot,0,1)$. The term "second-order arithmetic" designates either the theory of this structure, together with the second-order interpretation of the set variables (i.e. the set variables are to range over <u>all</u> subsets of the domain of the model, in this case ω), or a convenient first-order axiomatisation in a two-sorted language with variables ranging over individuals (i.e. numbers) and sets (of numbers). If one takes the first option, one has no second-order nonstandard models of arithmetic; the proof if Dedekind's categoricity result carries over: Let $\underset{\sim}{M}_2 = (M;P(M),\in,+_M,\cdot_M,0_M,1_M)$ be a model of second order arithmetic and define $F : \omega \to M$ by

$$F(0) = 0_M$$

$$F(n + 1) = F(n) +_M 1_M \ .$$

Induction on ω tells us that this is a one-one map. If it were not onto, there would be a number $a \in M$ not in the range of F. But

$$X_F = \{a \in M : a \notin \mathrm{Ran}(F)\} \ .$$

is an element of $P(M)$ and induction applies. Let $a_0 \in X_F$ be minimum. Since $a_0 \neq 0_M$, $a_0 -_M 1 \in M$ exists and $a_0 -_M 1 \notin X_F$, say $a_0 -_M 1 = F(n)$. Then we get the contradiction: $a_0 = F(n + 1)$. (This shows the map is one-one and onto. A little extra work shows it preserves the structure.)

Dedekind's categoricity argument does not work if

$$\underset{\sim}{M} = (M, \mathfrak{X} ; \in,+,\cdot,0,1) \qquad (*)$$

is a model of arithmetic in which $\mathfrak{X}$ is not the full power set of M-for, unless the set X_F defined above is in $\mathfrak{X}$, we cannot appeal to the induction axiom

$$\forall X[0 \in X \wedge \forall x(x \in X \to x + 1 \in X) \to \forall x(x \in X]$$

to minimise X_F. Hence, if we wish to speak of nonstandard models of second-order arithmetic, we must think of models of the form (*) of a two-sorted first-order theory. To help emphasise that one is not imposing the strong second-order restriction that $\mathfrak{X}$ be the power set of M, one does not refer to "second-order arithmetic". Instead, it has become customary to allude to the homeomorphism of Baire space with the set of irrationals in the unit interval, call the elements of $\mathfrak{X}$ _real numbers_ or _reals_, and refer to one's two-sorted first-order theory of arithmetic as _analysis_. This is a bit misleading, but rather firmly entrenched in the literature and, in any event, it is easier to refer to "analysis" than to "two-sorted first-order arithmetic".

Since the model $\underset{\sim}{M}$ of (*) has two major parameters--the set M of integers and the st $\mathfrak{X}$ of reals--there is immediately a greater variety of non-standard models of analysis than of nonstandard models of arithmetic. Moreover, the two-sorted language is more expressive (and more immediately so) than the single-sorted one and "second-order" restrictions of set theoretic principles can be studied, e.g.

$$\underline{AC}: \qquad \forall x \exists Y \phi(x,Y) \rightarrow \exists X \forall x \phi(x,X_x) ,$$

where

$$X_x = \{y : \langle x,y \rangle \in X\} ,$$

and $\langle\cdot,\cdot\rangle$ is a convenient pairing function. In the late 1960s and early 1970s, the use of sets to code trees was made much of because well-founded trees code sets of higher orders--the sets of naive set theory. The study of nonstandard models of analysis (not to be confused with nonstandard analysis) has largely been a study of subsystems and interrelations between schemata like AC and its variants. The typical result is an independence result and is proven by the exhibition of a model.

As already remarked, there is a greater variety, superficially at least, of nonstandard models of analysis than of nonstandard models of aritmetic. There are the totally nonstandard ones which contain infinite integers, but these have been least studied of all. The ω-_models_--those models in which all the integers are standard--have been most popular; they are, in a sense, the conceptually simplest: An ω-model

$$\underset{\sim}{M} = (\omega, \mathfrak{X}; \in, +, \cdot, 0, 1)$$

is determined completely by the class $\mathfrak{X}$ of sets of natural numbers, whence it can be identified with $\mathfrak{X}$. When this is done, the study of such models becomes a study of closure conditions on the classes $\mathfrak{X}$ and tools from Recursion Theory and Set Theory can be brought into play--or, if one prefers, one is studying curious families of sets of natural numbers rather than nonstandard models of analysis, so close are ω-models to being "standard". Yet, even though ω-models are very nearly standard models, they are not standard enough. The use of sets to code trees and the represnetation of sets by well-founded trees brings one to the expression of well-foundedness: Assuming R linear to avoid questions about representing functions as sets and how much choice holds, one would normally express the well-ordering of R by

$$\forall X[\exists x(x \in X)] \rightarrow \exists x(x \in X \wedge \forall y(yRx \rightarrow \neg y \in X))] .$$

I say "normally" because in $\underset{\sim}{M}$ the variable X is not ranging over all of $P(\omega)$, but only over $\mathfrak{X}$; in an ω-model the statement that a relation is well-founded can be deemed true of non-well-founded relations. The good ω-models in which this fails are called β-models (from bon ordre). In short, nonstandard models of analysis can be roughly trichotomised into the totally nonstandard ones, the semi-standard ω-models, and the most nicely behaved β-models.

For set theory, unlike arithmetic or analysis, there is no single canonical model all of whose isomorphs are to be declared "standard": the criterion for standardness is, thus, different: For a model of arithmetic,

$$\mathfrak{M} = (M,+,\cdot,0,1)$$

the condition on standardness is that M, $+$, $\cdot$, 0, 1 all be correct; for a model of analysis,

$$\mathfrak{M} = (M, \mathfrak{X}; \in,+,\cdot,0,1) ,$$

$\mathfrak{X}$ must also be correct; but for a model of set theory,

$$\mathfrak{M} = (M,E) ,$$

there being no single correct M, we must settle for E's being correct, i.e. we insist that E be the actual membership relation $\in$ restricted to M. Now, an old result of Andrzej Mostowski tells us that, assuming the universe of all sets to be well-founded, a model $\underset{\sim}{M} = (M,E)$ is isomorphic to one with the correct E iff the relation E is well-founded. Hence, a standard model of set theory is a well-founded one; a non-well-founded model of set theory is labelled "nonstandard".

The emphasis on well-foundedness ties in nicely with the emphasis on well-foundededness among nonstandard models of analysis. If we use the well-founded trees of a β-model of analysis to code sets, we will obtain a well-founded (hence: standard) model of (a possibly very weak) set theory; if we use the trees declared well-founded in a non-β-model, we will get a nonstandard model of (again, a weak) set theory. Moreover, if we think of natural numbers as finite ordinals, nonstandard models of arithmetic are nonstandard in the set theoretic sense as well: The ordering of the integers in a nonstandard model of arithmetic is non-well-founded. For, take an infinite integer a and consider the sequence

$$a , a - 1 , a - 2 , \ldots .$$

Since a is infinite, $a - n$ is never 0 and one can continue. Finally, let me remark that Feferman's use of ordinal notations, alluded to in the last section, was nonstandard in this sense: Some of the notations denoted non-well-founded, nonstandard ordinals.

Standard models of set theory have been widely used and studied, nonstandard models less so--so much less so that many logicians would have to stop and think a bit before citing any paper or result in which the use of nonstandard models of set theory is central. One example in Friedman's paper, "Countable models of set theory", which will be lightly discussed in the next-section.

We now have definitions of the notion of nonstandardness for models of arithmetic, analysis, and set theory, as well as the merest of indications of what is done with them. While these indications are not sufficient to allow the reader who is unfamiliar with nonstandard models to say anything definite about the study of any one type of nonstandard model, they should be broad enough to suggest that the studies of these three types of nonstandard models are three distinct disciplines, or, less grandiloquently put, follow at least three different paths. In a work of this sort, the obvious space limitations prevent me from following more than one such path, however short the distance and whether I restrict myself to Friedman's work or not. There are, to my knowledge, no published expositions of the basics of nonstandard model theory of analysis or set theory, but there is such for nonstandard models of arithmetic. For this reason, except for some discussion in the next section, I shall restrict my attention to nonstandard models of arithmetic. The background references are Smoryński, 1984 for an exposition of the fundamentals and McAloon et al. for exposition of the more recent research.

Citations of references, such as just made, generally offer a clear sign that the author has said his piece on the given topic and intends to move on to something else. In the present case, this is a false sign as I wish yet to make a couple of remarks about nonstandard models of arithmetic.

First, suppose

$$\underset{\sim}{M} = (M, +, \cdot, 0, 1)$$

is a nonstandard model of arithmetic. There can be no formula $\phi(x)$, with or without constants naming nonstandard elements of M, that defines the set ω of standard numbers. For, if $\phi(x)$ defined ω, one would have

$$\underset{\sim}{M} \models \phi(\bar{0}) \wedge \forall x(\phi(x) \rightarrow \phi(x + \bar{1})) .$$

but not the inductive conclusion,

$$\underset{\sim}{M} \models \forall x \phi(x) .$$

An immediate corollary to this observation is the following Overspill Principle, due (I believe) independently to Michael Rabin and Abraham Robinson:

Overspill Theorem. Let $\underset{\sim}{M}$ be a nonstandard model of arithmetic, and let $\phi(x)$ be a formula in the language of arithmetic such that

$$\underset{\sim}{M} \models \phi(\bar{n}) \text{, for all standard } n .$$

Then: There is an infinite integer $a \in M$ such that

$$\underset{\sim}{M} \models \phi(\bar{b})$$

for all $b \leqslant a$.

Proof: If the conclusion failed, the formula $\forall y \leqslant x\phi(y)$ would define ω .
Q.E.D.

In simple words, the Overspill Theorem says that anything holding of all standard numbers in a nonstandard model must carry over to some infinite integers. Note that Feferman's solution to Gaifman's problem, discussed in the introduction, is essentialy an instance of an Overspill Principle: The definability of the candidates for ordinal notations for which the iteration in question could be carried out, the inclusion of the genuine ordinal notations (i.e. standard ordinals) in this set, and the undefinability of the set of genuine ordinal notations all combine to yield similarly the existence of nonstandard ordinals out to which the iteration can be carried.

Second, suppose $\underset{\sim}{M}$ as given above is nonstandard and $\phi(x)$ is any formula defining a subset of M . Unless it is finite, the restriction of the set to ω ,

$$X = \{n \in \omega : \underset{\sim}{M} \models \phi(\bar{n})\} ,$$

is not definable in the language of $\underset{\sim}{M}$. Nonetheless, it is at least partially definable in the sense that it is the restriction of a definable set to ω and ought, thus, to be of some interest. The collection $\mathfrak{X}$ of all such partially definable set is, following Friedman, called the <u>standard system</u> of the model $\underset{\sim}{M}$. (The name comes from the fact that the sets $X \in \mathfrak{X}$ are the <u>standard</u> parts of the definable subsets of M .) Standard sets were first used by Stanley Tennenbaum around 1960 and the full standard system first introduced by him and Dana Scott, the latter publishing a nice characterisation of those countable families $\mathfrak{X}$ that can be the standard system of a countable nonstandard model of arithmetic. A glimpse of this importance can be had by observing (down a path we will not follow in the sequel) that, if $\mathfrak{X}$ is the standard system of a nonstandard model) $\underset{\sim}{M}$, $(\omega, \mathfrak{X}; \in,+,\cdot,0,1)$ is an ω-model of a (generally very) weak theory of analysis.

2. Friedman's Use of Nonstandard Models.

I fear that Friedman's use of nonstandard models and techniques so pervades his work that I must perforce fail to do him justice in attempting to describe it. That being the case, I shall only attempt the most superficial treatment in the present section. To remedy this deficiency, I offer the ensuing three sections, wherein some of Friedman's work will be discussed in greater depth, and I also note that nonstandard models make their presence felt in other contributions to this volume, e.g. in the chapters by Simpson, Smith, and Stanley.

No discussion of Friedman's work on nonstandard models, however superficial, would be complete without reference to his "Countable models of set theories". Here Friedman introduced the _standard part_ of a nonstandard model of a weak set theory along with the _standard system_ thereof and proved a number of theorems relating the original model to these standard bits. The standard part of a model of a set theory is just its well-founded part and bears to the original model a relation analogous to that of ω to a nonstandard model of arithmetic. The standard system of a nonstandard model of set theory, like the standard system of a nonstandard model of arithmetic, consists of the restrictions to the standard part of all sets definable in the model. Some of the theorems proven in this paper bear on the extent to which the standard part and standard system do or do not determine the original model. We will look at such results in the arithmetic case (roughly: set theory without the axiom of infinity) in the next section and I need not say anything further about them here. Let me simply note that Friedman does characterise those countable pairs $(A,\mathfrak{X})$ which are the standard part and standard system, respectively, of a certain weak set theory in terms of the structure $(A,\mathfrak{X};\in)$. In the arithmetic case, this translates to: A countable set $\mathfrak{X}$ is the standard system of a nonstandard model of arithmetic iff $(\omega,\mathfrak{X};\in,+,\cdot,0,1)$ models a weak theory of analysis, as already hinted at the end of the previous section.

The theme of looking at the standard parts of nonstandard models of set theory occurs again in "On the necessary use of abstract set theory", but here, as in the sequel, "Unary Borel functions and second-order arithmetic", the chief nonstandard emphasis is in constructing nonstandard models of analysis. In these papers, as well as in "Bar induction and Π_1^1 - CA" and "Iterated inductive definitions and Σ_2^1 - AC" , Friedman establishes metamathematical independence results and even conservation results by the outright construction of models. In so saying, I present the most philistine description of these papers possible: The latter two are old (1970s) and are devoted to traditional proof theoretic topics--proofs of the equivalence of various theories of analysis. The former two papers, however, are newer and more sophisticated in intent. But I need not say anything more about them as the exposition of their contents is taken up in Stanley's contribution to this volume.

Finally, before proceeding to the more detailed discussion of special topics, let me cite just one more paper on nonstandard models of set theory. This is the paper, "Categoricity with respect to ordinals", and deals directly with nonstandard models. In ordinary Zermelo-Fraenkel set theory with choice, ZFC , the connexion between the ordinals and the rest of the universe of sets is not completely determined. One new axiom, the Axiom of Constructibility, usually written $V = L$ (V being the univrse of sets and L the class of constructible sets), reduces the set theoretic universe explicitly to the ordinals. Friedman proves that this is the only axiom that does this in the following sense: Let T be a theory of set theory extending ZFC and consider the condition,

(+) for any models $\underset{\sim}{M}_1$, $\underset{\sim}{M}_2$ having the same ordinals, if $(\underset{\sim}{M}_1, \underset{\sim}{M}_2)$ satisfies transfinite induction on ordinals, then there is an isomorphism of $\underset{\sim}{M}_1$ onto $\underset{\sim}{M}_2$ which is the identity on ordinals.

Friedman's Theorem asserts that T satisfies (+) iff T proves $V = L$. In "Countable models of set theories", Friedman had already proven an analogous result for standard models of certain recursive set theories, including ZFC and and its recursive extensions.

Writing at this level of vague generality is boring and I assume the reading of it equally unpleasant. I have said enough to fulfill my duty to the editors and we can now get down to specific issues, and do mathematics instead of merely talk around it.

3. Back-and-Forth.

Friedman's paper "Countable models of set theories" did not deal with nonstandard models of arithmetic, but with nonstandard models of set theories, including the theory of hereditarily finite sets, a set theoretic equivalent to PA. It is the translation to arithmetic of the restriction of the results of this paper to this finite set theory for which it is chiefly known. Indeed, this paper stimulated the renascence of the study of nonstandard models of arithmetic in the 1970s. It did this by means of the following two results:

Theorem 1. Let $\underset{\sim}{M}_1$, $\underset{\sim}{M}_2$ be countable nonstandard models of PA. The following are equivalent:

i. $\underset{\sim}{M}_1$ is embeddable into $\underset{\sim}{M}_2$

ii. $SSy(\underset{\sim}{M}_1) = SSy(\underset{\sim}{M}_2)$ and $Th_{\exists}(\underset{\sim}{M}_1) \subseteq Th_{\exists}(\underset{\sim}{M}_2)$.

Theorem 2. Under the hypotheses of Theorem 1, the following are equivalent:

i. $\underset{\sim}{M}_1$ is isomorphic to an initial segment of $\underset{\sim}{M}_2$

ii. $SSy(\underset{\sim}{M}_1) = SSy(\underset{\sim}{M}_2)$ and $Th_{\Sigma_1}(\underset{\sim}{M}_1) \subseteq Th_{\Sigma_1}(\underset{\sim}{M}_2)$.

The notation "$SSy(\underset{\sim}{M})$" obviously denotes the standard system of $\underset{\sim}{M}$. For any set Γ of sentences, $Th_{\Gamma}(\underset{\sim}{M})$ denotes the Γ-theory of the model $\underset{\sim}{M}$, i.e. the set of all sentences of Γ true in $\underset{\sim}{M}$. It only remains to explain the two Γ's occurring in these Theorems. The set $\exists$ is easy to explain; it consists of all sentences of the form,

$$\exists x_1 \ldots x_n \phi ,$$

where ϕ contains no quantifiers. The set Σ_1 is a more delicate matter. Roughly, the set of Σ_1-formulae can be viewed as the closure of the set of $\exists$-formulae under $\wedge$, $\vee$, $\exists$ and bounded universal quantification, $\forall x < y$. The class of Σ_1-formulae arises most naturally in Recursion Theory, where they yield the natural arithmetic definitions of the recursively enumerable relations. The model theoretic significance of this class was established by Feferman and Georg Kreisel, who proved that the Σ_1-sentences are, up to provable equivalence, the sentences always preserved under the embeddability of a given model onto an initial segment of another. (In the arithmetic case, this follows directly from Theorem 2--cf. Smoryński 1984.) Finally, the closure conditions on the class Σ_1 arise most naturally in the course of proving the implication ii $\Rightarrow$ i of Theorem 2. Having said all of this, I note that Jurii Matijasevic's theorem on the Diophantine nature of recursively enumerable relations can be proven in PA and, if we restrict our attention to the arithmetic language, Σ_1 coincides (up to provable equivalence) with $\exists$ and we can ignore all of my remarks.

In each of Theorems 1 and 2, the difficult implication to prove is ii $\Rightarrow$ i . The overall structure of these proofs is best conveyed by first proving an infinitely simpler result and then discussing the specifics of the case at hand. This overall structure is called a back-and-forth argument and was introduced by Georg Cantor in his proof that any two countable dense linear orderings without endpoints are isomorphic. Actually, this is the structure of the construction underlying the proof of Theorem 2; the construction in the proof of Theorem 1 is more analogous to the proof that the countable dense linear orderings are universal for countable orderings. Let us first consider this last cited proof.

Let $(A,<)$ be a countable ordered set and let $(Q,<)$ denote the set of rational numbers with their ordering. Let $a_1, a_2, \ldots$ be an enumeration of A . To embedd $(A,<)$ into $(Q,<)$, map a_1 to any $q_1 \in Q$. Now, a_2 is either greater than a_1 or less than a_1 . In the former case map a_2 to some $q_2 > q_1$ (e.g. $q_1 + 1$) and in the latter map it to some $q_2 < q_1$ (e.g $q_1 - 1$) . Next, look at a_3 . Either it exceeds a_1, a_2 and we can map it to something larger than q_1, q_2 , or it lies between a_1, a_2 and we can map it to something between q_1, q_2 (e.g. their midpoint), or it is less than both a_1, a_2 and we must map it below q_1, q_2 . In the general case, suppose we have mapped $a_1,\ldots,a_n$ to $q_1,\ldots,q_n$ in an order preserving way and must decide what to do with a_{n+1} . If a_{n+1} exceeds all of the a_i's , map it to (say) $\max\{q_1,\ldots,q_n\} + 1$; if a_{n+1} is less than all the a_i's , map it (say) to $\min\{q_1,\ldots,q_n\} - 1$. Otherwise, there is a largest a_i and smallest a_j such that

$$a_i < a_{n+1} < a_j .$$

Then map a_{n+1} to (say) $(q_i + q_j)/2$. By step ω, all of A has been embedded into $(Q,<)$ in an order preserving way.

If $(A,<)$ is dense, without endpoints we can get an isomorphism by alternating the steps just described with steps mapping back from Q into A to guarantee the map is onto. (It is this alternation that gives the method the name "back-and-forth argument".)

Back-and-forth arguments are quite common in logic, both in theoretical generality where some abstract considerations presuppose the ability to carry them out and in concrete case where models of specific theories are involved. Generally, however, they have not been applied to <u>complex</u> specific theories because it is just too hard to decide where to map a_n to. Friedman's Theorems break with this tradition because he can use the theory of the model and the standard system to make this decision.

Let $\underset{\sim}{M}$ be a model of arithmetic, L the language of arithmetic, $a_1,\ldots,a_n$ elements of M and $\bar{a}_1,\ldots,\bar{a}_n$ constants naming them. The <u>Γ-type</u> of $a_1,\ldots,a_n$, for Γ a set of formulae of L with only the variables $x_1,\ldots,x_n$ free, is the set

$$\tau^{\Gamma}_{a_1\ldots a_n} x_1\ldots x_n = \{\phi \in \Gamma : \underset{\sim}{M} \models \phi\bar{a}_1\ldots\bar{a}_n\} ,$$

of formulae holding for $a_1,\ldots,a_n$. But for retaining the variables $x_1,\ldots,x_n$, we could identify τ with

$$Th_{\Gamma}(\underset{\sim}{M} ; a_1,\ldots,a_n) = \{\phi\bar{a}_1\ldots\bar{a}_n : \phi \in \Gamma \text{ and } \underset{\sim}{M} \models \phi\bar{a}_1,\ldots,\bar{a}_n\} ,$$

the Γ-theory for $\underset{\sim}{M}$ with constants $\bar{a}_1,\ldots,\bar{a}_n$ designating the elements $a_1,\ldots,a_n$.

The argument given to embedd $(A , <)$ into $(Q , <)$ used only those Γ-types for Γ consisting of formulae $x_i < x_j$. In the theory of dense linear order, every formula is equivalent to a propositional combination of such inequalities and these become the most general types. For algebraically closed fields, the types can merely assert algebraic dependencies and mutual transcendence--whence are again quite simple. In arithmetic, even the $\exists$-types can be quite complicated. For example, the $\exists$-type of a will tell which primes divide

a , which means that quite complicated sets can be coded up.

Now, a basic property of PA is that it has truth definitions for many restricted classes Γ of formulae, including $\Gamma = \exists, \Sigma_1$. Thus, for each n and each choice of Γ form $\exists, \Sigma_1$, there is a formula $Sat_n^\Gamma(x ; x_1,\ldots,x_n)$ such that

$$PA \vdash \phi x_1,\ldots x_n \leftrightarrow Sat_n^\Gamma(\ulcorner\phi\urcorner ; x_1,\ldots,x_n) ,$$

where $\ulcorner\phi\urcorner$ is some numerical code for the formula ϕ . Friedman combined this observations, Overspill, the standard system, and one more coding trick to determine what to map a_n onto during the n-th stage of the construction of the desired embedding.

First, let me mention the last cited coding trick, which goes back to Tennenbaum. There are several natural ways in which a number can code a finite chunk of information. One of the simplest is to wirte the given number in base 2: If

$$n = 2^{n_1} + 2^{n_2} + \ldots + 2^{n_k} , n_1 > n_2 > \ldots > n_k ,$$

we say n codes the set $D_n = \{n_1,\ldots,n_k\}$. This coding can be described in the language of arithmetic and with it, a Bounded Comprehension axiom schema can be proved:

$$PA \vdash \forall x \exists y \forall z < x\, [z \in Dy \leftrightarrow z < x \wedge \phi z] ,$$

where y does not occur in ϕz .

Friedman proved essentially the following:

<u>Lemma</u>. Let $\underset{\sim}{M}$ be a nonstandard model and let Γ be $\exists$ or Σ_1 .

i. if $a_1,\ldots,a_n \in M$, then $\tau^\Gamma_{a_1\ldots a_n} \in SSy(\underset{\sim}{M})$, i.e.

$$\{\ulcorner\phi\urcorner : \phi \in t^\Gamma_{a_1\ldots a_n}\} \in SSy(\underset{\sim}{M})$$

ii. if $\tau x_1,\ldots,x_n \subseteq \Gamma$ is consistent with $Th(\underset{\sim}{M})$, i.e. if

$$\underset{\sim}{M} \models \exists x_1 \ldots x_n \bigwedge\!\!\bigwedge \{\phi_{x_1\ldots x_n} : \phi \in \tau \wedge \ulcorner\phi\urcorner < k\}$$

for all $k \in \omega$, and if $\tau \in SSy(\underset{\sim}{M})$, then τ is included in the Γ-type of some

$a_1,\ldots a_n \in M$.

Proof: i. Observe that, for each k , the following holds in $\underset{\sim}{M}$:

$$\exists\, y\, \forall z(z \in D_y \leftrightarrow z < \bar{k} \wedge Sat_n^\Gamma(z\ ;\ \bar{a}_1,\ldots,\bar{a}_n))\ ,$$

i.e. for each k the finite set

$$D_y = \{\ulcorner\phi\urcorner : \ulcorner\phi\urcorner < k \wedge \underset{\sim}{M} \models \phi\bar{a}_1\ldots\bar{a}_n\}$$

exists. By Ovrspill, this holds for some infinite b in place of k and there is some a such that

$$\forall z(z \in D_{\bar{a}} \leftrightarrow z < \bar{b} \wedge Sat_n^\Gamma(z\ ;\ \bar{a}_1,\ldots,\bar{a}_n))\ ,$$

i.e. for each $\phi \in \Gamma$,

$$\underset{\sim}{M} \models \ulcorner\phi\urcorner \in D_{\bar{a}} \leftrightarrow Sat_n^\Gamma(\ulcorner\phi\urcorner\ ;\ \bar{a}_1,\ldots,\bar{a}_n)\ ,$$

i.e. $\underset{\sim}{M} \models \ulcorner\phi\urcorner \in D_{\bar{a}} \leftrightarrow \phi\bar{a}_1,\ldots,\bar{a}_n$.

Thus, the Γ-type of $a_1,\ldots a_n$ is $D_a \cap \omega \in SSy(\underset{\sim}{M})$.

ii. Let $D_a \cap \omega = \{\ulcorner\phi\urcorner : \phi \in \tau\}$ and observe that, for any k ,

$$\underset{\sim}{M} \models \exists\, y_1\ldots y_n\, \forall z < \bar{k}(z \in D_{\bar{a}} \rightarrow Sat_n^\Gamma(z\ ;\ y_1,\ldots,y_n))\ .$$

Choose b by Overspill so that

$$\underset{\sim}{M} \models \exists y_1\ldots y_n\, \forall z < \bar{b}(z \in D_{\bar{a}} \rightarrow Sat_n^\Gamma(z\ ;\ y_1,\ldots,y_n))\ .$$

Choose $a_1,\ldots,a_n$ instantiating this,

$$\underset{\sim}{M} \models \forall z < \bar{b}(z \in D_{\bar{a}} \rightarrow Sat_n^\Gamma(z\ ;\ \bar{a}_1,\ldots,\bar{a}_n))\ .$$

Let ϕ be any formula in τ : Since $\ulcorner\phi\urcorner \in D_a$,

$$\underset{\sim}{M} \models Sat_n^\Gamma(\ulcorner\phi\urcorner\ ;\ \bar{a}_1,\ldots,\bar{a}_n)\ ,$$

i.e. $\underset{\sim}{M} \models \phi \bar{a}_1 \ldots \bar{a}_n$. QED

With this lemma, the proof of Theorem 1 is fairly easy: Let $a_1, a_2, \ldots$ enumerate M_1 . One has $\tau^{\exists}_{a_1} x_1 \in SSy(\underset{\sim}{M}_1)$ by the Lemma. But $SSy(\underset{\sim}{M}_1) \subseteq SSy(\underset{\sim}{M}_2)$ by assumption ii of the Theorem. (We only prove ii $\Rightarrow$ i.) Since $Th_{\exists}(\underset{\sim}{M}_1) \subseteq Th_{\exists}(\underset{\sim}{M}_2)$, we see that $\tau^{\exists}_{a_1} x_1$, is consistent over $\underset{\sim}{M}_2$: For $\phi_1, \ldots \phi_k \in \tau^{\exists}_{a_1} x_1$,

$$\underset{\sim}{M}_1 \models \bigwedge\!\!\!\bigwedge \phi_i(\bar{a}_1) \Rightarrow \underset{\sim}{M}_1 \models \exists x_1 \bigwedge\!\!\!\bigwedge \phi_i(x)$$
$$\Rightarrow \underset{\sim}{M}_2 \models \exists x_1 \bigwedge\!\!\!\bigwedge \phi_i(x) .$$

By the Lemma, it follows that $\tau^{\exists}_{a_1} \subseteq \tau^{\exists}_{b_1}$ for some $b_1 \in M_2$. Map a_1 to b_1 . The induction step, given $a_1, \ldots, a_n$ mapped to $b_1, \ldots, b_n$, respectively, to map a_{n+1} onto some b_{n+1} is handled similarly.

The proof of Theorem 2 is more complicated and I omit it; it can be found in excessive detail in the general survey Smoryński 1984 (or, of course, in the set theoretic version in Friedman's original paper).

These two Theorems of Friedman launched a number of sequels by others, which begat further sequels. Papers for which these results are significant ancestors were written by Petr Hajek and Pável Pudlák, Don Jensen and Andrzej Ehrenfeucht, Laurie Kirby and Jeff Paris, Julia Knight and Mark Nadel, Roman Kossak, Henryk Kotlarski, Leonard Lipshitz, Pavel Pudlak and Antonin Sochor, C. Smoryński, Alec Wilkie, and George Wilmers, and others whom I have unjustly slighted. Results that are more-or-less immediatley related to Friedman's two Theorems, whence are more readily stated here, include the following:

Theorem. (Friedman). Let $\underset{\sim}{M}$ be a countable nonstandard model of arithmetic. $\underset{\sim}{M}$ is isomorphic to a proper initial segment of itself.

This result already appeared in Friedman's paper as a Corollary to Theroem 2, but it is worth mentioning.

Theorem. (Wilkie). Let $\underset{\sim}{M}_1$, $\underset{\sim}{M}_2$ be countable nonstandard models of arithmetic. The following are equivalent:

i. $\underset{\sim}{M}_1$ is isomorphic to arbitrarily large initial segments of $\underset{\sim}{M}_2$

ii. $SSy(\underset{\sim}{M}_1) = SSy(\underset{\sim}{M}_2)$ and $Th_{\Pi_2}(\underset{\sim}{M}_1) \subseteq Th_{\Pi_2}(\underset{\sim}{M}_2)$.

The class of Π_2-formulae consists, basically, of formulae $\forall x_1 \dots x_n \phi$, where $\phi \in \Sigma_1$. This result is usable--cf. Smorynski 1984 for the reference.

Theorem. (Lipshitz). Let $\underset{\sim}{M}$ be a countable nonstandard model of arithmetic. $\underset{\sim}{M}$ is isomorphic to arbitrarily small initial segments of itself (i.e. segments below any infinite integer) iff $\underset{\sim}{M}$ is Diophantine correct (i.e. $\underset{\sim}{M}$ satisfies only sentences in $\exists$ true already in the standard model).

The crucial Lemma that the Γ-types over $\underset{\sim}{M}$ were precisely those coded in $SSy(\underset{\sim}{M})$ depended on the existence of a truth definition for Γ within PA. It tells us nothing about the types with respect to all formulae in the language. The discovery of the concept of recursive saturation a couple of years after Friedman's paper appeared led quickly to the realisation that recursively saturated models were precisely those in which the Lemma holds for Γ as the set of all formulae. One result for these models which everyone recognised was:

Theorem. Let $\underset{\sim}{M}_1$, $\underset{\sim}{M}_2$ be countable recursively saturated models of arithmetic. The following are equivalent:

i. $\underset{\sim}{M}_1$ is isomorphic to $\underset{\sim}{M}_2$

ii. $SSy(\underset{\sim}{M}_1) = SSy(\underset{\sim}{M}_2)$ and $Th(\underset{\sim}{M}_1) = Th(\underset{\sim}{M}_2)$.

This resut is almost trivial. Far more sophisticated, and a favorite of mine, is the following:

Theorem. (Smoryński). Let $\underset{\sim}{M}$ be a countable recursively saturated model of arithmetic. Let $I \subseteq M$ be an initial segment. The following are equivalent:

i. I is closed under exponentiation

ii. There is an automorphism F of $\underset{\sim}{M}$ such that

a. $F|I$ is the identity

b. F moves elements of M that are arbitrarily small above I .

These results ought to give some flavour of the logically more immediate successors to Friedman's paper. A fairly complete picture of this subject as it was in 1980 can be found in Smoryński 1981. Laurie Kirby and Jeff Paris saw something completely different in Friedman's paper and followed another line. This led to the Paris-Harrington Theorem, discussed in a number of contributions to this volume, and this result led to more work on models of arithmetic on the one hand (cf. McAloon et. al.) and Friedman's Finite Form of Kruskal's Theorem (cf. at least half the contributions to this volume!).

4. Epistemic Arithmetic

There are numerous variants of logic. As we saw earlier, there are first- and second-order logics. One can study logic with and without quantifiers, purely monadic logic, or even logic with or without equality. These variants are primarily linguistic, the underlying true-false distinction being left untouched; there are logics which tamper with truth: intuitionistic logic, many-valued logics, modal logics, etc. Some of these logics are useful: Many-valued logics have applictions in computer science, as do modal logics (e.g. Dynamic Logic), which also have applications in other parts of logic (I am quite taken with Provability Logic and its use in understanding self-reference.) Of all these non-classical logics, however, intuitionistic logic has ruled supreme. There are applications--for example there are schemes afoot to translate intuitionistic proofs directly into machine programs--but it is not the usefulness of intuitionistic logic that has attracted so many logicians to it. What seems to give intuitionistic logic its drawing power is the deep philosophy of mathematics behind it. Intuitionistic logic is the tool of intuitionism and is thus part of a programme for the development of mathematics. It leads to a brave new mathematics which deviates from the classical--a fact that both attracts and repels at the same time. Modal and many-valued logics have not done this to date. They have merely developed mathematically, like lattice theory, as restricted branches of universal algebra with no philosophical hold on the mathematician. Epistemic Arithmetic may be a modest break with this tradition.

Epistemic arithmetic has been studied by Stewart Shapiro and William Reinhardt and their colleagues in what can be viewed as an attempt to replace intuitionistic logic by something more palatable to the classicist, yet something strong enough to allow the intuitionist's deviation in an acceptable reformulation

Brouwer's intuitionistic mathematics rejected the notion of a completed infinite. To Brouwer, any infinity, like that of the set of natural numbers, was a potentiality never actually completed. It followed that ordinary two-valued logic did not apply. One could not declare

$$\exists x\phi x \vee \neg \exists x\phi x$$

to be true automatically at any given time unless one had already found an n for which $\phi\bar{n}$ was true or one had a proof that the existence of such an n was contradictory. For, without the latter, the fact one had not yet found n would not make $\exists x\phi x$ false, because one might later come across such an n. To Brouwer, one could only assert a disjunction

$$\phi \vee \psi$$

if one could assert ϕ or one could assert ψ. The net result is a reinterpretation of the logical connectives and quantifiers in terms of proof--aiming at subjective knowledge rather than truth. This leads to a powerful logic and a novel mathematics, but it can be disconcerting to see the intuitionist calmly asserting

$$\neg\forall f[\forall x(\delta(x) = 0) \vee \neg\forall x(\delta(x) = 0)] ,$$

which is patently absurd under the usual interpretation of the connectives and quantifiers. The epistemicist would like to have the best of both worlds: He wants to embrace Brouwer's epistemological conservatism and yet preserve the meaning of the logical operators. He does this by referring directly to his own knowledge of mathematics (and his knowledge of his knowledge thereof, and so on) by adding to the basic language a new operator $\Box$. The senctence $\Box\phi$ can be read "I know that ϕ" or "it is known that ϕ" or "ϕ has been proven".

It can be argued that there is a vagueness to the box. This hardly an argument against it: The notion of proof on which intuitionism is supposedly based is vague, as is the notion of set on which classical mathematics is supposedly based. (Indeed, the geometric line is vague; It may not seem so because, after several millenia, it was replaced by the real number line, which only took--how long?--to clarify.) As I mentioned in the introduction, lack of a complete determination can be an adantage. The vagueness of the box allows a variety of interpretations: ordinary truth, provability, etc.

Enough said about philosophy! Let us look at Epistemic Arithmetic, EA. EA is the theory extending PA by the addition of the operator $\Box$ and some axioms, specifically:

i. full induction in the extended language

ii. some schemata concerning $\Box$:

a. $\Box\phi \wedge \Box(\phi \rightarrow \Psi) \rightarrow \Box\psi$

b. $\Box\phi \rightarrow \phi$

c. $\Box\phi \rightarrow \Box\Box\phi$.

One also assumes closure under the rule of inference: From ϕ conclude $\Box\phi$. (Note that ii.c. is a sort of reflexion of this rule.)

The axioms of $\Box$ are neutral. One can take $\Box\phi$ to mean ϕ and get a valid interpretation. It follows that one will not derive anything not already derivable in PA, whence nothing contrary to classical reasoning. On the other hand, one could add new principles about $\Box$ and, though not contradicting classical logic, derive some analogues to contradictory-looking assertions. For example, for the right axioms about $\Box$ and the right choice of ϕ , one could conceivably derive

$$\neg\forall x(\Box\phi x \vee \Box\neg\phi x) .$$

Thus, even without new axioms, one expects to show

$$\forall x(\Box\phi x \vee \Box\neg\phi x)$$

to be underivable. But, I digress...

Where do nonstandard models of arithmetic come in? The answer is: In joint work with Robert Flagg, Friedman used them to settle a moderately interesting axiomatic question, namely, what is the modal logic of EA? If we transfer the axiom schemata and rule of inference concerning $\Box$ to a purely propositional setting, i.e. if we add the box, these axioms, and this rule to the axioms and rules of the propositional calculus, we get a theory of modal logic known as S4.

We can then think of EA as PA + S4. We have already remarked that EA is <u>conservative</u> over PA, i.e. no new theorems in the language of arithmetic are provable. The question is: Is EA also conservative over S4, i.e. does S4 axiomatise all modal schemata valid in EA or do new schemata pop up? Friedman has shown that the answer is "no"; S4 is the modal logic of EA.

To explain anything about this proof, I must go back a decade or so ago and consider intuitionistic arithmetic, HA (for Heyting Arithmetic). HA is formulated in the same language as PA and has the same arithmetic axioms cited in the introduction, but, in place of the usual axioms of logic, it has those of intuitionistic logic. In the later 1960s, Dick deJongh showed that HA proves no new propositional axioms, i.e if $A(p_1,\dots p_n)$ is any propositional formula with propositional atoms $p_1,\dots p_n$, then $A(p_1,\dots,p_n)$ is provable in the intuitionistic propositional calculus iff, for all arithmetic sentences $\phi_1,\dots,\phi_n$, $HA \vdash A(\phi_1,\dots,\phi_n)$.

De Jongh's original proof of this result, now called de Jongh's Theorem, was rather esoteric and he never published it. Around 1970, C. Smoryński discovered a much simpler proof. (This proof, with background material, was expounded upon in <u>Smoryński 1973</u>. While I'm at it: Friedman's "Some applications of Kleene's methods for intuitionistic systems" included a more sophisticated proof of a very nice refinement. In <u>Smoryński 1973</u> one can find both another refinement and another proof of Friedman's refinement, both proofs being modifications of Smoryński's simple proof of de Jongh's Theorem.) This proof, as well as Friedman's analogous result for EA and S4, proceeds by simulating Kripke models for propositional logic (S4 in Friedman's case) by Kripke models for Heyting (respectively: Epistemic) Arithmetic.

A Kripke model for intuitionistic propositional logic is given by a triple $(K, \leq, \Vdash)$, where K is a nonempty set of nodes, $\leq$ is a partial ordering of K, and $\Vdash$ is a sort of truth valuation, $\alpha \Vdash A$ declaring <u>A is true at node α</u> (or, as one also says, A is <u>forced</u> at α_j the relation $\Vdash$ is called a <u>forcing</u> relation). Heuristically, one should think of the nodes α as representing states of knowledge, with $\alpha < \beta$ just in case β is a larger state of knowledge than α, i.e. β possesses more information than α. A truth valuation, or forcing relation, on propositional formulae is defined firstly by specifying which atoms are true at which nodes, subject to the condition

$$\alpha \Vdash p \text{ and } \alpha \leqslant \beta \Rightarrow \beta \Vdash p$$

(since knowledge grows), and secondly by inductively extending this valuation according to the clauses:

$$\alpha \Vdash A \wedge B \text{ iff } \alpha \Vdash A \text{ and } \alpha \Vdash B$$
$$\alpha \Vdash A \vee B \text{ iff } \alpha \Vdash A \text{ or } \alpha \Vdash B$$
$$\alpha \Vdash A \rightarrow B \text{ iff } \underset{\sim}{\forall}\beta \geqslant \alpha(\beta \Vdash A \Rightarrow \beta \Vdash B)$$
$$\alpha \Vdash \neg A \text{ iff } \underset{\sim}{\forall}\beta \geqslant \alpha(\beta \nVdash A) .$$

The latter two clauses might require a little thought. The growth of knowledge certainly requires

$$\alpha \Vdash A \rightarrow B \Rightarrow \underset{\sim}{\forall}\beta \geqslant \alpha(\beta \Vdash A \rightarrow B)$$
$$\Rightarrow \underset{\sim}{\forall}\beta \geqslant \alpha(\beta \Vdash A \Rightarrow \beta \Vdash B).$$

One accepts the converse for the sake of having a convenient criterion for forcing an implication. The same reasoning dictates the clause for negation.

Odd as they are, Kripke models offer a complete semantics for the intutionistic propositional calculus. In fact, not only is a propositional formula A derivable in Heyting's system iff it is valid in all Kripke models, but, if it is underivable, there is a finite <u>tree</u> model $(K, \leqslant, \Vdash)$ and node $\alpha_0 \in K$ (which can be taken to be the root of the tree) such that $\alpha_0 \nVdash A$. If A has atoms $p_1,\ldots,p_n$, one can derive de Jongh's Theorem by sticking nonstandard models of arithmetic at the nodes of K to obtain a Kripke model of HA, doing so in such a way that there are arithmetic sentences $\phi_1,\ldots,\phi_n$ forced at precisely the same nodes as $p_1,\ldots,p_n$, respectively. $A(\phi_1,\ldots,\phi_n)$ will then fail to be forced at α_0, just as $A(p_1,\ldots,p_n)$ failed to be so, and we see that $A(p_1,\ldots,p_n)$ is not a new valid propositional schema of HA.

As is implicit in this last observation, a Kripke model for an intuitionistic first-order theory is a partially ordered collection of classical structures. If $\alpha < \beta$, the domain, $D\alpha$, of the structure at α must be contained in the domain, $D\beta$, of the structure at β. Moreover, one has

$$\alpha \Vdash \exists x \phi x \text{ iff } \underset{\sim}{\exists} a \in D\alpha(\alpha \Vdash \phi \bar{a})$$
$$\alpha \Vdash \forall x \phi x \text{ iff } \underset{\sim}{\forall}\beta \geq \alpha\ \underset{\sim}{\forall} b \in D\beta(\beta \Vdash \phi \bar{b}) .$$

It is not a priori clear that the classical structures placed at the nodes must be nonstandard models of PA in order that the Kripke model be a model of HA. Indeed, it is probably false in general that the classical structures at the nodes of Kripke models of HA are models of PA. However, when the Kripke model is based on a finite tree, these structures are models of PA. This was first observed by Albert Visser, who applied the Friedman Translation ued by Friedman in "Classically and intuitionistically provably recursive functions" (cf. the contribution by Leivant for the definition and original application): If α is a node in such a finite tree model of HA, there is a sentence Λ such that

$$\beta \Vdash \Lambda \text{ if } \beta > \alpha ,$$

for all β in the model. Letting $\underset{\sim}{K}_{\Lambda}$ be the Kripke model obtained by deleting these nodes, and letting ϕ^{Λ} be the Friedman translation of ϕ based on Λ, it is not hard to show that, for any arithmetic sentence ϕ,

$$\alpha \Vdash_{\Lambda} \phi \text{ iff } \alpha \Vdash \phi^{\Lambda} .$$

In particular, induction at α in the new model reduces to that in the old, whence induction holds. Since α is terminal in the new model, the logic at α is classical (look at the clauses in the definition of $\Vdash$) and forcing at α agrees with truth in the classical structure placed at α. Thus, the classical structure at α is a model of PA.

From the above, models of HA constructed on finite trees must be constructed from nonstandard models of PA. However, some care is requied in fitting together such models to get Kripke models of HA: The truth of a sentence ϕ at any node α depends on the truth of its subformulae not only at α, but also at all $\beta > \alpha$. Thus, the classical truth or falsity of ϕ in the structure at α will usually not agree with the acceptance or rejection of the assertion $\alpha \Vdash \phi$ and it is conceivable that, for example, a bad fit of models will not result in a

Kripke model in which induction holds. Consider, e.g., the simple tree with three nodes $\alpha_1 < \alpha_2 < \alpha_3$. We want models $\underset{\sim}{M}_1, \underset{\sim}{M}_2, \underset{\sim}{M}_3$ of PA so that placing $\underset{\sim}{M}_1 < \underset{\sim}{M}_2 < \underset{\sim}{M}_3$ will yield a Kripke model of HA. Induction will be forced at α_3 for the simple reason that forcing at terminal nodes coincides with classical truth. A trivial way to get induction to hold at α_1, α_2 is to choose $\underset{\sim}{M}_1$ and $\underset{\sim}{M}_2$ to be the standard model. It is rather unimaginative, but not without use: Applied to fatter trees, this trick of putting the standard model at all but the terminal nodes will yield de Jongh's Theorem, if not any of the interesting refinements.

A second way of guaranteeing induction to hold is to insist that each model defines the basic operations of its successors: $\underset{\sim}{M}_1$ defines $\underset{\sim}{M}_2$ and $\underset{\sim}{M}_3$, while $\underset{\sim}{M}_2$ defines $\underset{\sim}{M}_3$. Questions about forcing at α_i reduce to questions about truth in $\underset{\sim}{M}_i$. In particular, induction in the Kripke model reduces to induction in the classical models, whence the Kripke model is one of HA.

The remarks on the successful construction of Kripke models of HA from models of PA carryover _mutatis mutandis_ to the construction of Kripke models of EA from models of PA--provided one has a finite tree underlying one's Kripke model of S4. Unfortunately, S4 is not complete for Kripke models on finite trees.

The Kripke models for S4 are triples $(K,R,\Vdash)$, where K is a nonempty set of nodes set of nodes, R is a reflexive, transitive binary relation on K , and the forcing relation satisifes

$$a \Vdash A \wedge B \quad \text{iff} \quad \alpha \Vdash A \text{ and } \alpha \Vdash B$$
$$a \Vdash A \vee B \quad \text{iff} \quad \alpha \Vdash A \text{ or } \alpha \Vdash B$$
$$a \Vdash A \rightarrow B \quad \text{iff} \quad \alpha \nVdash A \text{ or } \alpha \Vdash B$$
$$\alpha \Vdash \neg A \quad \text{iff} \quad \alpha \nVdash A$$
$$a \Vdash \Box A \quad \text{iif} \quad \underset{\sim}{\forall} \beta(\alpha R \beta \Rightarrow \beta \Vdash A) .$$

In other words, the usual connectives behave classically at each node and $\Box A$ is true at α iff A is true in "all possible worlds" accessible to α via R . There is no restriction on the forcing of propositional atoms as there was for intuitionistc logic. Thus, any assignment of truth values to atoms at the nodes of K extends inductively by the above clauses to a full forcing relation $\Vdash$ on (K,R) . S4 is valid in such a model and, in fact, completeness holds: For any modal sentence A , S4 $\vdash$ A iff A is valid in all Kripke models of this form. (An exposition of this and other background material can be found in Segerberg 1984.)

The completeness theorem for S4 can be improved. S4 is complete with respect to finite models, but not, as noted two paragraphs back, with respect to finite trees. The accessibility relation R need not be antisymmetric and model of, say, the form

β —— γ
α ,

where $\alpha R\beta$, $\alpha R\gamma$, $\beta R\gamma$, and $\gamma R\beta$, cannot be untangled to make a finite tree. The usual methods of splitting a structure like this into a tree would require a copy of β with a copy of γ above it, a copy of β above that, ... In short, one would get an infinite tree:

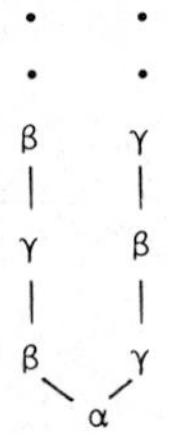

The situation is like a good-news - bad-news joke: The good news is that S4 is complete with respect to models on trees. The bad news is that the trees are not finite. The good news...the really good news will have to wait.

One nice thing about the completeness of S4 with respect to models on trees is that the intuitionistic propositional calculus is also complete with respect to such models. This allows a quick verification of a fundamental relationship, first noticed by Gödel, between the intuitionistic propositional calculus and S4: For any intuitionistic propositional formula A , define $A^{\Box}$ inductively by:

$$p^{\square} = \square p$$
$$(A \wedge B)^{\square} = A^{\square} \wedge B^{\square}$$
$$(A \vee B)^{\square} = A^{\square} \vee B^{\square}$$
$$(A \to B)^{\square} = \square(A^{\square} \to B^{\square})$$
$$(\neg A)^{\square} = \square\neg A^{\square} .$$

Then: A is an intuitionistic tautology iff $S4 \vdash A^{\square}$. One sees this simply by comparing the definitions of forcing in the two languages. The result extends: If we define

$$(\exists x \phi x)^{\square} = \exists x (\phi x)^{\square}$$
$$(\forall x \phi x)^{\square} = \square \forall x (\phi x)^{\square} ,$$

we get

$$HA \vdash \phi \quad \text{iff} \quad EA \vdash \phi^{\square} .$$

As for the analogue of de Jongh's Theorem, the necessary use of infinite trees poses a major problem. The two methods discussed of assigning models of PA to the nodes of a Kripke model do not work for infinite models. Take for example an S4 model on the infinite linear tree,

$$\alpha_0 < \alpha_1 < \ldots ,$$

with $\alpha_i \Vdash p$ iff i is even. We cannot affix the standard model to these nodes and hope for an arithmetic sentence ϕ exhibiting the same pattern: All the nodes look alike, whence either they all force ϕ or they all force $\neg\phi$. The definability trick will not work either: One can get a sequence $\underset{\sim}{M}_1, \underset{\sim}{M}_2, \ldots$ with each $\underset{\sim}{M}_i$ defining all of its successors, with even some degree of uniformity. But, when $\underset{\sim}{M}_1$ is deciding whether or not to force $\square A$, how, if it deems some α_n failing to force A to exist, does it know that n is <u>standard</u> (and one of the actual $\underset{\sim}{M}_i$'s occurrring)? Yet, one really must use the infinite trees to give a proof analogous to Smoryński's: Mutually accessible, nonidentical nodes α, β would require models $\underset{\sim}{M}_\alpha, \underset{\sim}{M}_\beta$ more-or-less contained in each other, but not identical. Friedman's embeddability criteria of section 3 allow this, but not in any way in which we can guarantee induction will hold.

Fortunately, as I hinted a moment ago, there is some good news. The result of splitting a finite S4 model into a tree model, although not finite, does have a great deal of periodicity built into it, a periodicity that can be exploited.

Let $\underline{K} = (K, <, \Vdash)$ be a tree Kripke model for S4. For $\alpha \in K$, let $\underline{K}_\alpha$ be the truncation of $\underline{K}$ to those nodes $\beta \geq \alpha$. If $\underline{K}$ arose from a finite S4-model by a canonical splitting of the nodes, then there are only finitely many isomorphism types among the truncations. The desired model of EA will have to imitate this repetition. In classical model theory, there is a name for a set of individuals repeating a common behaviour: such elements are called indiscernibles.

Formally, an ordered set $(A, <)$ in a model $\underset{\sim}{M}$ of some classical theory is called a set of indiscernibles relative to a set Γ of formulae if, for any $\phi x_1 \ldots x_n \in \Gamma$ and any

$$a_1 < \ldots < a_n, \ b_1 < \ldots < b_n \quad \text{in } A,$$

one has

$$\underset{\sim}{M} \models \phi \bar{a}_1 \ldots \bar{a}_n \leftrightarrow \phi \bar{b}_1 \ldots \bar{b}_n .$$

Obviously, the formulae of Γ cannot mention arbitrary parameters in M --for, otherwise, a_1, b_1 would be distinguished by the formula $\phi x_1 : x_1 = \bar{a}_1$. Yet, some parameters are often necessary. In Friedman's construction, elements of $\underset{\sim}{M}_\alpha$ and $\underset{\sim}{M}_\beta$ must be indiscernible relative to parameters from $\underset{\sim}{M}_\gamma$'s for γ below both α and β. Since the useful embeddings of nonstandard models are initial embeddings, $\underset{\sim}{M}_\gamma$ can be taken to be an initial segment of $\underset{\sim}{M}_\alpha$ and $\underset{\sim}{M}_\beta$. Supposing also that the ordering of A agrees with that of $\underset{\sim}{M}_\alpha$, $\underset{\sim}{M}_\beta$, and $\underset{\sim}{M}_\gamma$, the sort of indiscernibles one wants are Paris-Harrington indiscernibles satisfying: For

$$a \begin{array}{l} \nearrow a_1 < \ldots < a_n \\ \searrow b_1 < \ldots < b_n \end{array},$$

all in A, for $\phi_{x_1 \ldots x_n y_1 \ldots y_m} \in \Gamma$ with no parameters, and for $c_1, \ldots, c_m < a$,

$$\mathfrak{M} \models \phi_{\bar{a}_1 \ldots \bar{a}_n \bar{c}_1 \ldots \bar{c}_m} \leftrightarrow \phi_{\bar{b}_1 \ldots \bar{b}_n \bar{c}_1 \ldots \bar{c}_m} .$$

Paris-Harrington indiscernibles popped up in arithmetic in a different context, namely, that of combinatorial incompleteness theorems in arithmetic. I discussed their result in another contribution to this volume ("Some rapidly growing functions"), but did not discuss their indiscernibles much. Briefly, with the indiscernibles I described in that article (under the name "homogeneous set"), Jeff Paris and Leo Harrington constructed a model of a fragment of arithmetic with a lot of indiscernibles of the above type. With this minimal fragment, the typical arithmetic formula,

$$\forall x_1 \exists y_1 \ldots \forall x_n \exists y_n \phi , \ \phi \ \text{quantitier-free},$$

is equivalent to

$$\forall x_1 < a_1 \exists y_1 < a_2 \ldots \forall x_n < a_{2n-1} \exists y_n < a_{2n} \phi ,$$

where $a_1 < \ldots < a_{2n}$ are indiscernibles greater than any parameter occurring in ϕ. Thus, full induction follows from induction on Δ_0-formulae, i.e. formulae in which all quantifiers are bounded, and their model of a fragment of arithmetic which includes Δ_0-induction is a model of full induction.

Friedman's construction, given an S4 model, of a simulating Kripke model of EA similarly relies on the existence of Paris-Harrington indiscernibles. Unfortunately, after all this building up to this revelation, I cannot offer any thing further in the way of proof--this proof just isn't simple enough to expound upon here. I can merely make the anticlimatic remark that Friedman uses the Paris-Harrington indiscernibles as the key to reducing induction in the Kripke model to induction in the individual classical models at the nodes.

Friedman's proof will appear in a paper being written jointly, with Robert Flagg.

5. Translatability and Relative Consistency

The well-known proof of the consistency of non-Euclidean geometries proceeds by offering Euclidean re-interpretations of the geometric notions and observing that the non-Euclidean axioms hold under these interpretations. To those whose confidence in Euclidean geometry is shaken by such a process, this offers *relative* consistency proofs: Various non-Euclidean geometries are consistent relative to Euclidean geometry, i.e. they are consistent *if* Euclidean geometry is. Similary, when Hilbert presented an axiomatic theory of Euclidean geometry and proved its consistency, he did so relative to the arithmetic of the reals by translating the geometric theory into the real-arithmetic one. Finally, Gödel's famous proof of the consistency of the Axiom of Choice, i.e. the consistency of set theory with choice relative to set theory without choice, proceeded by defining within set theory a subuniverse of the universe of sets in which choice held, i.e. the proof was accomplished by an interpretation.

It would appear that every well-known relative consistency proof is established by an interpretation. The one counterexample that comes to mind is Gödel's Second Incompleteness Theorem: $T + \neg Con(T)$ is consistent relative to T. Yet, even here, if only for the restricted case of GB set theory, Peter Vopenka gave a proof of the relative consistency result by an explicit interpretation. For reasonably powerful theories T (such as PA, ZF, or GB), the combined formalisation of the Completeness Theorem (every consistent theory has a model.) and the Second Incompleteness Theorem (the consistency of $T + \neg Con(T)$ relative to T) yields an interpretation of $T + \neg Con(T)$ within T.

The last observation cited was first made by Feferman and Steven Orey around 1960. Feferman and Orey initiated the modern study of interpretability by proving such results as the following:

Theorem. (Feferman). Let T contain PA and be consistent. Then: $T + Con(T)$ is not interpretable in T.

Theorem. (Orey). A theory T is interpretable in PA iff every finite subtheory $T_0 \subseteq T$ is interpretable in PA.

Orey's Theorem, also known as the Orey Compactness Theorem, depends on a crucial property of PA: PA proves the consistency of all of its finite subtheories. If T is such that each finite subtheory $T_0 \subseteq T$ is interpretable in PA, then PA proves the consistency of a finite subtheory in which T_0 can be interpreted, whence PA proves the consistency of T_0. Bh a sort of Overspill, PA proves the consistency of all of T, but doesn't recognise this; PA thinks it proves the consistency of a large (nonstandard) finite chunk of T. By formalising the proof of the Completness Theorem in PA and applying it to this (imagined) chunk, PA interrpets T. (cf. Feferman 1960 for the results of Feferman and Orey; something of Orey's proof can also be gleaned from the discussion of the Arithmetised Completeness Theorem in Smoryński 1984.)

This reflexive property of PA, the ability to prove the consistency of its finite subsystems, is shared by ZF, but, of course, not by any finitely axiomatised theory like GB or the analogous so-called predicative extension ACA_0 of PA. Hence, a theorem similar to Orey's cannot be expected to hold for, say, GB. The picture of interpretability for such finitely axiomatised theories ought, thus, to be different from that for PA. In the early 1970s, Hájek began the study of the difference between relative interpretability in ZF and GB, where a sentence ϕ is said to be relatively interpretable in a theory T if $T + \phi$ has an interpretation in T. Hájek showed that, although ZF and GB prove the same set theoretic formulae, whence, for set theoretic ϕ, $ZF + \phi$ and $GB + \phi$ are mutually consistent, i.e. $ZF + \phi$ is consistent relative to ZF iff $GB + \phi$ is consistent relative to GB, there are set theoretic sentences ϕ such that ϕ is relatively interpretable in ZF, but not in GB. He also conjectured the opposite to hold, that there are set theoretic sentences ϕ such that ϕ is relatively interpretable in GB and not ZF (whence $GB + \phi$ is not interpretable in ZF); and he was able to draw some conclusions about the form ϕ could take, but it was only half a decade later that Solovay first constructed such.

The upshot of all this is that the notions of interpretability and relative consistency do not coincide and the exact relation between the two concepts stands in need of clarification. Moreover, we are not speaking here of obscure theories, but of natural theoreis like ZF and GB. On the one hand, the Orey Compactness Theorem offers, for reflexive theories like PA or ZF, a good characterisation of interpretability; on the other hand, it reveals an unintended ability to interpret theories--unintended insofar as the interpretation constructed via Orey's Compactness Theorem are not recognisably interpretations and do not entail provable relative consistency as a corollary. Moreover, there is the problem of non-reflexive theories, particularly, finitely axiomatised ones like GB and ACA_0, for which Orey's Compactness Theorem offers nothing.

When one sorts things out, two questions emerge. First, can one offer any decent characterisation of interpretability within theories like GB? Second, can one clarify the relation between interpretability and relative consistency?

Hájek and Viteslav Svejdar gave an answer to these quesitons for the case of interpreting theories GBL + ϕ (where GBL is GB plus Gödel's Axiom of Constructibility, V = L) in theories GB + ψ, for set theoretic sentences ϕ, ψ. (Their result applies to ACA_0 in place of GBL and GB, but, as their work developed in the set theoretic context, I shall refer to GBL and GB. Besides, the presence of the Axiom of Constructibility ought not to be overlooked; it indicates something of what is needed of the theories T to which the result applies.) This result is the completion of Solovay's work briefly alluded to a few paragraphs back and requires a word or two in preliminary explanation: Since GB is finitely axiomatised, it does not prove full arithmetic induction. Hence, there are formulae $I(x)$ in the language of GB which define initial segments of ω closed under successor. Because there is no guarantee of defining such an initial segment closed under exponentation (closure under addition and multiplication is unproblematic), and because the usual arithmetic coding requires the use of exponentiation, Solovay devised an alternate "proof predicate", $Prf^S_{ZFL}(x,y)$, asserting x to witness the provability of (the sentence with code) y in ZFL (= ZF + V = L).

<u>Theorem</u>. (Hajek-Svejdar). Let ϕ, ψ be set theoretic sentences. The following are equivalent:

i. GBL + ϕ is interpretable in GB + ψ

ii. For some formula $I(x)$, defining a subset of ω,

a. $GB + \psi \vdash I(\bar{0}) \wedge \forall x[I(x) \to I(x + \bar{1})]$

b. $GB + \psi \vdash \forall xy[I(x) \wedge y \leqslant x \to I(y)]$

c. $GB + \psi \vdash \forall x[I(x) \to \neg Prf^S_{ZFL}(x, \ulcorner \neg\phi \urcorner)]$.

In words: GBL + ϕ is interpretable in GB + ψ iff GB + ψ proves that any proof of the inconsistency of $ZFL + \phi$ (hence of $GBL + \phi$) must be nonstandard.

The Hájek-Svejdar Theorem appeared in 1983 (in Svejdar 1983, to be exact).

In a separate development, Friedman approached the problem syntactically and proved, for finitely axiomatised theories, the equivalence of interpretability and effective relative consistency. Although he proved this result in 1976, and although he even wrote up the proof in 1980, Friedman never published the result. Because of increasing interest in the subject, the Editors have politely requested that I include the proof here. In doing so, I follow Friedman's own exposition.

The first thing to note is that, where most people have spoken of "interpretability", Friedman refers to "translatability". Essentially, a translation of a theory S into a theory T is given by defining within the language of T a special set to be the domain of a model of S and additional relations interpreting the primitives of S in this model. Thus, PA is translated into ZF by specifying the set of finite ordinals as domain and the restriction of the recursively defined ordinal operations of addition and multiplication to be the arithmetic operations. A formal definition of the notion of translatability would not be too hard to present here, but would be rather detailed, particularly in view of the fact that we must deal with many-sorted theories (like GB and ACA_0). Only one fine point needs be attended to: It doesn't matter whether equality gets interpreted as equality or some congruence relation; the proof to come will show the two versions of translatability to be equivalent.

The second point is relative consistency: Where should relative consistency be proven? As Kreisel has pointed out, if one is actually interested in the consistency of S relative to T, one is fairly confiident about T and it would suffice to prove the relative consistency result in T:

$$T \vdash Con(T) \rightarrow Con(S) \ . \qquad (*)$$

If, however, one is not interested in the result for doctrinaire reasons or if one considers meaningless the provability within T of such an implication, as is the case for $T = PA + \neg Con(PA)$ where $T \vdash \neg Con(T)$, one wants the relative consistency result proven in a sound base theory. As Kreisel further noted, if the proof (*) of relative consistency in T is effective, say

$$T \vdash \forall x[\mathrm{Prov}_S(x,\ulcorner\bar{0} = \bar{1}\urcorner) \rightarrow \mathrm{Prov}_T(f(x),\ulcorner\bar{0} = \bar{1}\urcorner)] ,$$

where f is primitive recursive (and we assume a modicum of arithmetic in S , T) , and where PRA denotes Primitive Recursive Arithmetic, a weak subtheory of PA (of the same strength as Friedman's RCA_0--cf. 2nd contribution to this volume by S. Simpson), then

$$\mathrm{PRA} + \mathrm{Con}(T) \vdash \forall x[\mathrm{Prov}_S(x,\ulcorner\bar{0} = \bar{1}\urcorner) \rightarrow \mathrm{Prov}_T(f(x),\ulcorner\bar{0} = \bar{1}\urcorner)] , \qquad (**)$$

as can be shown by appeal to the equivalence of Con(T) to a weak form of Reflexion: $\mathrm{Pr}_T(\ulcorner\phi\urcorner) \rightarrow \phi$, for $\phi \in \Pi_1$. By (**),

$$\mathrm{PRA} \vdash \mathrm{Con}(T) \rightarrow \mathrm{Con}(S) .$$

As I already announced, Friedman's characterisation of translatability is in terms of an effective relative consistency and it will suffice to consider the implication provable over a weak base theory like PRA. Actually, Friedman considers several variants of effectiveness and an even weaker base theory than PRA.

The theory EFA of Elementary Function Arithmetic (deemed Exponential Function Arithmetic in the contribution by Harrington and Nerode to this volume) is the weak subtheory of PA with primitives $\bar{0}$, $\bar{1}$, + , $\cdot$ as before, as well as < and exp (exponentiation with base 2), and, along with basic axioms for these primitives, the schema of Δ_0-induction, i.e. induction restricted to Δ_0-formula (which, we recall, are formulae in which all quantifiers are bounded, i.e. of the form $\exists x < t$ or $\forall x < t$, for some term t). Jeff Paris has shown that EFA is finitely axiomatisable.

The crucial notion of effectiveness of a consistency proof for S relative to T is not the effective transformation of a proof of a contradiction in S to one in T ; effective bounds on the complexity of the latter proof from bounds on the former suffice. There are two relevant measures of the complexity of a proof--quantifier complexity and size complexity.

Size complexity is the easier of the two notions to define. The size of a derivation is simply the total number of symbols present in the derivation. We assume, however, that the alphabet of the language contains only finitely many symbols and that, e.g., the infinitely many variables $x_1, x_2, \ldots$ are each written as an x followed by the subscript written in binary. With such a convention, there will only be finitely many derivations of a given size.

The quantifier complexity of a derivation, like that of any finite set of formulae, is the maximum quantifier complexity of formulae occurring in it. To define the quantifier complexity of a formula ϕ, first define $qc(\phi) \leq 0$ for ϕ atomic and $qc(\phi) \leq k+1$ for ϕ a propositional combination of formulae of the forms $\forall x_1 \ldots \forall x_n \psi$ and $\exists x_1 \ldots \exists x_n \psi$ with $qc(\psi) \leq k$. Then define

$$qc(\phi) = k \quad \text{iff} \quad qc(\phi) \leq k \quad \text{and} \quad qc(\phi) \not\leq k-1 .$$

For a theory T and formula ϕ, we write

$$T \vdash \phi[m,n] \tag{*}$$

if T has a derivation of ϕ of quantifier complexity at most m and size at most n. We can code and discuss the syntax of reasonable theories T within EFA. For such a theory, we let

$$\text{"}T \vdash \phi[m,n]\text{"}$$

denote the EFA expression of (*). When we do not wish to assert any particular bound, we replace "m" or "n" by "∞".

The effective bounds I spoke of will be expressed in terms of the functions $2^{[a]}(b)$ and $2^{[a]}$ defined by

$$\begin{cases} 2^{[1]}(b) = 2^b \\ 2^{[a+1]}(b) = 2^{2^{[a]}(b)} \\ 2^{[a]} = 2^{[a]}(1) . \end{cases}$$

Finally, let us call a theory T adequate if it conains EFA and a reasonable theory of finite sequences of objects. (N.B. EFA has a theory of finite sequences of numbers. Theories like ACA_0, ZF, and GB have other sorts of objects and they admit special codings of finite sequences of such. For example, in ACA_0, a sequence $X_1,\ldots,X_n$ of sets would be coded by

$$X = \{1\} \times X_1 \cup \ldots \cup \{n\} \times X_n.)$$

With all of this notation, we can state Friedman's result for theories like GB and ACA_0:

Theorem 1. Let S, T be finitely axiomatised adequate theories. The following are equivlent:

i. S is translatable into T

ii. for some c,

$$EFA \vdash \forall k[\text{"}S \vdash \bot[k,\infty]\text{"} \to \text{"}T \vdash \bot[k + \bar{c},\infty]\text{"}]$$

iii. for some m,

$$EFA \vdash \text{"}S \vdash \bot[qc(S),\infty]\text{"} \to \text{"}T \vdash \bot[\bar{m},\infty]\text{"}$$

iv. for some c, n

$$T \vdash \forall qk[\text{"}S \vdash \bot[q,k]\text{"} \to \text{"}T \vdash \bot[q + \bar{c},2^{[\bar{n}]}(k)]\text{"}]$$

v. for some m, n,

$$T \vdash \forall k[\text{"}S \vdash \bot qc(S),k]\text{"} \to \text{"}T \vdash \bot[\bar{m},2^{[\bar{n}]}(k)]\text{"}] ,$$

where $\bot$ is any convenient contradictory statement (e.g. $\bar{0} = \bar{1}$) and $qc(S)$ is the quantifier complexity of the set of axioms of S.

For single-sorted theories S, let S' denote its predicative extension obtained by adding class variables and the relative comprehension axiom schema,

$$\exists X \forall x[x \in X \leftrightarrow \phi(x)] ,$$

where X does not occur in ϕ and no class quantifiers occur in ϕ, but free class variables can so occur. In place of any schemata of S, individual axioms about classes are added. Thus, the induction schema,

$$\phi\bar{0} \wedge \forall x[\phi(x) \to \phi(x + \bar{1})] \to \forall x\phi(x) ,$$

of PA is replaced by the induction axiom,

$$\bar{0} \in X \wedge \forall x(x \in X \rightarrow x + \bar{1} \in X) \rightarrow \forall x(x \in X) ,$$

of PA' . I remark that PA' is just ACA_0 , that ZF' is GB, and that, if S has only finitely many axiom schemata, then S' is finitely axiomatised.

Using proof theory, Friedman derived the following from Theorem 1:

<u>Theorem 2</u>. Let S , T be adequate single-sorted theories with only finitely many axiom schemata. The following are equivalent:

i. S' is translatable into T'

ii. EFA $\vdash$ Con(T) $\rightarrow$ Con(S) , i.e.

$$EFA \vdash \text{"}S \vdash \bot[\infty,\infty]\text{"} \rightarrow \text{"}T \vdash \bot[\infty,\infty]\text{"}$$

iii. for some n ,

$$T \vdash \forall k[\text{"}S \vdash \bot[\infty,k]\text{"} \rightarrow \text{"}T \vdash \bot[\infty,2^{[\bar{n}]}(k)]\text{"}] .$$

Most of the rest of the present section is devoted to an exposition of a sketch of the proof of Theorem 1. Implications i $\Rightarrow$ ii and i $\Rightarrow$ iv are fairly unremarkable: Via the interpretation of S in T , an inconsistency proof of S translates back to one of T . The bounds cited are mere matters of book-keeping. Implications ii $\Rightarrow$ iii and iv $\Rightarrow$ v are trivial.

The nontrivial implications are v $\Rightarrow$ iii, iii $\Rightarrow$ v, and v $\Rightarrow$ i. Two of these can be disposed of on general principles.

v $\Rightarrow$iii. By Kreisel's remark cited earlier, v immediately yields

$$\begin{aligned} EFA &\vdash \forall k[\text{"}S \vdash [qc(S),k]\text{"} \rightarrow \text{"}T \vdash \bot[\bar{m},2^{[\bar{n}]}(k)]\text{"}] \\ &\vdash \exists k\text{"}S \vdash \bot[qc(S),k]\text{"} \rightarrow \exists k\text{"}T \vdash \bot[\bar{m},2^{[\bar{n}]}(k)]\text{"} \\ &\vdash \text{"}S \vdash \bot[qc(S),\infty]\text{"} \rightarrow \text{"}T \vdash \bot[\bar{m},\infty]\text{"} . \end{aligned}$$

iii $\Rightarrow$ v. Assertion iii yields

$$EFA \vdash \forall k \exists p[\text{"}S \vdash \bot[qc(S),k]\text{"} \rightarrow \text{"}T \vdash \bot[qc(S),p]\text{"}] .$$

This is of the form $\forall k \exists p\, \phi(k,p)$ with $\phi \in \Delta_0$. If one could not find any finite n such that p could be bounded by $2^{[n]}(k)$, then by ordinary model theory there would be a model $\underset{\sim}{M}$ of EFA and a (possibly nonstandard) integer $k_0 \in M$ such that

$$\underset{\sim}{M} \models \exists p \phi(\bar{k}_0,p)$$

but, for each finite n ,

$$\underset{\sim}{M} \models \forall p[\phi(\bar{k}_0,p) \rightarrow p > 2^{[\bar{n}]}(k_0)] .$$

Starting with k_0 construct, á la Ryll-Nardzewski, a new model $\underset{\sim}{M}_0$ by taking as the domain of $\underset{\sim}{M}_0$ the set

$$M_0 = \{a \in M : a < 2^{[n]}(k_0) \text{ for some } n \in \omega\} .$$

The model $\underset{\sim}{M}_0$ is one of EFA: The trickiest thing to verify is Δ_0-induction, but this is also a general fact: An initial segment of a model of Δ_0-induction satisfies Δ_0-induction.

Now we quickly get a contradiction: $\text{EFA} \vdash \forall k \exists p\, \phi(k,p)$, whence

$$\underset{\sim}{M}_0 \models \exists p\, \phi(\bar{k}_0,p) .$$

But, to be in M_0 , p must be less than $2^{[n]}(k_0)$ for some finite n , contrary to assumption.

All that is left is the implication v $\Rightarrow$ i. Before discussing this proof, let me quickly note that the implications iii $\Rightarrow$ ii and v $\Rightarrow$ iv both follow from the formalisation within EFA of Herbrand's Theorem. Thus, ii-v of Theorem 1 are equivalent on general principles and the crucial step in proving Theorem 1 is the outright construction of a translation of S into T on assumption of v.

In the ensuing lemmas, T^* will denote the theory $T + "S \vdash \perp[qc(S),\infty]"$.

Lemma 1. If S is translatable into T^* , then S is translatable into T .

T^* is, basically, $T + \neg \text{Con}(S)$. If S were translatable into both $T + \text{Con}(S)$ and $T + \neg\text{Con}(S)$, it would be translatable into T . For, one could take the translation to be defined by the former if $\text{Con}(S)$ and the latter if

$\neg$Con(S) . Were T strong enough, the full completeness Theorem would be provable in T + Con(S) and a model of S would result. This model would yield the translation. T is not, however, assumed to be very strong, and it certainly does not have full induction (by Ryll-Nardzewski's Theorem). Nonetheless, this is almost the approach Friedman follows to construct the translation.

One style of completeness proof for the predictate calculus consists of trying to construct a countermodel to a finite set of formulae. Occasionally in the attempt one must branch to two attempts, as when, for example, one tries to make $\phi \wedge \psi$ false by making ϕ false or ψ false. The result of such an attempt is a tree-called a refutation tree. The attempt fails if on each path through the tree one encounters both ψ and $\neg\psi$ at the same node, for some ψ . If, however, there is a path through the tree in which this never happens (and one has made an honest attempt at the construction), the path can be used to construct a countermodel to the given set of formulae. Moreover, things can be arranged neatly: First, we can agree to stop working on a path as soon as some ψ , $\neg\psi$ occur simultaneously. Second, if the sentence to be countermodelled has only infinite countermodels (say, it entails a minimal amount of arithmetic), it will require an infinite path through the tree. Thus, the tree yields a countermodel iff there is an infinite path through it. Finally, the actual construction of the tree is fairly canonical and can be described in EFA.

The describability of the canonical refutation tree in EFA has its uses. For one thing, a failed construction--one in which every path is blocked by a pair ψ , $\neg\psi$--flips over to a derivation of the sentence one cannot refute. Thus, assuming Con(ϕ) , there is an infinite path modelling ϕ in the canonical refutation tree for $\neg\phi$. As I already said, however, without induction we cannot prove this. But we can <u>define</u> such a path: Take the set of leftmost nodes all of whose predecessors are leftmost nodes with <u>consistent</u> sets of formulae. This offers a definition of the leftmost infinite path and gives a model of ϕ . Applying this to $\phi = \bigwedge\!\!\bigwedge S$ and working inside T , we get the desired interpretation of S in T + Con(S) and therewith Lemma 1.

With Lemma 1, the task of constructing a translation of S into T reduces to one of translating S into T^* . As with Solovay's work and the resultant

Hájek-Svejar Theorem, this is accomplished by using the nonstandard integer proving the inconsistency of S as a large upper bound on the size of a proof. The argument is, however, a bit more refined.

In T^*, let x be the least k such that $S \vdash \bot[qc(S),k]$. The number x is infinite if S is consistent (as we assume).

Lemma 2. $T^* \vdash$ "$T \vdash \bot[\bar{m},2^{[\bar{n}]}(x)]$".

This follows directly by v, where m, n are as in v and x is as just chosen.

Let $I(\phi,y)$ denote induction on ϕ up to y :

$$\phi(\bar{0}) \wedge \forall x(\phi(x) \to \phi(x+\bar{1})) \to \phi y .$$

Lemma 3. There is a formula ϕ such that

$$T^* \vdash \neg I(\phi,2^{[\bar{n}]}(x)) .$$

Proof: The proof that PA or ZF is reflexive can be invoked: Using the theory of finite sequences, one can give a truth definition for formulae of quantifier-complexity at most m (m is standard). Using this, an induction up to $2^{[n]}(x)$ shows anything provable in complexity $[m,2^{[n]}(x)]$ is true. Thus, if ϕ is the formula of this induction,

$$T + I(\phi,2^{[\bar{n}]}(x)) \vdash \neg\text{"}T \vdash \bot[\bar{m},2^{[\bar{n}]}(x)]\text{"} ,$$

whence

$$T^* \vdash \neg I(\phi,2^{[\bar{n}]}(x)) . \qquad \text{Q.E.D.}$$

Lemma 4. For some formula ψ, $T^* \vdash \neg I(\psi,x)$.

Standard proof theory reduces $I(\phi,2^{[\bar{n}]}(x))$ to $I(\psi,x)$ for some ψ of higher quantifier complexity.

Lemma 5. $T^* \vdash$ "the height of the canonical refutation tree for S is at least logloglogx".

This is a matter of careful construction of the canonical refutation tree and careful bookkeeping.

Lemma 6. For some formula ρ , $T^{*}\vdash$ "if y is the height of the canonical refutation tree for S , then $\neg I(\rho,y)$" .

This follows from Lemma 5 and the proof of lemma 4.

One now finishes the proof of Theorem 1 as follows: The formula ρ of Lemma 6 can be used to define an infinite initial segment of the leftmost path of maximum height y in the canonical refutation tree for $\neg S$. Although T^{*} cannot prove it, this path has enough logical closure to devise a model of S . This offers the translation of S into T . This completes our outline of the proof of Theorem 1.

Although there really were no nonstandard models in the above proof, there certainly was the element of nonstandardness--the infinite size of x and y . Similarly, ρ (or, perhaps better: $\forall y \leqslant x\, \rho(y)$) defines an infinite initial segment of the integers closed under successor. In a nonstandard model, it could extend beyond ω . Now, as Andrzej Grzgorczyk first noticed, in Ryll-Nardzewski's proof that induction cannot be finitely axiomatised, the set of natural numbers is definable in the nonstandard model constructed: For the domain of the model was of the form

$$I = \{x \in \underset{\sim}{M} : \text{for some } n \in \omega ,\ x \leqslant F^{(n)}(a)\} ,$$

for some model $\underset{\sim}{M}$, $a \in M$, and function F ; ω is defined simply by

$$\omega = \{x \in I : F^{(x)}(a) \text{ exists}\} .$$

Refining this and combining it with a translation of S into T , Friedman proved the following:

Theorem 3. Let S , T be as before. The following are equivalent:

i. S is translatable into T

ii. S is faithfully translatable into T .

Here, I should explain, a translation τ of S into T is faithful if one has

$$S \vdash \phi \Leftrightarrow T \vdash \tau(\phi)$$

for all sentences ϕ of the language of S. The earliest result on faithful translatability was a variant of the Orey Compactness Theorem and, thus, required reflexiveness. It also required soundness:

Theorem. (Feferman, Kreisel, Orey). Let S, T be recursive theories, with $T \supseteq PA$ reflexive and Σ_1-sound (i.e. T proves only true Σ_1-sentences). Then: The following are equivalent:

i. S is translatable into T

ii. S is faithfully translatable into T.

Actually, Feferman, Kreisel, and Orey didn't offer their Theorem in this form. What they did was to prove the Orey Compactness Theorem anew, using the Σ_1-soundness of T during the construction of the interpretation of S to guarantee faithfulness. The Σ_1-soundness condition cannot be dropped, but it can be weakened. In 1980, Per Lindström proved the following:

Theorem. (Lindström). Let S, T be recursive theories, T containing PA and reflexive. The following are equivalent:

i. S is faithfully translatable into T

ii. S is translatable into T and every sentence of the language of S that T proves is derivable in the predicate calculus is provable in S.

These two Theorems, along with the Orey Compactness Theorem, characterise those recursive theories faithfully interpretable in a theory like PA and ZF; Friedman's Theorem 3 characterises those finite (actually, the result extends to recursive) S which are faithfully interpretable in theories like ACA_0 and GB.

Common to the proofs of all three Theorems are the use of independent formulae and the Arithmetised Completeness Theorem. Friedman's proof can briefly be described thus: Let π be an interpretation of S in T and let t be a term for which $T + t = \bar{n}$ is consistent for each natural number n. Because the complexity of the sentences $t = \bar{n}$ is uniform, Ryll-Nardzewski's proof provides a single formula $\sigma(x)$ which, for each n, will define the standard integers in some model of $T + t = \bar{n}$. A new interpretation τ is defined by examining t.

If t equals the code of a sentence ϕ, try to construct a Henkin model of $S + \phi$ using the integers satisfying σ. If this results in a model, we use the model for a translation; otherwise, we fall back on π. The translation τ must be faithful because, if $S \nvdash \phi$, in the model of $T + t = \ulcorner \neg\phi \urcorner$, τ interprets ϕ by its falsity in the Henkin model constructed (successfully because $S + \neg\phi$ is consistent and we are carrying out the standard construction in this model), whence $T \nvdash \tau(\phi)$.

It is probably worth remarking at this point on the origin of Friedman's work on translatability and his philosophical preference for finitely axiomatised theories. In the early 1970's, Friedman worked on a programme he called the Calibration of Mathematical Texts. In real mathematics, one rarely uses the full power of any axiom system. Friedman sought to see how much actually was used through case studies of mathematical textbooks: He chose a textbook, formulated a weak, finitely axiomatised base theory for it, and proceeded through the book to measure the relative strengths of the various results by their proof-theoretic consistency strengths, i.e. by their provable relative consistencies. Eventually, this programme matured into Reverse Mathematics (for which see one of Simpson's contributions to the present volume). A dissatisfaction with relative consistency in this context led Friedman to consider translatability. Because of this context of comparing single sentences over finite base theories, it is the translatability of finite theories that is significant to him.

However, I digress. What is of especial interest is not where Friedman's results came from (They, like the Hájek-Svejdar Theorem, could have come from curiosity about the difference between ZF and GB.), but where they led to. A natural corollary to the quantitative refinement of relative consistency is a quantitative view of consistency. Among the Finite Forms at Gödel's Incompleteness Theorems, Friedman proved the following:

Theorem. Let T be finitely axiomatised, adequate, and consistent. There is a real number $c > 0$ such that, for all sufficiently large n,

$$T \nvdash \text{"}T \nvdash \bot[\infty,n]\text{"}\ [\infty,n^c],$$

i.e. T cannot prove in n^c steps that T cannot prove an absurdity in n steps.

I shall not prove this here. A stronger result of Friedman's will appear in a joint paper with Pavel Pudlák, who, spurred on by the result of Hájek and Svejdar, also began to study the relation between Gödel's Theorems and the lengths of proofs. I will, however, cite one consequence of this Theorem for translatability:

Theorem. If T is finitely axiomatised, adequate, and consistent, then T' is not translatable into T.

The corresponding result for reflexive T is an easy consequence of the Orey Compactness Theorem. Consider, e.g. GB and ZF: For GB to be translatable into ZF, every finite subtheory of GB would have to be provably consistent in ZF. But this includes GB, the consistency of which is equivalent to that of ZF. Gödel's Second Incompleteness Theorem tells us, however, that ZF cannot prove its own consistency and we see that such a translation is precluded.

The appeal to Orey does not work for finite theories T, like ACA_0 or GB, and their predicative expansions (predicative third-order arithmetic and predicative third-order set theory, respectively); Friedman's Theorem covers these.

The proof of this Theorem proceeds roughly as follows: If T' were translatable into T, we would get a relative consistency proof. Now, it happens that T' proves "$T \nvdash \perp[\infty,n]$" very quicky--with approximately $\log n$ symbols in the language of T'. (This proof proceeds by a metamathematical induction on n. I don't have space to say more.) Composing the translation of T' into T, one can show that T' proves "$T' \nvdash \perp[\infty,dn]$" for some $0 < d < 1$. But this contradicts the quantitative refinement of Godel's Second Incompleteness Theorem applied to T'; Relabelling, the translation gives

$$T' \vdash \text{"}T' \nvdash \perp[\infty,n]\text{"}\ [\infty,O(\log n)] ,$$

while the incompleteness result asserts

$$T' \nvdash \text{"}T' \nvdash \perp[\infty,n]\text{"}\ [\infty,O(n^c)] ,$$

and yet $n^c > \log n$ for sufficiently large n.

I shall have to leave the reader with this inadequate exposition of these matters. The publication of the Friedman-Pudlák paper will, it is hoped, provide the ground work for such quantitative results, a foundation on which an exposition can rest.

Bibliography

Papers of Friedman cited:

1969 Bar induction and Π^1_1- CA , JSL 34, pp. 353-362.

1970 Iterated inductive definition and Σ^1_2-AC, in: Kino, Myhill, and Vesley, eds., Intuitionism and Proof Theory, North-Holland, Amsterdam.

1973A Countable models of set theories, in: Mathias and Rogers, eds., Cambridge Summer School in Mathematical Logic, Springer-Verlag, Heidelberg.

1973B Some applications of Kleene's methods for intuitionistic systems, in: Mathias and Rogers, eds., Cambridge Summer School in Mathematical Logic, Springer-Verlag, Heidelberg.

1978 Classically and intuitionistically provably recursive functions, in: Müller and Scott, eds., Higher Set Theory, Springer-Verlag, Heidelberg.

1981 On the necessary use of abstract set theory, Adv. in Math 41, pp. 209-280.

1983 Unary Borel functions and second-order arithmetic, Adv. in Math 50, pp. 155-159.

Expositions of background material:

Ken McAloon, et al.

1980 Modèles de l'Arithmétique, Astérisque 73.

1981 Model Theory and Arithmetic, Springer-Verlag, Heidelberg.

Krister Segerberg and Robert Bull

1984 Basic modal logic in: Gabbay and Gwenthner, eds., Handbook of Philosophical Logic, II, Reidel, Dordrecht.

C. Smoryński

1973 Applications of Kripke models, in: Troelstra, ed., Metamathematical Investigation of Intuitionistic Arithmetic and Analysis, Springer-Verlag, Heidelberg.

1983 Recursively saturated nonstandard models of arithmetic, JSL 46, pp. 259-286.

1984 Lectures on nonstandard models of arithmetic, in: Lolli, Longo, and Marcja, eds., Logic Colloquium '82, North-Holland, Amsterdam.

Other papers cited:

Sol Feferman

1960 Arithmetization of metamathematics in a general setting, Fund. Math. 49, pp. 35-92.

Sol Feferman, Georg Kreisel, and Steven Orey

1962 1-consistency and faithful interpretations, Arch. Math. Logik 6, pp. 52-63.

Petr Hájek

1971/1972 On interpretability in set theories I, II, Comment. Math. Univ. Carol. 12, pp. 73-79; 13, pp. 445-455.

Per Lindström

A On faithful interpretability, to appear.

Vitezslav Svejdar

1983 Modal analysis of generalized Rosser sentences, JSL 48, pp. 986-999.

HARVEY FRIEDMAN'S RESEARCH ON THE FOUNDATIONS
OF MATHEMATICS, L.A. Harrington et al. (editors)

Intuitionistic formal systems

Daniel Leivant

Department of Computer Science
Carnegie-Mellon University

This chapter surveys Harvey Friedman's work on intuitionistic logic and intuitionistic arithmetic, most of which dates from 1971 through 1977. It was also around 1971 that Friedman took interest in set theories based on intuitionistic logic, a topic covered in the chapter by Andrej Scedrov. Proofs are outlined for the most important of Friedman's published theorems, as well as for some (but not all) of Friedman's unpublished results, available to date only in sketchy memoranda.

Contents.

1. Intuitionism in Friedman's Calibration Programme.

1.1. Calibrating levels of abstraction in Mathematics.

A leitmotif in Friedman's work has been the calibration and assessment of levels of mathematical abstraction, on the background of the incompleteness results and the failure of Hilbert's Programme in its original form. Given a principle P that introduces a new level of abstraction, one compares the consequences of admitting P, as opposed to confining Mathematics to its less speculative portion that avoids P. Intuitionism fits well into this scheme, since it is rooted in the rejection of certain forms of mathematical and of logical construction, precisely because of what is perceived as their speculative nature.

Given a formalism M for a part of Mathematics that is confined to a particular level of abstraction, several properties of M are important for the kind of calibration considered. The focus changes, however, from level to level and from formalism to formalism. For instance, an abstract principle P which Friedman has studied extensively is the existence of uncountable sets. The most interesting advance there is the discovery of cases where it is impossible to avoid P (i.e. the existence of uncountable sets) in deriving certain natural theorems about objects of Analysis [Fri81], or even combinatorial statements about natural numbers. For Intuitionism, however, the important issues are the strength and conceptual coherence of Intuitionistic Mathematics itself. The restrictive aspects of intuitionistic formalisms are rather self-evident.

1.2. The role of formalization.

The formalization of a level M of Mathematics is an essential precondition to assessing its strength: A comparison between principles can be carried out only in a framework where the notion of provability is given a precise formal meaning, and where the abstract principles whose strength is being investigated are formally coded by certain axioms and axiom schemas. Choosing the appropriate formal systems and axioms is, however, not always a trivial or obvious task.

Take again as an example the strength of $P \equiv$ "*the uncountable exists*". One natural formalism for mathematics that avoids P is ZFC minus the Power-set Axiom [Fri81], a theory consistent with the formula asserting "*all sets are countable*". An alternative is to avoid not the presence of uncountable sets, but essential uses of their uncountability[1]. [Fri81, Appendix].

The formalization of Intuitionism is particularly interesting in the framework of the Calibration Programme, as the formalization process has moved here dialectically between ontological and formalistic concerns. The inception of Intuitionism was tinted by Brouwer's wholesale opposition to formal reasoning. The formalization of Intuitionism was subsequently accepted by Brouwer as an *a posteriori* process. It then turned out, however, that the study of intuitionistic formal theories led to a better understanding of the constructive nature of the subject, its connections with classical mathematics and with classical recursive mathematics, and the status of various axioms and rules therein. This success motivated Friedman and Myhill to extend the use of intuitionistic logic into ZF set theory, whose underlying ontology is completely alien to Brouwerian Intuitionism. In the next few paragraphs we briefly expand these historical comments. For an introduction to Intuitionism and to intuitionistic formal systems see for example [Tro77,vDa73,DT8?,Hey66]. An outline of formalisms for intuitionistic logic and arithmetic is given in §2 below.

1.3. Brouwerian Intuitionism and its formalization.

Probably the most fundamental aspect of Brouwer's thought was the acceptance of *time*, understood as a discrete progression of individual moments, as the only *a priori* notion[2]. Brouwer's insistence on this unique *a priori* governed both his ontological and epistemological choices. Its main consequence for mathematical ontology is an acceptance of infinite sets only as constructions in progress, not as completed totalities. The consequence for epistemology is even more radical: the notion of "*truth*" may be applied only to propositions that can be perceived and decided by the human mind within a discrete progression of time, which naturally must be *finite*. There results a fundamental gap between mathematical objects, which can be infinite, and mathematical perception, which is always finite. It follows then that the Aristotelian notion of "*truth*" can not be applied in Mathematics, in particular, the rule of *excluded third*, $\varphi \vee \neg\varphi$ is not valid for a proposition φ that refers to infinite totalities, unless one of φ and $\neg\varphi$ is perceived as true [Bro23].

The rejection of Aristotelian logic was so audacious and spectacular, that Brouwer and others associated with it a complete expulsion of logic from mathematics[3]. Principal components of formalization, in the common, Fregean, perception, were initially an anathema to Brouwer, such as the primality of formal language, the precedence of Logic to Mathematics, and the axiomatic method. Summing up a lifelong position that had already mellowed considerably [vSt79,82], Brouwer declared in 1951 the *First Act of Intuitionism* to be "Completely separating mathematics from mathematical language, and hence from the phenomena of language described by theoretical logic, recognizing that intuitionistic mathematics is an essentially languageless activity of the mind" [Bro81]. Brouwer also stated that reasoning by logical rules may lead "from scientifically accepted premisses to inadmissible conclusions" [Bro08].

However, of this wholesale repudiation of mathematical practice the necessary part was merely the rejection of *particular* mathematical and logical principles, primarily the *excluded third*. Brouwer himself showed acceptence of formal language by inventing terms to denote new kinds of mathematical constructions. And although he initially admitted logical principles "not to guide arguments... but to describe regularities" [Bro08], Brouwer proceeded to accept as valid virtually all propositional principles that do not imply the *excluded third* [Bro81, p.9; Hey30]. The formalization in earnest of Intuitionistic Mathematics *and Logic* was in fact begun by Kolmogorov and Heyting in the mid 20's, with Brouwer's blessing [Kol25, Hey30]. "The difference", wrote Heyting, "is that intuitionism proceeds independently of the formalization, which can but follow after the mathematical construction" [Hey66, p.5]. The same thing could be said, however, of most all mathematical research. Brouwer and Heyting had simply set out to delineate the very notions and principles proper to their novel way of thought, and were rightfully cautious not to restrain their thinking by premature formalization[4].

1.4. *The metamathematics of Intuitionism.*

One of Brouwer's most original ideas was the self-inspection of mathematical thinking [Bro25], and he took pride in predating the inception of similar ideas in Hilbert's metamathematics. However, a principal component of the latter, which was alien to Brouwer's intentions, is the need for a fixed formalization of a given body of mathematics, to which one applies metamathematical reasoning "from the exterior". It is therefore all the more interesting that Hilbert's approach to metamathematics has had such a glowing success when applied to intuitionistic formal systems.

Friedman has made contributions in four areas in which the metamathematics of Intuitionism has been particularly successful. A first area is the reduction of classical theories to their intuitionistic analogues. One form of such a reduction is the *translation* of classical theories into their intuitionistic counterparts. Friedman has obtained such translations for set theories[5], with, as a consequence, the consistency of classical ZF relative to intuitionistic ZF [Fri73a]. Another type of reduction asserts that, for certain kinds of formulas, provability in a given classical theory implies provability already in the corresponding intuitionistic theory. Friedman has proved results of this kind for formulas that assert the totality of partial recursive functions.

A second research direction addresses those properties of intuitionistic theories which reveal their constructive character, thereby linking Intuitionism to Recursion Theory. Important among such properties are instantiation properties, constructive interpretations, and consistency with Church's Thesis.

Yet another area consists of the relations between intuitionistic logic and intuitionistic mathematics. An example of such a relation is the proof-theoretic maximality of intuitionistic propositional logic, which was proved independently by D.H.J. de Jongh [dJo70b,73] and Friedman [Fri73b]. Related to this is a theorem of Friedman asserting the existence of universal test-case for the provability of transfinite induction over a given recursive ordering.

Finally, we mention the characterization of intuitionistic logic. Most important is the characterization by completeness relative to an adequate notion of semantics. Semantic characterizations of this kind have linked Intuitionism to classical notions of non-standard semantics and to Category Theory. Within an intuitionistic metamathematics completeness with respect to set-theoretic (=Tarskian) semantics also makes sense, and Friedman has refined and expanded previous results in this area. Another type of characterization is obtained by showing that extensions of intuitionistic logic fail to have some syntactic constructive properties of intuitionistic logic. Friedman obtained such a characterization for Intuitionistic Propositional Logic (independently of a similar discovery by D.H.J. de Jongh).

2. Formalisms for intuitionistic logic and arithmetic.

Of the many styles of logical calculi Gentzen's natural deduction is particularly suitable for the formalization of intuitionistic logic: it most closely parallels the constructive explication of logical constants, and it agrees with the separation of logic from mathematics advocated by Brower. The canonical text on natural deduction is [Pra65]. The following is a brief summary.

A *line* in a natural deduction proof is not a formula, but a statement that a formula can be derived from certain assumptions[6]. Write Σ / φ for "φ is derived from the finite set of formulas Σ". *Initial* lines of a natural deduction proof are of the form $\{\varphi\} / \varphi$. Lines that are not initial must be derived from previous lines by one of the *inference rules*, which generally consist of an *introduction rule* and an *elimination rule* for each one of the logical constants. Use Σ_{12} to abbreviate $\Sigma_1 \cup \Sigma_2$, and Σ, φ to stand for either $\Sigma \cup \{\varphi\}$ or Σ.

Σ_1/φ_1 and $\Sigma_2/\varphi_2 \Rightarrow \Sigma_{12}/\varphi_1\wedge\varphi_2$. $\qquad \Sigma/\varphi_1\wedge\varphi_2 \Rightarrow \Sigma/\varphi_i \quad (i=1,2)$

$\Sigma,\varphi/\psi \Rightarrow \Sigma/\varphi\rightarrow\psi \qquad \Sigma_1/\varphi\rightarrow\psi$ and $\Sigma_2/\varphi \Rightarrow \Sigma_{12}/\psi$

$\Sigma/\varphi \Rightarrow \Sigma/\forall x\varphi$
(x not free in Σ) $\qquad \Sigma/\forall x\varphi \Rightarrow \Sigma/\varphi[t/x]$
(t free for x in φ)

$\Sigma/\varphi_i \Rightarrow \Sigma/\varphi_1\vee\varphi_2 \quad (i=1,2) \qquad \Sigma_0/\varphi_1\vee\varphi_2$, $\Sigma_1,\varphi_1/\psi$ and $\Sigma_2,\varphi_2/\psi \Rightarrow \Sigma_{012}/\psi$

$\Sigma/\varphi[t/x] \Rightarrow \Sigma/\exists x\varphi$
(t free for x in φ) $\qquad \Sigma_0/\exists x\varphi$ and $\Sigma_1,\varphi/\psi \Rightarrow \Sigma_{01}/\psi$
(x not free in Σ_{01},ψ)

A useful convention is to regard the negation $\neg\varphi$ of a formula φ as an abbreviation for $\varphi\rightarrow\bot$, where $\bot$ is a logical constant denoting "false" or "absurd". In *minimal logic* (denoted $\mathbf{ML}_1$ or M) there is no special rule for $\bot$, and the rules for negation are special cases of the rules for implication. In *intuitionistic logic* ($\mathbf{IL}_1$ or I) there is, in addition, a rule $\bot_I$ that permits to infer Σ/φ from $\Sigma/\bot$ for any formula φ. *Classical logic* ($\mathbf{CL}_1$ or C) is obtained by generalizing that to a rule $\bot_C$, that permits to infer Σ/φ from $\Sigma,\neg\varphi/\bot$.

The terms t permitted in instances of $\forall$-introduction and $\exists$-elimination depend on the formalism. If function letters are used then terms are defined inductively as usual. In many-sorted formalisms the sort of the term must agrees with the sort of the bound variable it replaces. Second-Order Logic (with full comprehension) is obtained simply by allowing second order abstraction terms, of the form $\lambda\bar{x}.\varphi$, where φ is any formula, and $(\lambda\bar{x}.\psi)(\bar{t})$ is read as an abbreviation for $\psi[\bar{t}/\bar{x}]$.

Intuitionistic first order arithmetic is usually referred to as *Heyting's Arithmetic* (HA), and is defined exactly like Peano Arithmetic, except that the underlying logic is the First-order Intuitionistic Logic. Second-Order Heyting Arithmetic $\mathbf{HA}^2$, or HAS, is similarly based on (full) Intuitionistic Second Order Logic. It is useful to formulate HA and $\mathbf{HA}_2$ with an Induction Rule, rather than an axiom schema,

$$\Sigma_0/\varphi[0] \text{ and } \Sigma_1/\varphi[n]\rightarrow\varphi[Sn] \ \Rightarrow\ \Sigma_{01}/\varphi[t]$$
(x not free in Σ_{01})

The notion of a *normal proof* is one that renders transparent a good number of results about intuitionistic formal system. A proof is *normal* if the first premise of an instance of an elimination rule is never derived by one of the introduction rules or one of the elimination rules for $\vee$, $\exists$. Any proof can be effectively converted, with last line unchanged, into a normal proof. For First-Order Logic this is due to Prawitz [Pra65], and for Second-Order Logic to Girard [Gir71,72,Pra72,Mar72,Tai75,FLO83]. (A less effective variant, stating only that every provable formula has a normal proof, was proved for second-order logic in [Tai66,Tak67,Pra70]).

The notion of a normal proof is extended to the formalism described above for arithmetic. One requires that the rule of induction be used only when it cannot be eliminated by simple syntactic conversions. In particular, the eigen-term t should not be closed. The normalization proofs mentioned above can be applied to first and second order arithmetic, in spite of the additional conditions on normality [Pra71,Mar71,Tro73].

3. Reduction of classical proofs to intuitionistic proofs.

3.1. *Translations of classical into intuitionistic theories.*

The translation of classical first order logic and arithmetic into their intuitionistic counterparts to intuitionistic logic is due to Kolmogorov and Godel [Kol25, God32].[7]. Godel's translation is as follows. If φ is a first order formula, let φ^- be defined inductively by: $\alpha^- \equiv \neg\neg\alpha$ for atomic α ($\alpha^- \equiv \alpha$ in first order arithmetic); $(\varphi o \psi)^- \equiv (\varphi^- o \psi^-)$ for $o \equiv \rightarrow, \wedge$; $(\varphi \vee \psi)^- \equiv \neg(\neg\varphi^- \wedge \neg\psi^-)$; $(\forall x.\varphi)^- \equiv \forall x(\varphi^-)$; $(\exists x.\varphi)^- \equiv \neg\forall x \neg(\varphi^-)$. Then $\vdash_C \varphi$ iff $\vdash_I \varphi^-$, where C is Classical First-Order Logic (or Arithmetic) and I is Intuitionistic First-Order Logic (Arithmetic, respectively).

This simple translation shows that narrowing the logic underlying a mathematical theory from classical to intuitionistic generates in fact a richer theory with finer distinctions, with no loss of proof theoretic power. It also establishes the consistency of classical theories relative to their intuitionistic counterparts. This is a significant reduction, because the ontological and epistemological principles underlying Intuitionistic (Heyting's) Arithmetic are definitely more restrictive than the ones underlying Classical (Peano's) Arithmetic.

If Γ is any set of first-order axioms, then $\Gamma \vdash_C \varphi$ implies $\Gamma^- \vdash_I \varphi^-$ by Godel's translation for First-Order Logic. This is not quite as satisfactory a reduction as the one above for arithmetic, because Γ^- does not convey the same "existential content" as Γ, when the context is intuitionistic (consider for example AC^- where AC is the Axiom of Choice). A genuine reduction of classical Γ to intuitionistic Γ is obtained using a classical variant $\hat{\Gamma}$ of Γ (conveying the underlying informal ideas in a sufficiently direct manner), for which one shows that $\hat{\Gamma}^- \vdash_I \gamma$ for each $\gamma \in \hat{\Gamma}$. This establishes the desired reduction: $\hat{\Gamma} \vdash_C \varphi$ implies $\hat{\Gamma} \vdash_I \varphi^-$ for all formulas φ. This is the kind of reduction obtained by Friedman for set theory based on intuitionistic logic [Sce85]. In particular, it follows then that classical ZF is consistent relative to intuitionistic ZF.

3.2. *Provably recursive functions.*

Reducing the classical provability of a formula φ to the intuitionistic provability of φ^- is not always a satisfactory conservation result. A case in hand are formulas φ that are Π_2 in the arithmetical hierarchy, or, equivalently, formulas that express the convergence of a recursive function for all arguments. Consider a formula $\varphi \equiv \forall x\, \exists y \alpha$, where α is a primitive recursive formula expressing that "*y codes a complete computation of algorithm* A *on input* x". The Godel translation of φ is $\forall x \neg \forall y \neg \alpha$, conveying the convergence of algorithm A for all input only in a weak sense.

The missing link is provided by *Markov's Rule*: if $\neg \forall \neg x.\alpha$ (or, equivalently, $\neg\neg \exists x \alpha$) is a theorem, where x is a numeric variable and α is a primitive recursive relation, then $\exists x.\alpha$ is a theorem. Kreisel [Kre58a] showed that Markov's Rule holds for Intuitionistic (Heyting's) Arithmetic. Composing this with Godel's translation from classical to intuitionistic arithmetic, it follows that the functions provably recursive in Classical Arithmetic are provably recursive already in Intuitionistic Arithmetic. This conservation result is important in that it guarantees that no information about the convergence of recursive functions is lost when proofs are restricted to constructive logic, thus removing a potential objection to the use of constructive logic in reasoning about programs (see [CO78] for example). Conversely, no objection can be raised by intuitionists to proofs of Π_2 formulas that use classical reasoning, because such proofs can be converted to constructive proofs (this has been exploited extensively; see [Smo82]).

In [Fri78] a purely syntactic proof of Markov's rule is given for any formal theory that satisfy a simple syntactic criterion. (A similar translation was discovered independently by A.G. Dragalin [Dra79].) Proofs of Markov's Rule for intuitionistic theories had required, until Friedman's proof, a relatively sophisticated mathematical apparatus. The chief method used Godel's "Dialectica" interpretation (see [Tro73b,§3]). Other proofs used normalization of proofs, provable reflection for subsystem [Gir73], and Kripke models [Smo73]. Moreover, to adapt these proofs to new theories had required that the underlying metamathematical techniques be adapted first, usually a non-trivial step.

Friedman's proof has often been viewed as a hard to motivate trick. In fact it is an immediate consequence of a trivial translation from intuitionistic to minimal logic. For generalizations, details, and an alternative syntactic method, see [Lei85].

Given formulas ψ and ψ, let φ^ψ result from simultaneously replacing in φ each occurrence of $\bot$ by ψ. In general, if $\varphi \Rightarrow \varphi^\#$ is a transformation of formulas, Γ a set of formulas, then $\Gamma^\#$ will denote $\{ \varphi^\# \mid \varphi \in \Gamma \}$.

Lemma 3.1. Assume that no variable free in ψ is bound in Γ or φ. If $\Gamma \vdash_M \varphi$ then $\Gamma^\psi \vdash_M \varphi^\psi$.

Proof. Induction on the proof of $\Gamma \vdash_M \varphi$ in M. ■

Let φ be a formula. Let φ^F arise by replacing every non-$\bot$ atomic subformula α of φ by $\alpha \vee \bot$.

Lemma 3.2. (F is a translation from I into M). If $\Gamma \vdash_I \varphi$ then $\Gamma^F \vdash_M \varphi^F$.

Proof. By induction on the length of the proof in I for $\Gamma \vdash \varphi$. The only non-trivial case is when the last inference rule of the proof is the absurdity rule $\bot_I$, deriving φ from $\bot$. We assume that $\Gamma^F \vdash_M \bot$, and we have to show that $\Gamma^F \vdash_M \varphi^F$. Prawitz [Pra65] observed that, without loss of generality, $\bot_I$ may be used only when the conclusion, φ in our case, is atomic. But, for φ atomic, $\bot \vdash_M \varphi^F$ trivially.■

A theory Γ is *closed under* F if $\Gamma\vdash_M\gamma^F$ for each axiom γ of Γ. It can be easily shown that Peano's axioms, as well as several formulations of set theory, are closed under F [Fri78]. There are simple syntactical conditions on the axioms and schemas of a theory Γ which guarantee closure under F [Fri78,Lei85].

Theorem 3.3. Suppose that Γ is a theory closed under F, and $\Gamma\vdash_I\neg\neg\varphi$, where $\varphi\equiv\exists x\alpha$, with α atomic.
(i) If $\perp$ does not occur in Γ, then $\Gamma\vdash_M\varphi$.
(ii) If $\perp$ occurs in Γ only positively, then $\Gamma\vdash_I\varphi$.

Proof. Suppose $\Gamma\vdash_I\neg\neg\varphi$. Then by 3.2 $\Gamma^F\vdash_M(\neg\neg\varphi)^F$. But Γ is closed under F, and $(\neg\neg\exists x\alpha)^F\vdash_M\neg\neg\exists x\alpha$. Hence $\Gamma\vdash_M\neg\neg\varphi$. By 3.1 this implies $\Gamma^\varphi\vdash_M(\neg\neg\varphi)^\varphi$. Now, $(\neg\neg\varphi)^\varphi\equiv(\varphi\to\varphi)\to\varphi$, and if $\perp$ does not occur in Γ, then $\Gamma^\varphi\equiv\Gamma$, so $\Gamma\vdash_M\varphi$, concluding (i). If $\perp$ occurs only positively in γ then $\gamma\vdash_I\gamma^\varphi$ for any φ, and (ii) follows. ■

Corollary 3.4. The classical and intuitionistic variants of the following theories have exactly the same provably recursive functions: First-Order Arithmetic, Second-order Arithmetic, Type Theory, ZF Set Theory (suitably axiomatized).

Proof. For Arithmetic in any finite type, as well as type theory, this follows from theorem 3.3 since the set of axioms in question constitute a theory closed under F. For ZF one uses a version of ZF, which is closed under F [Fri78], and although formally weaker than ZF is just as strong for deriving arithmetic sentences [Fri78].■

4. Constructive properties of intuitionistic formal theories.

The simplest constructive properties of intuitionistic theories are the *instantiation properties*, such as numeric existential instantiation: if $\exists x\varphi$ is a theorem, then $\varphi[\bar{n}/x]$ is a theorem for some numeral $\bar{n}$. More general are *constructive interpretations*, such as the various realizability interpretations and Godel's "Dialectica" translation. Constructive interpretations are one important method for establishing the *consistency with Church's Thesis* of an intuitionistic formal system.

Excluding his work on intuitionistic set theories, discussed by Scedrov, Friedman's main contributions in this area have been on instantiation properties.

Under the constructive interpretation of the logical constants a formula $\varphi\vee\psi$ is recognized as *true* iff either φ or ψ is recognized as true. This excludes the validity of the *excluded third* schema $p\vee\neg p$, which would require a uniform method for establishing the truth of all formulas $\varphi\vee\neg\varphi$. From this interpretation it follows that an adequate formalism I for intuitionistic logic must satisfy the *disjunction instantiation* property (DI): if $\varphi\vee\psi$ is a theorem, then either φ or ψ must be a theorem: For the premise implies that either φ or ψ is constructively true, and assuming completeness for constructively true schemas that formula must be provable. That the usual formalization of intuitionistic logic indeed satisfies this property has been known since [Gen35]. This is easy to see from the formalism of §2: if the last line of a normal proof is $\Rightarrow\varphi\vee\psi$, then the penultimate line must be $\Rightarrow\varphi$ or $\Rightarrow\psi$.

Similarly, formalisms for intuitionistic logic satisfy the *existential instantiation* property (EI): if $\exists x\varphi$ is a theorem, where x ranges over a given sort s, then there is a term t of sort s such that $\varphi[t/x]$ is a theorem.

The instantiation properties hold also for intuitionistic mathematical theories, such as Heyting's Arithmetic. However, because predicate letters are absent in first order theories, they do not capture entirely the constructive nature of these theories: for example, the set of all classically true sentences of arithmetic also satisfies the instantiation properties. Thus, in the absence of (free) predicate letters these properties do not convey as much of the constructive character of the formalism as they do for first order intuitionistic logic.

4.1. Characterization of constructive logic by instantiation properties.

Lukasiewicz [Luk52] conjectured that the disjunction instantiation property DI characterizes Intuitionistic Propositional Logic I_0 from above: no proper extension of I_0 which is closed under substitution obeys the property. This was refuted by Kreisel and Putnam [KP57], turning the issue of characterizing I_0 by syntactically coded constructive properties into a non-trivial problem.

A natural characterization of this kind, conjectured by Kleene [Kle62], was established by D.H.J de Jongh in 1968 [deJ70a]. An alternative characterization is given in [Fri73b], with an interesting direct syntactic proof.

A characterization of Intuitionistic First-Order Logic $\mathbf{IL}_1$ of a similar syntactic nature has not yet been found.

4.2. Proof theoretic calibration of disjunction instantiation and numeric existential instantiation.

All proofs of the disjunction instantiation property for a formalism T use methods that cannot be formalized in T. Myhill proved [Myh73] that this is no accident: no consistent r.e. extension T of **HA** proves DI for itself. In fact, it is easy to exhibit simple counterexamples. Assume T is a consistent r.e. extension of **HA** that satisfies DI. Let $Prf[z,a]$ be a primitive recursive relation asserting that z is a code of a proof in T of the formula coded by a. (We use a subscript T when in danger of confusion.) If α is a formula with code z, let $\overline{\alpha}\equiv\overline{z}$ = the z'th numeral. Set $Pr(\varphi)\equiv\exists zPrf(z,\overline{\varphi})$.

Theorem 4.2.1.[Fri75][7]. Let T be an r.e. extension of **HA**. There are Σ^0_1 sentences ρ, σ, such that if

(*) $Pr(\rho\vee\sigma)\ \rightarrow.\ Pr(\rho)\vee Pr(\sigma)$ is provable in T, then T proves its own inconsistency

Proof. [Fri76] (attributed to Leivant). Let ρ be equivalent in T to "*there exists a proof of* $\neg\rho$ in T *with no shorter proof of* ρ". Let $\sigma\equiv$ "*there is a proof of* ρ *with no shorter proof of* $\neg\rho$". Then $Pr(Pr(0=1))\rightarrow Pr(\rho\vee\sigma)$. As in Rosser's incompleteness argument, $Pr(\rho)\rightarrow Pr(0=1)$, and similarly

$Pr(\sigma) \rightarrow Pr(0=1)$. These implications are all provable in **HA**. Thus, assuming (*) is provable in T,

$$Pr(Pr(0=1)) \rightarrow Pr(0=1)).$$

is a theorem of T. By Lob's Theorem [Lob55,Smo77] it follows that $Pr(0=1)$ is a theorem.■

The numeric EI property is a kind of infinitary DI, which it trivially implies, since a disjunction $\varphi \vee \psi$ is equivalent to $\exists x(x=0 \rightarrow \varphi. \wedge .x \neq 0 \rightarrow \psi)$ by very little arithmetic. Thus, no consistent r.e. extension of **HA** proves numeric EI for itself either.

An intersting question is whether numeric EI is strictly stronger than DI for a given theory. It is rather surprising that the answer is always negative:

Theorem 4.2.2 [Fri74, Fri75]. Every r.e. extension T of **HA** that obeys the disjunction instantiation property also obeys numeric existential instantiation.

Proof. ([Lei74], reproduced in [Fri75]). Let $\varphi \equiv \varphi[x]$ be a formula with free variable x. By the standard diagonalization theorem [KL68,Smo77], there is a formula α provably equivalent to $\exists x\beta$, where

$$\beta \equiv \beta[x] \equiv Prf(x, \overline{\neg\alpha}) \vee \varphi. \wedge .\forall y < x \neg Prf(y, \overline{\alpha}).$$

Lemma 1. (1) If $\vdash_T \alpha$ then $\vdash_T \varphi[\overline{z}]$ for some z.

(2) If $\vdash_T \neg\alpha$ then $\vdash_T 0=1$.

(3) $\varphi[x] \rightarrow \alpha \vee Pr(\alpha)$, provably in **HA**.

(4) If φ is decidable then $\varphi[x] \rightarrow \alpha \vee \neg\alpha$, provably in **HA**.

Proof. (1) Assume $\vdash_T \exists x\beta$, by a proof with code n say. By the definition of β we have $Prf(n, \overline{\alpha}) \wedge \beta[z] \rightarrow z < n$. So $\vdash_T \vee_{z<n} \beta[\overline{z}]$. By DI $\vdash_T \beta[\overline{z}]$ for some z, so $\vdash_T Prf(\overline{z}, \overline{\neg\alpha}) \vee \varphi[\overline{z}]$. If the first disjunct is true then $\vdash_T 0=1$, and if it is false then $\vdash_T \varphi[\overline{z}]$.

(2) Assume $\vdash_T \neg\alpha$, by a proof with code n say. If $\forall y < n \neg Prf(y, \overline{\alpha})$ then $\vdash_{\mathrm{HA}} Prf(\overline{n}, \overline{\neg\alpha}). \wedge . \forall y < \overline{n} \neg Prf(y, \overline{\alpha})$. So $\vdash_T \alpha$. Else, then again $\vdash_T \alpha$ (by a proof coded by some $z < n$).

(3) Assume φ. If $\forall y < x \neg Prf(y, \overline{\alpha})$ then $\beta[x]$, so α. Else $\vdash_T \alpha$ (by a proof coded by some $y < x$).

(4) Consider the "else" case in (3). If $Prf(y, \overline{\alpha})$ for some $y < x$ then $\beta[z]$ can be true only with $z < y$. But if φ is decidable then so is β, and α can be decided by checking $\beta[\overline{z}]$ for $z < y$. ■ **lemma 1.**

Lemma 2. If φ is decidable in T and $\vdash_T \exists x\varphi$ then $\vdash_T \varphi[\overline{n}]$ for some n.

Proof. Assume $\vdash_T \exists x\varphi$. Then $\vdash_T \alpha \vee \neg\alpha$ by (4), and so $\vdash_T \alpha$ or $\vdash_T \neg\alpha$ by DI. This implies $\vdash_T \varphi[\overline{n}]$ for some n by (1) and (2).■ **lemma 2.**

Proof of theorem 4.2.2 concluded. Assume $\vdash_T \exists x\varphi$. Then $\vdash_T \alpha \vee Pr(\alpha)$ by (3), and so $\vdash_T \alpha$ or $\vdash_T Pr(\alpha)$ by DI. The latter case implies $\vdash_T Prf(\overline{n}, \overline{\alpha})$ for some n, by lemma 2, and so $\vdash_T \alpha$ in either case. By (1), $\vdash_T \varphi[\overline{n}]$ for some n. ■

The requirement that T be r.e. is essential [Fri76]. Let φ be a false Σ_1^0 sentence independent over **HA**. Expand **HA**$+\varphi$ into a complete Δ_2^0 set T of formulas (as in Hilbert-Bernays' completeness proof [HB39,Kl52]). Then T satisfies DI, but fails to obey EI even for the Σ_1^0 formula φ.

Corollary 4.2.3. [Fri76]. The following are provably equivalent in **HA**.

(a) **HA** satisfies DI for disjunctions of Σ_1^0 sentences.

(b) **HA** satisfies numeric EI for Σ_1^0 sentences.

(c) **HA** is inconsistent or is Σ_1^0-sound.

(d) **HA** satisfies DI.

(e) **HA** satisfies numeric EI.

Proof. (e)→(d)→(a) is trivial. (a)→(b) is lemma 2 above. (b)←→(c) is straightforward. We prove (b)→(e), using **q**-realizability[8]. [Tro71] proves that if φ is a theorem of **HA**, then $\bar{n}\mathbf{q}\varphi$ is a theorem, for some n. Troelstra's proof, which proceeds by induction on the given proof of φ, is in fact formalizable in **HA** + (b): for example, if $\bar{n}\mathbf{q}(\alpha\rightarrow\beta)$ and $\bar{m}\mathbf{q}\alpha$ are theorems, then so is $\exists x(T(\bar{n},\bar{m},x) \wedge Ux\mathbf{q}\beta)$. By (b), then, $\overline{Ux}\mathbf{q}\beta$ is a theorem for some x.

Towards proving (e), assume now that $\exists x\varphi$ is provable. Given (b), it then follows that $\bar{n}\mathbf{q}(\exists x\varphi)$ is a theorem for some n. From the definition of **q**-realizability it follows that $\varphi[\bar{m}]$ is a theorem for some m.■

4.3. *Existential instantiation in Second-Order Arithmetic.*

Instantiation properties for Intuitionistic Second-Order Arithmetic $\mathbf{HA}_2$ fall out from the Normal Form Theorem for $\mathbf{HA}_2$ (cf. §2):

DI: Disjunction instantiation: if $\varphi\vee\psi$ is a theorem, with no numeric variable free, then one of the disjuncts is a theorem. (Note that here $\varphi\vee\psi$ may have free set variables.)

NEI: Numeric existential instantiation: if $\exists n\varphi$ is a theorem, with no numeric variable free, then $\varphi[\bar{n}]$ is a theorem for some n.

SEI: Set existential instantiation: if $\exists R\,\varphi[R]$ is a theorem, with no numeric variable free, then $\varphi[\lambda\bar{x}\psi]$ is a theorem for some formula ψ where R is not free.
(Recall from §2 that $\varphi[\lambda\bar{x}\psi]$ can be rewritten as $\exists R(\varphi \wedge \forall\bar{x}(R(\bar{x}) \leftrightarrow \psi))$.)

These proofs of the instantiation properties are presented in [Tro73a,73b]. Alternative proofs, using Kripke models, are in [JS74] (using a trick from [Fri73b]). [Fri73b] proves these properties by combining Kleene's slash method [Kle62] with ideas similar to those used in Girard's normalization proof [Gir71,72], and discovered independently by Friedman.

In [Fri76] Friedman tackles the proof-theoretic calibration of the instantiation properties listed above. The most remarkable result is the following, which one should contrast with Myhill's [Myh73] theorem on the underivability of numeric instantiation, mentioned above.

Theorem 4.3.1. [Fri76] The set existential instantiation property of $\mathbf{HA}_2$ is provable in $\mathbf{HA}_1$. ■

[Fri76] also extends the work described in §4.2 above to Second-Order Arithmetic. The main technique is a variant of **q**-realizability [Tro73b], along lines suggested in broad terms in [KT70]. Thus, we have:

Theorem 4.3.2. The following are provably equivalent in **HA**

(a) HA_2 satisfies DI for disjunctions of Σ_1^0 sentences.

(b) HA_2 satisfies NEI for Σ_1^0 sentences.

(c) HA_2 satisfies DI.

(d) HA_2 satisfies NEI.

Finally, some independence relations between variants of the instantiation properties are obtained in [Fri76], by building on the proof of theorem 4.2.1. above. In particular,

Theorem 4.3.3.

(a). There is a true arithmetic sentence φ such that $HA_2+\varphi$ does not obey SEI, even for sentences.

(b). There is a true arithmetic sentence φ such that $HA_2+\varphi$ obeys SEI, but not even DI for sentences. ■

4.4. Recursive instantiation of countable choice functions in Second Order Arithmetic.

The constructive character of a second order theory is sometimes apparent in a recursive instantiation of a function-existence theorem. A simple case is: if $\exists f\varphi$ is a theorem then there is a recursive function f such that $\varphi[\mathrm{f}/f]$ is true (or even provable). This is indeed the case, for example, for second order intuitionistic arithmetic HA_2. Here we show, following [Fri77d], that a similar property holds for the schema of countable choice:

$$\forall x\,\exists y\varphi[x,y]\rightarrow\ \exists f\forall x\varphi[x,f(x)].$$

Lemma 4.4.1. Let Γ be a set of first order arithmetic formulas. Suppose that for each (strictly positive) subformula $\exists x\varphi$ of a formula in Γ, if $\vdash_{HA^2}\exists x\varphi$, then $\vdash_{HA^2}\varphi[\bar{n}/x]$ for some numeral $\bar{n}$. Supposes that an analogous statement holds for disjunctive formulas. Then $HA^2+\Gamma$ has the DI property and the EI for numbers.

Proof. This is a corollary of the application made in [Fri73b] of Kleene's | method [Kle62]. ([Sca72,Lei73] prove the lemma directly, as an alternative to Kleene's |).■

Theorem 4.4.2.[Fri77d]. Let $\varphi[x,y]$ be a Π_1^0 formula such that $\forall x\,\exists y{<}g(x)\,\varphi[x,y]$, where g is a total recursive function. Suppose that

(1) $\vdash_{HA^2}\forall x\,\exists y{<}g(x)\,\varphi[x,y]\rightarrow(\exists f\forall x)(f(x){<}g(x)\wedge\varphi[x,f(x)])$.

Then $(\exists\text{recursive } f\forall x)(f(x){<}g(x)\wedge\varphi[x,f(x)])$.

Proof (in *classical* second order arithmetic). Assume the premises. There exist then functions p such that

(2) (2) $\forall x\ (p(x){<}g(x)\wedge\varphi[x,p(x)])$.

For each such function define the theory

$$T_p\equiv HA^2\cup\{\varphi[\bar{n},\overline{p(n)}]\}_{n\geq 0}\cup\{\forall x\,\exists y{<}g(x)\varphi[x,y]\}.$$

By 4.4.1 each such theory T_p satisfies DI. Fix now p to p_0. The premise of (1) is an axiom of T_p, so the conclusion is a theorem, and applying DI (for species) yields a formula $\psi[u,v]$, such that

(3) $\forall x\, \exists! y\, \psi[x,y] \wedge \forall x\, \exists y(\psi[x,y] \wedge \varphi[x,y])$

is a theorem of T_{p_0}. Let σ be an initial segment of p_0 such that (3) is a theorem of

$$\mathrm{T}_\sigma \equiv \mathrm{HA}^2 \cup \{\varphi[\overline{n},\overline{s(n)}]\}_{n \in Dom(s)} \cup \{\forall x\, \exists y{<}g(x)\varphi[x,y]\}.$$

Then (3) is a theorem of T_p for any extension p of σ satisfying (2).

Lemma. $\psi[\overline{n},\overline{m}]$ holds exactly when provable in each T_p where p extends σ and satisfies (2).

Proof. Assume $\psi[\overline{n},\overline{m}]$. Suppose that p extends σ and satisfies (2). By definition, $\varphi[\overline{n},\overline{p(n)}]$ is an axiom of T_p. Since p extends σ, (3) is a theorem of T_p, and so $\psi[\overline{n},\overline{p(n)}]$ is a theorem. To see now that $\overline{m} \equiv \overline{p(n)}$ observe that, on the one hand, $\overline{m}$ is the only numeral x such that $\psi[\overline{n},x]$, since (3) is true; and on the other hand $\psi[\overline{n},\overline{p(n)}]$ is true, since T_p is sound.

Conversely, if $\psi[\overline{n},\overline{m}]$ is a theorem of T_p, then it is true, since T_p is sound. ■**lemma.**

To conclude the proof of the theorem, let f be defined by ψ, i.e. $f(n)=m$ iff $\psi[n,m]$. By the lemma, f is uniformly recursive in every p extending σ and satisfying (2). Any function p is represented by an infinite branch $\{\langle p(1),...,p(n)\rangle\}_n$ in the universal spread, and for the functions p extending σ and satisfying (2) these are precisely the infinite branches of the tree

$$\{\vec{x} \equiv \langle x_0, \ldots, x_k\rangle \mid \vec{x} \text{ extends } \sigma \;\&\; \varphi[x_i,x_{i+1}] \;\&\; x_i{<}g(i)\}.$$

Let φ be $\forall z \varphi_0$, where φ_0 is recursive; then the infinite branches of the tree above are precisely the infinite branches of the recursive tree

$$T \equiv \{\vec{x} \equiv \langle x_0, \ldots, x_k\rangle \mid \vec{x} \text{ extends } \sigma \;\&\; \forall z{<}k\, \varphi_0[x_i,x_{i+1}] \;\&\; x_i{<}g(i)\}.$$

Thus, f is uniformly recursive, via some recursive function h, in every infinite branch of a recursive tree, which does have infinite branches (by (1)), and whose fan-outs are recursively bounded (by g). Given $n \geq 0$, compute $f(n)$ as follows. For $k=1,2,...$ perform k steps in th calculation of h for n and for all $\langle x_1, \ldots, x_k\rangle \in T$. When, for a sufficiently large k, all such calculation converge and return a uniform value, that value is $f(n)$. ■**Theorem.**

Theorem 4.4.3. The axiom schema of countable choice is independent of HA^2.

Proof. Choose a Π^0_1 formula $\varphi[x,y]$ and a recursive function g such that $\forall x\, \exists y{<}g(x)\, \varphi[x,y]$ is true, but not recursively true. Then, by theorem 4.4.2, (1) must fail. ■

5. Transfinite induction in intuitionistic arithmetic

This topic is covered in [Fri77], which we render here. An alternative treatment is developed in [FS84a,84b]. The main theorem is

Theorem 5.1. Let ρ be an arithmetic binary relation, for which transfinite induction is provable in **HA** for all formulas β. Then ρ defines a well-ordering of type $< \varepsilon_0$.

In [Fri77] this is stated to be false for Peano Arithmetic **PA**, in that for every recursive ordinal α there is a provably linear primitive recursive ordering of type α for which transfinite induction is not provable in **PA**.

The precise formulation of these statements uses the following definitions. Let $\rho[x,y]$ be a formula with free variables x,y, $\varphi[x]$ a formula with free variable x. *Transfinite induction over* ρ *for* φ is then expressed by the formula

$$TI_R[\varphi] \equiv Prog_\rho[\varphi] \rightarrow \forall x\varphi[x],$$

where

$$Prog_\rho[\varphi] \equiv \forall x(\forall y(\rho[x,y] \rightarrow \varphi[y]) \rightarrow \varphi[x]).$$

Also, let

$$TI_\rho \equiv \forall P\ TI_\rho[P].$$

A direct formulation of the well-foundedness of ρ is given, using second order quantification, by

$$WF_\rho \equiv \forall R[\exists xR(x) \rightarrow \exists x(R(x) \wedge \forall y(\rho[y,x] \rightarrow \neg R(y)))].$$

The well-foundedness of ρ under n is given by

$$WF_\rho[n] \equiv \forall R[\exists x(\rho[x,n] \wedge R(x)) \rightarrow \exists x((\rho[x,n] \wedge R(x)) \wedge \forall y(\rho[y,x] \rightarrow \neg R(y)))].$$

The relation between TI and WF is discussed in [Kre68§8].

The principal tool used by Friedman in analyzing the provability of transfinite induction and well-foundedness in intuitionistic theories is the combinatorial analysis of recursive infinitary proof figures built using the ω-rule. The technique was developed in [Lei75,79a], and is applied here in much the same way.

The formalism $\mathbf{HA}^\infty$, Heyting Arithmetic with the recursive ω-rule, is most easy to use when it is based on a sequential representation. The language is that of first-order arithmetic, with no variables used free. (For $\mathbf{HA}^\infty[P]$, where P is a predicate letter, use also P in formulas.) The negation of a formula φ is defined as $\varphi \rightarrow 0=\bar{1}$. Nodes in a proof are *sequents*, which are pairs $\Gamma \Rightarrow \varphi$, where Γ is a finite set of sentences (the antecedent), and φ is a sentence (the succedent).

The *initial sequents* are those sequents $\Gamma \Rightarrow \varphi$ where $\varphi \in \Gamma$, or φ is a true (closed) equation, or some $\gamma \in \Gamma$ is a false equation.

The inference rules fall into ones operating on antecedents (L-rules), and those operating on succedents (R-rules). As in § 2, we use Γ,φ to stand for either $\Gamma \cup \{\varphi\}$ or Γ.

$$\frac{\Gamma,\varphi_1,\varphi_2 \Rightarrow \psi}{\Gamma,\varphi_1 \wedge \varphi_2 \Rightarrow \psi} \qquad \frac{\Gamma \Rightarrow \varphi_1 \quad \Gamma \Rightarrow \varphi_2}{\Gamma \Rightarrow \varphi_1 \wedge \varphi_2}.$$

$$\frac{\Gamma \Rightarrow \varphi \quad \Gamma, \chi \Rightarrow \psi}{\Gamma, \varphi \rightarrow \chi \Rightarrow \psi} \qquad \frac{\Gamma, \varphi \Rightarrow \psi}{\Gamma \Rightarrow \varphi \rightarrow \psi}$$

$$\frac{\Gamma, \forall x \varphi \Rightarrow \psi}{\Gamma, \varphi[t/x] \Rightarrow \psi} \qquad \frac{\{\Gamma \Rightarrow \varphi[\bar{n}/x]\}_{n \geq 0}}{\Gamma \Rightarrow \forall x \varphi}$$

$$\frac{\Gamma, \varphi_1 \Rightarrow \psi \quad \Gamma, \varphi_2 \Rightarrow \psi}{\Gamma, \varphi_1 \vee \varphi_2 \Rightarrow \psi} \qquad \frac{\Gamma \Rightarrow \varphi_i}{\Gamma \Rightarrow \varphi_1 \vee \varphi_2}$$

$$\frac{\{\Gamma \Rightarrow \psi\}_{n \geq 0} \Rightarrow \psi}{\Gamma, \exists x \varphi \Rightarrow \psi} \qquad \frac{\Gamma \Rightarrow \varphi[t/x]}{\Gamma \Rightarrow \exists x \varphi}$$

The *proof figures* are recursive, well-founded (= every branch is finite) trees of sequents that relate according to the inference rules, and where the leaves are initial. The resulting system is denotes $\mathbf{HA}^{\infty}$. If predicate letter P is used in the language, denote the system $\mathbf{HA}^{\infty}[P]$.

Lemma 5.2. Let P be a unary predicate variable. If WF_ρ is true then $TI_R[P]$ is derived in $\mathbf{HA}^{\infty}[P]$.

Proof. [Lei79a]. ■

Let $\alpha[x,y]$ be a Σ_1^0 formula with x,y as the only free variables, let $\bar{\alpha}$ be the Π_1^0 form of $\neg\alpha$. We think of α as enumerating the r.e. sequence of r.e. sets $A_n \equiv \{m \mid \alpha(n,m)\}$.
Let $\beta[x] \equiv \forall y(\alpha[x,y] \vee \bar{\alpha}[x,y])$.

Lemma 5.3. Suppose that $TI_\rho[\beta]$ is derived in $\mathbf{HA}^{\infty}$, where $\neg WF_\rho$. Then there is an n such that A_n is recursive in $\{A_k \mid k \neq n\}$.

Proof. The assumption implies that $Prog_\rho[\beta] \rightarrow \beta[\bar{m}]$ is derived in $\mathbf{HA}^{\infty}$ for all $m \geq 0$. Choosing an m such that $\neg WF_\rho[m]$, we get the lemma's conclusion from the following.

Suppose $\neg WF_\rho[m]$, and let δ_m be one of the following formulas.

(1) $\beta[\bar{m}]$;

(2) $\rho[\bar{m},\bar{p}] \rightarrow \beta[\bar{m}]$, where the premise is true;

(3) $\forall x(\rho[x,\bar{m}] \rightarrow \beta[x])$;

Suppose that each formula in Γ has one of the following forms, where p is such that $WF_\rho[p]$; $\forall x(\rho[x,\bar{p}] \rightarrow \beta[x]) \rightarrow \beta[\bar{p}]$; $\beta[\bar{p}]$; true subsentences of $\beta[\bar{p}]$; true sentences $\rho[\bar{x},\bar{y}]$. If $\Gamma \Rightarrow \delta_m$ is derived in $\mathbf{HA}^{\infty}$, then there is some $n \geq 0$ such that A_n is recursive in $\{A_k \mid k \neq n\}$.

The proof is by induction on the (well-founded) proof-tree π deriving $\Gamma \Rightarrow \delta_m$ in $\mathbf{HA}^\infty$. For the base case observe that this sequent can not be initial: no atomic sentence in Γ is false, δ_m is not atomic, and δ_m in Γ we must have both $WF_\rho[m]$ and $\neg WF_\rho[m]$.

The induction step is by cases on the terminal rule of π. The only interesting case for a left-rule is $\rightarrow L$, where the conclusion is $\Gamma', \forall x(\rho[x,\bar{p}] \rightarrow \beta[x]) \rightarrow \beta[\bar{p}] \Rightarrow \delta_m$, and the premises are $\Gamma' \Rightarrow \forall x(\rho[x,\bar{p}] \rightarrow \beta[x])$ and $\Gamma', \beta[\bar{p}] \Rightarrow \delta_m$. If $WF_\rho[p]$ then apply induction assumption to the second premise, otherwise apply induction assumption to the first premise.

The cases for a right rule are determined by the form of δ_m. If case (3) applies for δ_m, with $\forall R$ as the rule, choose q such that both $\neg WF_\rho[q]$ and $\rho[q,m]$, and apply induction assumption to the q'th premise. If case (2) applies (with $\rightarrow R$), apply induction assumption to the single premise. The core of the proof is case (1), deriving $\Gamma \Rightarrow \forall x(\alpha(\bar{m},x) \vee \bar{\alpha}(\bar{m},x))$ from $\Gamma \Rightarrow \alpha(\bar{m},\bar{x}) \vee \bar{\alpha}(\bar{m},\bar{x}))$ $(x \geq 0)$. This case is settled by the following

Sublemma. Let Γ be as above. If $\Gamma \Rightarrow \alpha[\bar{m},\bar{x}] \vee \bar{\alpha}[\bar{m},\bar{x}]$ is derived in $\mathbf{HA}^\infty$, then $\alpha[m,x]$ can be decided from sentences $\alpha[k,z]$ with $k \neq m$.

Proof of sublemma. By induction on the proof in $\mathbf{HA}^\infty$. As above for δ_m, the given sequent cannot be initial. Consider cases for the terminal inference. If the terminal rule is $\vee R$, with $\Gamma \Rightarrow \alpha[\bar{m},\bar{x}]$ or $\Gamma \Rightarrow \bar{\alpha}[\bar{m},\bar{x}]$ as the premise, then the succedent of the penultimate sequent is true, since all formulas in the antecedent are true, and the sublemma is proved. If the rule is $\rightarrow L$, with $\forall x(\rho[x,\bar{p}] \rightarrow \beta[x]) \rightarrow \beta[\bar{p}]$ as the active formula, then argue as above for this case, applying the sublemma's induction assumption to the second premise if ρ is well-founded under p, and the lemma's induction assumption to the first premise otherwise. If the terminal rule is $\vee L$, with $\alpha[\bar{p},\bar{y}] \vee \bar{\alpha}[\bar{p},\bar{y}]$ as the active formula, and with premises $\Gamma', \alpha[\bar{p},\bar{y}] \Rightarrow \alpha[\bar{m},\bar{x}] \vee \bar{\alpha}[\bar{m},\bar{x}]$ and $\Gamma', \bar{\alpha}[\bar{p},\bar{y}] \Rightarrow \alpha[\bar{m},\bar{x}] \vee \bar{\alpha}[\bar{m},\bar{x}]$, then apply induction assumption to the first or second premise, according to whether $\alpha[p,y]$ is true or false. Other cases are handled trivially. **■Sublemma ■Lemma.**

Theorem 5.4. There is a formula $\beta[x]$ such that for all primitive recursive ρ, $TI_\rho[\beta]$ is derived in $\mathbf{HA}^\infty$ iff WF_ρ.

Proof [Fri77c]. By [Sac63] there exists a Σ_1^0 formula $\alpha[x,y]$ such that the sequence of r.e. sets $A_n \equiv \{x \mid \alpha[n,x]\}$ is strongly independent: no A_n is recursive in $\{A_k\}_{k \neq n}$. Let β be defined from α as above. If WF_ρ then we even have $\mathbf{HA}^\infty[P]$ deriving $TI_\rho[P]$, by lemma 5.2. If $\neg WF_\rho$ then $TI_\rho[\beta]$ cannot be derived in $\mathbf{HA}^\infty$, by lemma 5.3.■

Theorem 5.5 [Fri77c]. Let ρ be a formula defining a binary relation. There is a formula $\beta[x]$ such that $TI_\rho[\beta]$ is derived in $\mathbf{HA}^\infty$ iff WF_ρ.

Proof. Relativize the proof of theorem 5.4. If ρ is classically Π_n^0, then $\beta[x]$ will have the form $\forall y(\alpha[x,y] \vee \bar{\alpha}[x,y])$, where α is Σ_{n+1}^0 and $\bar{\alpha}$ is Π_{n+1}^0.■

Lemma 5.6. The proof of theorem 5.5. is formalizable in $\mathbf{PPA}^2$ = Second-Order Peano Arithmetic, with comprehension restricted to formulas free of second-order quantifiers.

Proof. Inspection of the proof.■

Theorem 5.7 [Fri77c]. Let T be one of the following theories.

- HA (more precisely, $\mathbf{PHA}^2$ = Second Order Heyting Arithmetic, with comprehension restricted to formulas free of second order quantifiers, a theory conservative over HA [Tro73b]);
- Second-order Intuitionistic Arithmetic $\mathbf{HA}^2$; or
- Intuitionistic Theory of Types $\mathbf{HA}^\omega$.

Let ρ be an arithmetic formula defining a binary relation. $TI_\rho[\beta]$ is derived in T for all arithmetic β iff WF_ρ is derived in T.

Proof. Given ρ as above, let β be as in theorem 5.5. By [Lei75b], every theorem φ of T is derived in in $\mathbf{HA}^\infty$, provably in T. By lemma 5.6. WF_ρ is then provable in the classical version $\mathbf{T}^C$ of T. This implies that WF_ρ is provable in T, by theorem 3.5.■

Proof of theorem 5.1. Assume the theorem's premise. Then $TI_\rho[\beta]$ is derived in $\mathbf{HA}^\infty$, provably in **HA** [Sch60], for any β. Let now β be obtained from theorem 5.5. Since $TI_\rho[\beta]$ is derived in $\mathbf{HA}^\infty$, provably in PA, it follows by 5.6 that WF_ρ provably in $\mathbf{PPA}^2$. By [Sch60] (see [Kre60, Kre68§8]), ρ then defines a well-ordering of type $< \varepsilon_0$. ■

6. Completeness of intuitionistic logic.

Notions of validity.

Consider the statement of completeness for a first-order logic **L**:

$$\text{Comp(L)} \qquad \models \varphi \Rightarrow \vdash_L \varphi,$$

Here $\models$ is defined as Tarskian validity, in all set-theoretic structures. That is, if $\overline{P} \equiv \langle P_1, \ldots, P_k \rangle$ is the list of all predicate letters in φ, and $\mathbf{M} \equiv \langle \mathbf{D}, \overline{\mathbf{P}} \rangle$ is a structure where $\mathbf{P}_i$ is a relation over $\mathbf{D}$ whose arity is the arity of P_i, then

$$\models_\mathbf{M} \varphi \;\; \equiv \;\; \varphi^\mathbf{M}[\overline{\mathbf{P}} / \overline{P}],$$

($\varphi^\mathbf{M}$ is φ with quantifiers restricted to **D**), and

$$\models \varphi \;\; \equiv \;\; \forall \mathbf{M} \, (\models_\mathbf{M} \varphi).$$

Clearly, Comp is false under a classical reading, even for minimal propositional formulas: take $\varphi \equiv (p \to q) \vee (q \to p)$.

The completeness of $\mathbf{IL}_1$ can be studied, in classical metamathematics, if the notion of validity is based on non-standard semantics that attempt to capture explicitly a constructive interpretation of the logical constants. Semantical notions of this sort include pseudo-boolean algebras, topological models, Beth models, Kripke models, and Lauchli's abstract realizability (see [Tro75,§2] for review and references). An alternative approach is the study of Comp in intuitionistic metamathematics, where the constructive interpretation of the logical constants is implicit in the intuitionistic reading of Val. Since Comp is refutable in (a weak subsystem of) classical second order logic, any intuitionistic proof thereof must use principles that are incompatible with classical logic. This can be done by parametrizing Tarskian (set theoretic) models with functions (choice sequences), and using continuity principles for these [Kre58c]. The statement of validity becomes then

$$\models \varphi[\bar{P}] \equiv \forall \varepsilon \forall D \forall \bar{Q} \varphi^D[\overline{Q^\varepsilon} / \bar{P}].$$

Here ε ranges over choice sequences (of natural numbers, or of 0's and 1's) that satisfy certain continuity conditions, and each P_i in $\bar{P}$ is replaced by Q_i^ε defined by $Q_i^\varepsilon(\bar{x}) \equiv Q_i(\varepsilon, \bar{x})$.

This twist on Tarskian semantics may be viewed as another way of rendering explicit the constructive interpretation of the connectives. For example, suppose $Val(\varphi \vee \psi)$, where the only predicate letter in the formula is the unary P). A suitable continuity principle implies that $\forall P\ \varphi[P^\varepsilon] \vee \psi[P^\varepsilon]$ for any ε extending some finite function-segment σ. A variant of Konig's Lemma implies then that there is a finite set of sequences σ_i, such that given any ε and P^ε, $\varphi[P^\varepsilon] \vee \psi[P^\varepsilon]$ depends only on an initial segment $\sigma_i \subset \varepsilon$. Thus the truth of $\varphi \vee \psi$ in a model depends only on a finite amount of information about the model. In other words, there is a general decision procedure that, given a model, returns either φ or ψ as true in the model.

Variants of completeness.

The schema Comp has a number of classically equivalent variants, which are not intuitionistically equivalent. The ones of interest here are the following, where $\sim$ is used for negation at the metamathematical level[9].

Weak Completeness is the contra-positive of Comp, stating that unprovable formulas are not valid:

WComp(L) $\quad \not\vdash_L \varphi \Rightarrow \not\models \varphi.$

or, equivalently,

$$\models \varphi \Rightarrow \sim\sim \vdash_L \varphi.$$

A form of "counter-example-completeness" often used in Model Theory,

$$\not\vdash \varphi \Rightarrow \exists M \models_M \neg\varphi,$$

can hold intuitionistically only for small fragments [Lei72,Tro75]: for example, $p \vee \neg p$ is not a theorem, but $\neg\neg(p \vee \neg p)$ is. A related form is *Satisfyability-Completeness*, stating that every consistent formula has a model:

SatComp(L) $\quad \varphi \not\vdash_L \bot \Rightarrow \exists M \models_M \varphi.$

Both SatComp and Comp imply WComp, but SatComp and Comp are incomparable, as is clear from theorems 6.1, 6.2 and 6.3 below.

Parallel to these variants of completeness we have variants of strong completeness:

$$\text{StComp(L)} \qquad \Gamma \models \varphi \Rightarrow \Gamma \vdash_L \varphi,$$

Where Γ is a set of formulas, which for the results below we assume to be r.e. Similarly, we have a strong variant of WComp,

$$\text{StWComp(L)} \qquad \Gamma \nvdash_L \varphi \Rightarrow \Gamma \not\models \varphi,$$

for r.e. sets Γ. Likewise, for satisfyability completeness,

$$\text{StSatComp (L)} \qquad \Gamma \nvdash_L \bot \Rightarrow \exists M \models_M \Gamma,$$

for r.e. Γ.

Completeness theorems.

The theorems below are from [Fri77a,77b,77c,84].

A first theorem states that Comp holds for *First-Order Minimal Logic*, ML_1, as defined in §2. The theory in which $Comp(ML_1)$ is proved must be classically inconsistent, since ML_1 is classically incomplete. That theory, HL, is defined as follows. The language has first order variables, intended to range over natural numbers, set variables and unary function variables. The constants are those of HA, plus a predicate L over unary functions, where $L(f)$ is intended to render "f is a lawless 0-1 function". We write $\forall f_L \cdots$ for $\forall f(L(f) \rightarrow \cdots)$, and $\exists f_L \cdots$ for $\exists f(L(f) \wedge \cdots)$.

The axioms of HL are as follows.

1. The axioms of HA, with induction for all formulas in the language;
2. Arithmetical Comprehension for sets:

$$\exists S \forall x (S(x) \leftrightarrow \varphi), \ (S \text{ not free in } \varphi);$$

3.

$$\forall f_L . f(n)=0 \vee f(n)=1;$$

4. Existence of lawless functions:

$$\forall \sigma \exists f_L \ \sigma \subset f.$$

Here σ ranges over functions from some domain $\{0,...,n\}$ to $\{0,1\}$, and $\subset$ denotes function extension.

5. Continuity for arithmetical properties: for arithmetical φ with f as the only non-number variable,

$$\forall f_L \varphi[f] \rightarrow \exists \sigma \subset f \forall g_L \supset \sigma \ \varphi[g];$$

6. Continuity for primitive-recursive spreads: for primitive recursive ρ over numbers,

$$\forall f_L \exists n\, \rho(\bar{f}(n)) \to \exists k \forall f_L \exists n{<}k\, \rho(\bar{f}(n)).$$

Theorem 6.1. Comp(ML_1) and even StComp(ML_1) are provable in **HL**. ■

To derive completeness results for IL_1, i.e. when negation may be present in formulas, a theory **HC** richer than **HL** is used. The language of **HC** is the same as the language of **HL**, with the additional unary predicate constant $C(f)$ over functions, intended to render "f is a lawlike function." The axioms and rules of **HC** are as follows.

1. The axioms of **HA**, with induction for all formulas in the language;

2.

$$\forall f_L . f(n)=0 \vee f(n)=1;$$

3. Existence of lawless functions:

$$\forall \sigma \exists f_L\, \sigma \subset f.$$

4. Continuity for lawless functions and arithmetical properties: for arithmetical φ where all parameters other than f are numbers and lawlike functions,

$$\forall f_L \varphi[f] \to \exists \sigma \subset f \forall g_L \supset \sigma\, \varphi[g];$$

5. Continuity for lawless functions and primitive-recursive spreads: for primitive recursive ρ over numbers and lawlike functions,

$$\forall f_L \exists n\, \rho(\bar{f}(n)) \to \exists k \forall f_L \exists n{<}k\, \rho(\bar{f}(n)).$$

6. Lawlike choice for arithmetic properties: for arithmetic φ and ψ, with only number and lawlike parameters,

$$\forall n(\varphi[n] \to \exists m \psi[n,m]) \to \exists f_C \forall n(\varphi[n] \to \psi[n,f(n)]).$$

In addition, we consider the following two schemas.

1. Markov's Principle: for primitive recursive ρ, with number parameters only

$$\mathbf{MP}_0\text{:} \quad \neg\forall n \rho \to \exists n \neg\rho.$$

2. For primitive recursive ρ_1 and ρ_2, with number parameters only, and with n not free in ρ_2 and m not free in ρ_1,

$$\mathbf{Q}\text{:} \quad \forall n \forall m(\rho_1[n] \vee \rho_2[m]) \to (\forall n \rho_1[n]) \vee (\forall m \rho_2[m]).$$

Theorem 6.2. The following are equivalent in **HC**.

(a) SatComp ($\mathbf{IL}_1$);

(b) StSatComp ($\mathbf{IL}_1$);

(c) SatComp ($\mathbf{CL}_1$);

(d) StSatComp ($\mathbf{CL}_1$);

(e) **Q**. ∎

Theorem 6.3. The following are equivalent in **HC**.

(a) WComp ($\mathbf{IL}_1$);

(b) StWComp ($\mathbf{IL}_1$);

(c) WComp ($\mathbf{CL}_1$);

(d) StWComp ($\mathbf{CL}_1$);

(e) $\neg\neg\mathbf{Q}$. ∎

Theorem 6.4. The following are equivalent in **HC**.

(a) Comp ($\mathbf{IL}_1$);

(b) StComp ($\mathbf{IL}_1$);

(c) Comp ($\mathbf{CL}_1$);

(d) StComp ($\mathbf{CL}_1$);

(e) $\neg\neg\mathbf{Q} + \mathbf{MP}_0$. ∎

Footnotes.

1. This is somewhat analogous to Brouwer's objection not to the presence of infinite sets, but to using them as completed totalities.

2. Time and space were the two *a priori* notions for Kant. Brouwer's choice was dual to Frege's, who accepted space as *a priori*, but not time.

3. A fact which seems to have been mostly unnoticed is that Aristotle and the Epicureans knew about the failure of excluded third for future contingent events [Luk30].

4. The kind of "informal rigor" involved in this type of enterprise is discussed in [Kre67].

5. Similar work was pursued independently by W. Powell.

6. Assumptions have to be tagged when combinatorial properties of proofs are studied, with assumptions possibly repeated with different tags [Pra65, How81, Lei79b]. These are not needed for a general understanding of the style.

7. In [Fri76] Friedman notes that the proof in [Fri75] actually establishes a result weaker than the theorem stated.

8. An alternative proof: The implication (c)→(e) is an immediate consequence of the provability in **HA** of the implication

   (*) $Pr(\varphi\vee\psi) \rightarrow Pr(Pr(\varphi)\vee Pr(\psi))$

   [Lei75, Vis82]. The same proof establishes that (a)-(e) for $\mathbf{HA}^2$ are equivalent, provably in **HA**. This is because (*) for $\mathbf{HA}^2$ is provable in **HA**.

9. The terminology for variants of completeness has not been uniform. Comp is referred to as "Strong Completeness" in [Kre62,Lei72], *Comp* in [Tro75], Completeness in [Fri77abc]. WComp is "Weak Completeness" in [Kre62,Lei72,Fri77abc], *Comp'* in [Tro75]. Counter-example completeness is *Comp"* in [Tro75]. SatComp is "satisfyability of the irrefutable" in [Lei72], "Strong Completeness" in [Fri77abc].

References.

[Bar77] John Barwise (editor), **Handbook of Mathematical logic** North-Holland (Studies in Logic #90), Amsterdam (1977) xi+1165 pp.

[Bro08] L.E.J. Brouwer, *De onbetrouwbaarheid der logische principes*; **Tijdschrift voor Wijsbeggerte 2** (1908) 152-158. English translation: *The unreliability of logical principles*; in [Hey75] 107-111.

[Bro23] L.E.J. Brouwer, *Over de rol van het principium tertii exclusi in de wiskunde, in het bijzonder in de functiontheorie*; **Wis- en natuurkundig tijdschift 2** (1923) 1-7. German translation: **J. reine angew. Math. 154** (1924) 1-7. [Hey75] 268-274. English translation: *On the significance of the principle of excluded third in mathematics, especially in function theory*; in [vHe] 334-341.

[Bro25] L.E.J. Brouwer, *Zur Begrundung der intuitionistischen Mathematik I*; **Mathematische Annalen 93** (1925) 244-257. [Hey75] 301-314.

[Bro81] L.E.J. Brouwer, **Brouwer's Cambridge Lectures on Intuitionism** (D. van Dalen, Editor), Cambridge University Press, Cambridge (1981) xii+109.

[CO78] Robert Constable and Michael O'Donnell, **A Programming Logic**, Winthrop, Cambridge (MA) (1978) x+389pp.

[vDa73] D. Van Dalen, *Lectures on Intuitionism*; in [MR73] 1-94.

[DT8?] D. van Dalen and A.S. Troelstra, **Introduction to Intuitionism**; in preparation.

[Dra79] A. G. Dragalin, *New kinds of realizability*; Abstracts, Sixth International Congress for Logic, Methodology, and Philosophy of Science (Hanover, 1979).

[Fen71] Jens E. Fenstad (editor), **Proceedings of the Second Scandinavian Logic Symposium**; North-Holland, Amsterdam, 1972, vii+405pp.

[FLO83] Steven Fortune, Daniel Leivant and Michael O'Donnell, *The expressiveness of simple and second order type systems*; **Journal of the ACM 30**, 1983, 151-185.

[Fri73a] Harvey Friedman, *The consistency of classical set theory relative to a set theory with intuitionistic logic*; **Journal of Symbolic Logic 38** (1973) 315-319.

[Fri73b] Harvey Friedman, *Some applications of Kleene's methods for intuitionistic systems*; in [MR73] 113-170.

[Fri74] Harvey Friedman, *The disjunction property implies the existence property*; memo, June 1974.

[Fri75] Harvey Friedman, *The disjunction property implies the numeric existence property*; **Proc. Nat. Acad. Sci. 72** (1975) 2877-2878.

[Fri76] Harvey Friedman, *On the derivability of instantiation properties*; memo, May 1976.

[Fri77a] Harvey Friedman, *Intuitionistic completeness of Heyting's predicate calculus*; **Notices of the Amer. Math. Soc.**

[Fri77b] Harvey Friedman, *The intuitionistic completeness of intuitionistic logic under Tarskian semantics*; memo, March 1977.

[Fri77c] Harvey Friedman, *New and old results on completeness of HPC*; memo, June 1977.

[Fri77d] Harvey Friedman, *Note on arithmetic choice in the intuitionisitc theory of species*; memo, July 1977.

[Fri77e] Harvey Friedman, *Transfinite induction in intuitionistic arithmetic*; memo, July 1977.

[Fri78] Harvey Friedman, *Classically and intuitionistically provably recursive functions*; in **Higher Set Theory** (G.H. Muller and Dana S. Scott, editors), Springer-Verlag (LNM #669), Berlin (1978) 21-28.

[Fri81] Harvey Friedman, *On the necessary use of abstract set theory;* **Advances in Mathematics 41** (1981) 209-280.

[Fri84] Harvey Friedman, *Intuitionistic completeness*; handwritten notes, 1984.

[FS84a] Harvey Friedman and Andrej Scedrov, *Intuitionistically provable recursive well-orderings;* manuscript, 1984.

[FS84b] Harvey Friedman and Andrej Scedrov, *Arithmetic transfinite induction and recursive well-orderings;* manuscript, 1984; to appear in **Advances in Mathematics.**

[Gen35] Gerhard Gentzen, *Untersuchungen uber das logische Schlissen;* **Mathematische Zeitschrift 39** (1935) 176-210. English translation: [Sza69] 68-131.

[Gir71] Jean-Yves Girard, *Une extension de l'interpretation de Godel a l'analyse, et son application a l'elimination des coupures dans l'analyse et la theorie des types;* in [Fen71] 63-92.

[Gir72] Jean-Yves Girard, *Interpretation fonctionelle et elimination des coupures dans l'arithmetique d'ordre superieur*, These de Doctorat d'Etat, 1972, Paris.

[Gir73] Jean-Yves Girard, *Cloture par rapport a Markov*, handwritten memo, 1973.

[God32] Kurt Godel, *Zur intuitionistischen Arithmetik und Zahlentheorie*; **Ergebinisse eines mathematischen Kolloquiums 4** (1932-1933) 34-38. English translation in M. Davis (editor), **The Undecidable**, New York (1965) 75-81.

[HB39] David Hilbert and P. Bernays, **Die Grundlagen der Mathematischen Wissenschaften II**, Springer, Berlin, 1939.

[vHe67] Jean van Heijenoort (editor), **From Frege to Godel**; Harvard university Press, Cambridge (1967) xi+660 pp.

[Hey30] Arend Heyting, *Die formalen Regeln der intuitionistischen Logik*; **Sitzungsber. preuss. Akad. Wiss. Berlin**, 1930, 42-56.

[Hey66] Arend Heyting, **Intuitionism, an introduction**, second edition, North-Holland, Amsterdam (1966) ix+137pp.

[Hey75] Arend Heyting (editor), **L.E.J. Brouwer Collected Works**, North-Holland, Amsterdam (1975) xv+628I .

[How80] William A. Howard, *The formulae-as-types notion of construction*, in [SH80] 479-490.

[dJo70a] D.H.J. de Jongh, *A characterization of the intuitionistic propositional calculus*; in [MKV70] 211-217.

[dJo70b] D.H.J. de Jongh, *The maximality of intuitionistic predicate calculus with respect to Hayting's arithmetic;* Abstract, **Journal of Symbolic Logic 35**, 1970, 606.

[dJo73] D.H.J. de Jongh, *The maximality of intuitionistic predicate logic with respect to Hayting's arithmetic;* Technical Report, University of Amsterdam, 1973.

[JS74] D.H.J. de Jongh and Craig Smorynski, *Kripke models and the theory of species;* Report 74-03, University of Amsterdam, 1974.

[KL68] Georg Kreisel and Azriel Levy, *Reflection principles and their use for establishing the complexity of axiomatic systems*; **Zeitsch. fur math Logik und Grundlagen der Mathematik 14** (1968) 1-50.

[Kle52] S.C. Kleene, **Introduction to Metamathematics**, Wolters-Noordhoff, Groningen (1952) x+550 pp.

[Kle62] S.C. Kleene, *Disjunction and existence under implication in elementary intuitionistic formalisms*; **Journal of Symbolic Logic 27** (1962) 11-18.

[Kol25] Alexander Kolmogorov; *Sur le principe de tertium non datur*; **Recueil Math. de la Soc. Math. de Moscou** 32 (1925) 647-667. English translation: *On the principle of excluded middle*; in [vHe67] 414-437.

[KP57] Georg Kreisel and Hilary Putnam, *Eine Unableitbarkeitbeweismethode fur den intuitionistischen Aussagenkalkul*; **Archiv fur Math. Logik u. Grundl. Forsch. 3** (1957) 74-78.

[Kre58a] Georg Kreisel, *Mathematical significance of consistency proofs*; **Journal of Symbolic Logic 23** (1958) 155-182.

[Kre58b] Georg Kreisel, *Elementary completeness properties of intuitionistic logic, with a note on negations of prenex formulas*; **Journal of Symbolic Logic 23** (1958) 317-330.

[Kre58c] Georg Kreisel, *A remark on free choice sequences and the topological completeness proofs*; **Journal of Symbolic Logic 23** (1958) 369-388.

[Kre60] Georg Kreisel, *Status of the first epsilon-number in first order arithmetic;* (abstract) **Journal of Symbolic Logic 25** (1960) 390.

[Kre62] Georg Kreisel, *Weak completeness of intuitionistic predicate logic*; **Journal of Symbolic Logic** 27 (1962) 139-158.

[Kre67] Georg Kreisel, *Informal rigor and completeness proofs*; in: **Problems in the Philosophy of Mathematics** (I. Lakatos, ed.), North-Holland, Amsterdam (1967) 138-171.

[Kre68] Georg Kreisel, *A survey of proof theory I;* **Journal of Symbolic Logic** 33 (1968) 321-388.

[Kre70] Georg Kreisel, *Church's thesis: a kind of reducibility axiom inc onstructive mathematics*; in [KMV].

[KT70] Georg Kreisel and Anne S. Troelstra, *Formal systems for some branches of intuitionistic analysis;* **Annals of Mathemaatical Logic 1** (1970) 229-387.

[Lei72] Daniel Leivant, *Notes on the completeness of the intuitionistic predicate calculuss*; report ZW 40/72, Mathematisch Centrum, Amsterdam (1972) pp. 25.

[Lei73] Daniel Leivant, *Existentially-mute theories and existence under assumptions*; report ZN 53/73, Mathematisch Centrum, Amsterdam (1973).

[Lei74] Daniel Leivant, *Friedman's proof that disjunction instantiation implies existential instantiation*; memo, August 1974.

[Lei75] Daniel Leivant, *Two properties of Rosser's sentence*; memo, March 1975.

[Lei79a] Daniel Leivant, **Absoluteness of Intuitionistic Logic**; Mathematical Center Tract 73, Amsterdam, 1979, pp.ix+137. Revised version of a PhD dissertation under the same title, Amsterdam 1976.

[Lei79b] Daniel Leivant, *Assumption classes in natural deduction*; **Zeit. math. Logik 25**, 1979, 1-4.

[Lei85] Daniel Leivant, *Syntactic translations and provably recursive functions*; **Journal of Symbolic Logic 50**, 1985, 252-258.

[Lob55] M.H. Lob, *Solution of a problem of Leon Henkin*; **Journal of Symbolic Logic 20** (1955) 115-118.

[Luk30] J. Lukasiewicz, *Philosophische Bemerkungen zu mehrvertigen Systemen des Aussagenkalkuls*; **Comptes Rendus des seances de la societe des Sceinces et des letters de Varsovie, 23** (1930), Cl. III 75-77 ("Zur Geschichte des Zweiwertigkeitssatzes").

[Luk52] J. Lukasiewicz, *On the intuitionistic theory of deduction*; **Indag. Math. 14** (1952) 202-212.

[Mar72] Per Matin-Lof, *Hauptsatz for the theory of species;* in [Fen71] 217-234.

[MKV68] J. Myhill, A. Kino and R.E. Vesley (eds.), **Intuitionism and Proof Theory (Proceeding of the Summer Conference at Buffalo, 1968)**, North-Holland. Amsterdam (1970) viii+516pp.

[MR73] A.R.D. Mathias and H. Rogers (editors, **Cambridge Summer School in Mathematical Logic (1971)**, Spinger-Verlag (LNM #337), Berlin (1973) ix+660pp.

[Myh73] John Myhill, *A Note on indicator functions;* **Proceedings of the AMS 39** (1973) 181-183.

[Pra65] Dag Prawitz, **Natural Deduction**, Almqvist & Wiksell, Uppsala (1965) 113 pp.

[Pra70] Dag Prawitz, *Some results for intuitionistic logic with second order quantification rules*; in [MKV70] 259-269.

[Pra72] Dag Prawitz, *Ideas and results of Proof Theory*; in [Fen71] 235-308.

[Sac63] Gerald Sacks, **Degrees of Unsolvability;** *Annals of Mathematics Studies 55* Princeton University Press, Princeton (1963) 175pp.

[Sca72] Bruno Scarpellini, *Disjunctive properties of intuitionistic systems*; typewritten notes, 1972.

[Sce85] Andrej Scedrov, *Set theory based on Heyting's predicate calculus*; this volume.

[Sch60] Kurt Schutte, **Beweistheorie**; Springer, Berlin (1960).

[SH80] J.P. Seldin and J.R. Hindley (editors), **To H.B. Curry: Essays on Combinatory Logic, Lambda Calculus and Formalism** Academic Press, London (1980) 606pp.

[Smo73] Craig A. Smorynski, *Applications of Kripke models*; chapter V in [Tro73b].

[Smo77] Craig A. Smorynski, *The incompleteness theorems*; in [Bar77] 821-866.

[Smo82] Craig A. Smorynski, *Nonstandard models and constructivity*; in **The L.E.J. Brouwer Centenary Symposium** (A. Troelstra and D. van Dalen, editors), North-Holland, Amsterdam (1982) 459-464.

[vSt79] Walter P. van Stigt, *The rejected parts of Brouwer's dissertation of the Foundations of Mathematics*; **Historia Mathematicae 6** (1979) 385-404.

[vSt82] Walter P. van Stigt, *L.E.J. Brouwer, the signific interlude*; in **The L.E.J. Brouwer Centenary Symposium** (A.S. Troelstra and D. van Dalen, editors), North-Holland (Studies in Logic #110), Amsterdam (1982) 505-512.

[Sza69] M.E. Szabo (editor), **The Collected Papers of Gerhard Gentzen**, North-Holland, Amsterdam, 1969.

[Tai66] W.W. Tait, *A nonconstructive proof of Gentzen's Hauptsatz for second order predicate logic*; **Bull. Amer. Math. Soc. 72** (1966) 980-988.

[Tai75] W.W. Tait, *A realizability interpretation of the theory of species*, in R. Parikh (ed.) **Logic Colloquium**, Springer-Verlag (LNM #453), Berlin (1975) 240-251.

[Tak67] M. Takahashi, *A proof of cut-elimination theorem in simple type theory*; **Jour. Math. Soc. Japan 19** (1967) 399-410.

[Tro71] Anne S. Troelstra, *Notions of realizability for intuitionistic arithmetic and intuitionistic arithmetic in all finite types:* in [Fen71] 369-405.

[Tro73a] Anne S. Troelstra, *Notes on intuitionistic second order arithmetic;* in [MR73] 171-205.

[Tro73b] Anne S. Troelstra, **Metamathematical Investigation of Intuitionistic Arithmetic and Analysis**, Springer-Verlag (LNM #344), Berlin (1973) xvii+485pp.

[Tro75] Anne S. Troelstra, *Completeness and validity for intuitionistic predicate logic*; **Colloque Internationale de Logique (Clermont-Ferrand)**, (M. Guillaume, ed.), CNRS, Paris (1975) 39-58.

[Tro77] Anne S. Troelstra, *Aspects of constructive mathematics*; in [Bar77] 973-1052.

[Tro81] Anne S. Troelstra, Course Notes on Intuitionistic Mathematics, Rijksuniversiteit Utrecht, 1980-1981.

[Vis82] Albert Visser, *On the completeness principle: a study of provability in Heyting's Arithmetic and extensions*; **Annals of Mathematical Logic 22** (1982) 263-295.

HARVEY FRIEDMAN'S RESEARCH ON THE FOUNDATIONS
OF MATHEMATICS, L.A. Harrington et al. (editors)

INTUITIONISTIC SET THEORY*

Andrej Ščedrov

Department of Mathematics
University of Pennsylvania
Philadelphia, PA 19104
U.S.A.

0. Introduction

In 1930 Heyting proposed a formal logic describing intuitionistic reasoning advocated by Brouwer [Bw]. One simply deletes the Law of Excluded Middle from the axiom schemata of classical propositional calculus (e.g. as axiomatized in [K]). Brouwer himself was opposed to formalization in mathematics in general, so, properly speaking, the phrase "intuitionistic formal system" is a misnomer, although it is frequently used in describing the formal systems based on Heyting's predicate calculus. Nevertheless, these systems (in particular, first order arithmetic based on Heyting's predicate calculus, HA) have been quite successful in capturing constructive reasoning. Not only does HA suffice for intuitively constructive elementary number theory, but it also proves only intuitively constructively provable facts in the subject. Furthermore, its importance is enhanced by a result of Gödel (1932) which exhibits classical first order arithmetic (PA) as an equiconsistent special case of HA (negative interpretation).

In fact, Friedman proved [Fr2] that this result of Gödel extends even to ZF set theory. He formulated the relevant intuitionistic counterpart, the theory ZFI . It is conveniently formulated with the binary predicate symbol ε as the only non-logical symbol. The underlying logic is Heyting's predicate calculus, and the non-logical axioms are as follows:

* Related topics are discussed in Leivant's article.

Extensionality $\forall x(x \varepsilon u \leftrightarrow x \varepsilon v) \rightarrow (A(u) \leftrightarrow A(v))$,

Pairing $\exists x(u \varepsilon x \wedge v \varepsilon x)$,

Union $\exists x \forall y \forall z(z \varepsilon u \wedge y \varepsilon z \rightarrow y \varepsilon x)$,

Separation $\exists x \forall y(y \varepsilon x \leftrightarrow y \varepsilon u \wedge A(y))$,

Infinity $\exists x(\exists u . u \varepsilon x \wedge \forall y \varepsilon x . \exists z \varepsilon x . y \varepsilon z)$,

Power Set $\exists x \forall y(\forall z \varepsilon y . z \varepsilon u \rightarrow y \varepsilon x)$,

ε -induction $\forall x(\forall y \varepsilon x . A(y) \rightarrow A(x)) \rightarrow \forall x A(x)$,

Collection $\forall x \varepsilon u . \exists y A(x,y) \rightarrow \exists v . \forall x \varepsilon u . \exists y \varepsilon v . A(x,y)$,

with the usual restrictions on the occurence of free variables in schemata.

As D. Scott has often emphasized since, this theory is also suggested quite independently of intuitionistic considerations by a particularly rich variety of category-theoretic, topological, recursion-theoretic and forcing interpretations, many of which are mathematically interesting in their own right [Fo2, FoS, Sco, Gr, Fr1 (§6), Be, H1, H2, Mu, Mc, S1, S2]. At the same time, the state of affairs is not yet entirely satisfactory because of the present lack of a good mathematical characterization of these interpretations (implying a completeness theorem), e.g. of the kind available for intuitionistic type theory [Fo1, LS] in terms of elementary toposes.

A set theorist might also be interested in ZFI because it is bound to cast light on classical set theory for the following reason. As a result of the parsimonious nature of intuitionistic logic, the difficulties and the depth inherent in the study of intuitionistic systems are much greater than the problems related to the study of their classical counterparts. Intuitionistic propositional calculus is therefore at least as hard as classical predicate calculus, intuitionistic predicate calculus at least as hard as classical first order arithmetic (PA) , intuitionistic first order arithmetic (HA) at least as hard as classical second order arithmetic, intuitionistic second order arithmetic (HAS) at least as hard as classical ZF , and intuitionistic ZF set theory is at least as hard as low large cardinals ($\leq$ measurable) in the classical setting.

The difficulties arise already in the formulation of the axioms. Let ZFI_R be the system obtained from ZFI by replacing Collection with

<u>Replacement</u> $\forall x \varepsilon u. \exists! y A(x,y) \rightarrow \exists v \forall x \varepsilon u. \exists y \varepsilon v. A(x,y)$,

which is easily provable in ZFI . This system was first studied by Myhill [M1]. Classically, Replacement implies Collection by the use of the Least Ordinal Principle. On the other hand, Friedman proved (in 1983) that ZFI_R does not imply ZFI . He used the methods that rely on important theorems about classical set theory. These methods have already led to much stronger results in this direction [FS4], and are likely to open up a new field of investigation connecting ZFI_R and ZFI with models of ZFC obtained by indiscernibles and infinitary partition properties.

It is an old observation of Myhill that the usual form of Regularity, $\forall x \forall y(y \varepsilon x \rightarrow \exists z \varepsilon x. \forall u \varepsilon z. u \not\varepsilon x)$, trivially implies Excluded Middle. Indeed, the set $\{\{x \varepsilon \{0\} | A\}, 1\}$ contains 1 , but if 1 is ε -minimal A must fail, and if 0 is ε -minimal A must hold, so Regularity would decide A . The Least Ordinal Principle is thus replaced by ε -induction.

The Axiom of Choice implies Excluded Middle as well [D,GM]. It is interesting to note that Zorn's Lemma does not imply Excluded Middle, and it in fact holds in Heyting-valued models of ZFI , when assumed externally [Gr].

We will describe Friedman's contributions to intuitionistic set theory in some detail. We will also briefly mention some of the results related to or inspired by Friedman's work in this area.

Friedman's extension of Gödel's negative interpretation is discussed in section 1. Section 2 deals with Friedman's extension of Kleene's recursive realizability. In section 3, we present Friedman's short proof that many intuitionistic systems, in particular ZFI , prove the same Π^0_2 sentences as their counterparts based on classical logic. Section 4 deals with the disjunction and existence properties. In section 5 we present Friedman's recent use of Kripke models to prove that Replacement does not imply

Collection. In section 6 we discuss a proof theoretically weak fragment of ZFI, which Friedman showed to be conservative over HA and suitable for the foundations of constructive analysis. In section 7 we discuss several partially intuitionistic fragments of ZFC for which Excluded Middle holds for an important class of formulas.

This paper is entirely expository, and it is self-contained as far as the statements of theorems are concerned. Detailed proofs are rarely given, although the basic methods used are clearly outlined.

We would like to thank H. Friedman, G. Kreisel, J. Myhill, and D. Scott for very valuable and inspiring discussions of the subject.

1. Negative interpretation

One of the first significant results about intuitionistic systems was obtained in 1932 by Gödel (cf. [K, §81]) who gave a syntactical translation of classical predicate calculus (Peano arithmetic PA, resp.) into Heyting's predicate calculus (Heyting's arithmetic HA, resp.). Thus the consistency of a system with classical logic is reduced to the consistency of a system with intuitionistic logic, and furthermore the classical system can be viewed as a subsystem (or a special case) of an intuitionistic one! Gödel's interpretation came to be known as the _negative interpretation_, since it is built on $\neg\neg$-prefixing. More precisely, to each formula A one associates a formula A^-, defined inductively as follows:

A^-	is	$\neg\neg A$, for atomic A
$(A \wedge B)^-$	is	$A^- \wedge B^-$
$(A \vee B)^-$	is	$\neg\neg(A^- \vee B^-)$
$(A \to B)^-$	is	$A^- \to B^-$
$(\neg A)^-$	is	$\neg(A^-)$
$(\forall x A)^-$	is	$\forall x(A)^-$
$(\exists x A)^-$	is	$\neg\neg\exists x(A)^-$,

and it is readily checked that if A is provable classically, A^- is provable intuitionistically. (Another interpretation was given by Kolmogorov

[Ko]). Myhill [M2] gave a negative interpretation for higher-order arithmetic (type theory), and Friedman [Fr2] gave a negative interpretation for ZF. Thus ZF becomes an equiconsistent special case of ZFI.

A technical problem is that the translation, as given above, does not work outright, as the reader will be quickly convinced when trying to translate Extensionality, Power Set, and Collection. E.g., (Extensionality)$^-$ is:

$$\forall z(\neg\neg(z\varepsilon x) \leftrightarrow \neg\neg(z\varepsilon y)) \rightarrow (\neg\neg(x\varepsilon u) \rightarrow \neg\neg(y\varepsilon u)) \quad ,$$

and Extensionality in ZFI simply does not suffice to verify this. We will see in the other sections of this paper that Extensionality creates a nuissance with most metamathematical techniques for intuitionistic systems like ZFI (as it does, e.g., in Boolean-valued models of ZF as well). Thus Friedman [Fr2] disposes of Extensionality once and for all, by first proving a useful technical lemma, interpreting ZFI (ZF, resp.) into ZFI (ZF, resp.) without Extensionality. This auxiliary theory (with $A \vee \neg A$ added) is then negatively interpreted in ZFI. However, one needs to be careful with the formulation of the power set axiom. Consider the following sentence:

$$\forall u.\exists v.\forall x.\exists y\varepsilon v.\forall z(z\varepsilon y \leftrightarrow z\varepsilon x \wedge z\varepsilon u) \qquad \text{(Weak Power Set)} \ ,$$

intuitionistically equivalent to the Power Set in the presence of Extensionality.

Let S be ZFI-Extensionality, and with Weak Power Set instead of Power Set. Let S^c be S in classical logic. In S, "x contains an ordered pair of a,b" is given by a formula, which we write as "$\langle a,b\rangle \varepsilon x$". Let EQR(a) abbreviate:

$$\forall x\forall y(\langle x,y\rangle \varepsilon a \rightarrow (\forall z\varepsilon x)(\exists w\varepsilon y)(\langle z,w\rangle \varepsilon a) \wedge (\forall w\varepsilon y)(\exists z\varepsilon x)(\langle z,w\rangle\varepsilon a))) \ ,$$

and let $a \sim b$ abbreviate $\exists x(\mathrm{EQR}(x) \wedge \langle a,b\rangle \varepsilon x)$. Let $a \varepsilon^* b$ abbreviate $\exists x \varepsilon b \,.\, x \sim a$. For a formula A of ZFI, let A^* be the (formula of ZFI abbreviated by the) result of replacing each ε in A by ε^*.

Let RDC be the schema of Relativized Dependent Choice:

$$A(z) \wedge \forall x(A(x) \to \exists y(A(y) \wedge B(x,y)) \to$$
$$\to \exists \text{function } f.(\text{dom}(f) = N \wedge f(0) = z \wedge \forall n \in N.(A(f(n)) \wedge B(f(n),f(n+1)))) .$$

<u>Lemma 1.1.</u> If A is provable in ZFI (ZFI + RDC , ZF , resp.), then A^* is provable in S (S + RDC, S^c , resp.).

The proof first establishes that $\sim$ is an equivalence relation, and shows that $S \vdash (a \sim b) \leftrightarrow \forall x(x \varepsilon^* a \leftrightarrow x \varepsilon^* b)$. ε -induction and Collection are used essentially - actually the proof fails for weaker systems. Then one checks that the axioms translate.

<u>Remark.</u> This translation can be composed with any of the interpretations of S discussed below to give direct interpretations of ZFI . In fact, the usual Boolean-valued interpretation of "$x \varepsilon y$" and "$x=y$" is obtained this way from a simpler Boolean-valued interpretation of S^c .

The next step is to apply the negative interpretation:

<u>Lemma 1.2.</u> If $S^c \vdash A$, then $S \vdash A^-$.

For the proof, reformulate Collection as:

$$u. \; v. \; x. \; s[x \varepsilon u \quad A(x,s) \to \quad y \varepsilon v.A(x,y)] .$$

Thus we actually get:

<u>Theorem 1.3.</u> Let $ZF \vdash A$. Then ZFI-Extensionality $\vdash A^{*-}$.

2. <u>Recursive realizability</u>

In 1945, Kleene gave a recursive interpretation of the intuitive meaning of constructive connectives and quantifiers, cf. [K,§82] and [Tr1, §3.2]. Under this interpretation, natural numbers are recursive indices of a recursive information about (a proof of) a formula A : we say that a natural number n realizes A , and write $n \; \underset{\sim}{r} \; A$. This interpretation is known as 1945 (recursive) realizability. For HA , it was defined by Kleene as follows:

$$n \; \underset{\sim}{r} \; A \quad \text{is} \quad A \quad , \text{ for atomic } A$$

$$n \; \underset{\sim}{r} \; (A \wedge B) \quad \text{is} \quad \pi_1(n) \underset{\sim}{r} \; A \wedge \pi_2(n) \underset{\sim}{r} \; B$$

$$n \; \underset{\sim}{r} \; (A \vee B) \quad \text{is} \quad \begin{bmatrix} \pi_1(n) = 0 \rightarrow \pi_2(n) \underset{\sim}{r} \; A \\ \wedge \\ \pi_1(n) \neq 0 \rightarrow \pi_2(n) \underset{\sim}{r} \; B \end{bmatrix}$$

$$n \; \underset{\sim}{r} \; (A \rightarrow B) \quad \text{is} \quad \forall k (k \; \underset{\sim}{r} \; A \rightarrow \{n\}(k)\downarrow \text{ and } \{n\}(k) \underset{\sim}{r} \; B)$$

$$n \; \underset{\sim}{r} \; \forall x A(x) \quad \text{is} \quad \forall k (\{n\}(k)\downarrow \text{ and } \{n\}(k) \underset{\sim}{r} \; A(k))$$

$$n \; \underset{\sim}{r} \; \exists x A(x) \quad \text{is} \quad \pi_2(n) \underset{\sim}{r} \; A(\pi_1(n)) \quad ,$$

where π_1, π_2 are primitive recursive pairing functions.

Kleene showed that for each formula A provable in HA , there is a natural number n such that $n \; \underset{\sim}{r} \; A$. The most interesting feature of this interpretation is that it verifies Church's Thesis (CT), which in a constructive setting (since only computable functions are considered) claims:

$$\forall n \in \mathbf{N}. \exists m \in \mathbf{N}. A(n,m) \rightarrow \exists e \in \mathbf{N}. \forall n \in \mathbf{N}. (\{e\}(n)\downarrow \wedge A(n, \{e\}(n))) \; .$$

Indeed, $k \; \underset{\sim}{r} \; \forall n \exists m A(n,m)$ gives a number j such that for all $n, \pi_2(\{j\}(n)) r \; A(n, \pi_1(\{j\}(n)))$. Then one easily recursively computes a realizor of the consequent of CT .

Shortly afterwards, Nelson treated 1945-realizability as a syntactical translation: by formalizing Kleene's proof in HA , he obtained that HA + CT $\vdash$ A implies HA $\vdash \bar{n} \; \underset{\sim}{r} \; A$, for some numeral $\bar{n}$.

It is surprising that 1945-realizability would extend to impredicative systems, whose constructivity is far from being clear comparably to HA . Troelstra (cf. [Tr1,§3.2.31]) extended it to second order intuitionistic arithmetic HAS , from which one easily extends it to higher order intuitionistic arithmetic HAH . Hyland [Hy] gave a conceptually simpler treatment. Friedman [Fr1,§6] was the first to extend 1945-realizability to ZFI . Compare this with the result of the previous section: ZFI + CT is consistent up to ZFI , although ZFI "contains" ZF !

Technically, the problem is to formulate 1945-realizability for ZFI in such a way that it coincides with the original 1945-realizability for arithmetic formulas. The reader acquainted with Boolean-valued models of ZF may think of 1945-realizability analogously as taking $P(\mathbf{N})$ as an "algebra of truth-values", so $n \underset{\sim}{r} A$ can be thought of as $n \in \|A\|$. Complications in the definition of $\|y \in x\|$ are avoided if one defines 1945-realizability for ZFI - Extensionality, and then combining it with Lemma 1.1. Thus, let us define 1945-realizability as a syntactical translation of ZFI - Extensionality to ZFI as follows:

$n \underset{\sim}{r} \perp$ is $\perp$

$n \underset{\sim}{r} (y \in x)$ is $\langle n,y \rangle \in x$

$n \underset{\sim}{r} \forall x A$ is $\forall x. n \underset{\sim}{r} A$

$n \underset{\sim}{r} \exists x A$ is $\exists x. n \underset{\sim}{r} A$,

and the rest is as before. Notice that the (unbounded) set-quantifiers are translated trivially. One then has:

<u>Theorem 2.1</u>. Let A be provable in ZFI + CT (ZFI+CT+RDC, ZFI + CT + RDC + MP , resp.). Then for some numeral $\bar{n}$, $\bar{n} \underset{\sim}{r} A^*$ is provable in ZFI(ZFI+RDC , ZFI+RDC+MP , resp.).

Here MP is Markov's Principle:

$$\forall n \forall m (A(n,m) \vee \neg A(n,m)) \wedge \forall n. \neg\neg \exists m. A(n,m) \rightarrow \forall n \exists m. A(n,m)$$

(Analysis in ZFI+CT+MP is essentially the Markov-Šanin constructive analysis, cf. [Tr2], [Ab].)

The original 1945-realizability for numerical quantifiers is recaptured by the clause for implication in the consideration of restricted quantifiers. The internal "set of natural numbers" is given by the set $\mathbf{N}$

for which $n \mathrel{\underset{\sim}{r}} (x \in N)$ iff $x=n$.

Remark. These results were extended beyond ZFI in [FS1].

Remark. In [Fr1,§6], Friedman actually gives a set-theoretic version of 1945-realizability for the system $ZFC^{1/2}$, which goes beyond ZFI (it has Choice, and the Law of Excluded Middle for Δ_0 -formulas, but only Δ_0 - Separation), cf. section 7 below. Definitions in [Fr1, §6] are easily adapted to ZFI , as above.

3. Classically and intuitionistically provably recursive functions

Gödel's functional interpretation shows that PA and HA prove the same Π^0_2 sentences, i.e. they have the same provably recursive functions. This result was extended to higher-order arithmetic by the use of delicate and complicated proof-theoretic techniques of functional interpretation (Spector, Girard) and normalization of proofs (Prawitz). Friedman, on the other hand, gave short, elegant proofs of the following results:

Theorem 3.1. ZF and ZFI prove the same Π^0_2 sentences.

Theorem 3.2. ZF and ZFI have the same provable ordinals (Π^1_1 sentences).

More precisely, Theorem 3.2 states that if $ZF \vdash \forall f.\exists n.F(\tilde{f}(n)) = 0$, then $ZFI \vdash \forall f.\exists n.F(\tilde{f}(n)) = 0$, where F is a primitive recursive function symbol, and f ranges over subsets of $\mathbf{N} \times \mathbf{N}$ such that $\forall n \exists! m.\langle n,m\rangle \in f$ (i.e., f is a function internal to ZFI). Variables n,m range over $\mathbf{N}$. $\tilde{f}(n)$ stands for the finite sequence $\langle f(0),\ldots,f(n)\rangle$.

The proofs are given by a syntactical translation like the negative interpretation, they use no proof theory whatsoever, and they apply not only to ZF , but to first-order arithmetic (for Theorem 3.1), and higher-order arithmetic. They work under very general conditions [Fr9]. In fact, the proof of Theorem 3.1 establishes the closure of the relevant system under the

Primitive Recursive Markov Rule:

$$\frac{\neg\neg\exists m \in \mathbf{N}.A(n,m)}{\exists m \in \mathbf{N}.A(n,m)} \quad ,$$

where n is a numerical parameter and A is recursive. (The rest is left to the negative interpretation.)

To give the desired translation, let B be any formula, and let ϕ be any formula none of whose bound variables is free in B . Define the B-translation of ϕ to be the formula ϕ_B obtained by simultaneously replacing each atomic subformula ψ of ϕ by $\psi \vee B$.

Lemma 3.3. If $\phi \vdash \psi$ (in the intuitionistic logic), and ϕ_B and ψ_B are defined, then $\phi_B \vdash \psi_B$ and $B \vdash \phi_B$ (in the intuitionistic logic).

Proof. Straightforward verification.

Lemma 3.4. If $Q \vdash \phi$ and ϕ_B is defined, then $Q \vdash \phi_B$.

Here Q is the intuitionistic system under consideration (HA, higher-order intuitionistic arithmetic HAH , or intuitionistic set theory). For set theory, it is best formulated in two-sorted Heyting's predicate calculus, with numerical variables and set variables (for a precise description, cf. [Fr9]).

One now shows the closure of Q under the Primitive Recursive Markov Rule by suitably choosing B :

Lemma 3.5. Let $Q \vdash \neg\neg\exists mA(n,m)$, where n is a numerical variable, and A is recursive, Then $Q \vdash \exists mA(n,m)$.

Proof. We have $Q \vdash ((\exists mA(n,m)) \to \bot) \to \bot$. By Lemma 3.3, using $B = \exists mA(n,m)$, we have:

$$Q \vdash (\exists m.(A(n,m) \vee \exists mA(n,m)) \to \exists mA(n,m)) \to \exists mA(n,m) \quad ,$$

and hence

$$Q \vdash \exists mA(n,m) \quad ,$$

done.

Remark. Theorem 3.2 is proved by a suitable choice of a formula B as well.

Remark. These results were extended beyond ZFI in [FS1].

4. Disjunction and existence properties

In describing the intuitive meaning of intuitionistic connectives and quantifiers, the clauses for disjunction and for existential quantification are usually given as follows: one can prove $A \vee B$ iff one can prove A or prove B , one can prove $\exists xA(x)$ iff one can find an object a of the domain under consideration, and prove $A(a)$. It turns out that intuitionistic systems often have the corresponding metamathematical properties, stating the closure under the following rules:

$$\frac{\vdash A \vee B \quad \text{(closed)}}{\vdash A \ \text{ or } \vdash B} \qquad \text{Disjunction Property (DP)} ,$$

$$\frac{\vdash \exists xA(x) \quad \text{(closed)}}{\text{for some closed term } t\ ,\ \vdash A(t)} \qquad \text{Existence Property (EP)} ,$$

and its special case:

$$\frac{\vdash \exists x \varepsilon \mathbf{N}.A(x) \quad \text{(closed)}}{\text{for some numeral } \bar{n}\ , \vdash A(\bar{n})} \qquad \text{Numerical Existence Property } (EP_{\mathbf{N}})$$

As Kreisel has often pointed out [Kr1,2], one should not make a mistake of regarding these properties as inherent to all intuitionistic systems.

Remark. $EP_{\mathbf{N}}$ clearly implies DP for theories containing arithmetic. Remarkably, DP implies $EP_{\mathbf{N}}$ as well, under very general conditions on a theory [Fr4]. This contribution of Friedman is discussed in [L].

Kleene obtained DP and $EP_{\mathbf{N}}$ for HA in 1962 [Tr1,§3.1], by using the following interpretation of HA . We define $|A$ (read: slash A) by induction of the complexity of a sentence A :

$\mid A$	is	A , for A atomic
$\mid A \wedge B$	is	$\mid A$ and $\mid B$
$\mid A \vee B$	is	($\vdash A$ and $\mid A$) or ($\vdash B$ and $\mid B$)
$\mid A \to B$	is	($\vdash A$ and $\mid A$) implies $\mid B$
$\mid \forall x A(x)$	is	$\mid A(t)$ for all closed terms t
$\mid \exists x A(x)$	is	$\vdash A(t)$ and $\mid A(t)$, for some closed term t .

The soundness theorem gives DP and $EP_{\mathbb{N}}$ immediately: e.g. for $EP_{\mathbb{N}}$, say $\vdash \exists x A(x)$ (closed). Then $\mid \exists x A(x)$, so in particular $\vdash A(t)$ for some closed term t . Evaluating t gives the numeral $\bar{n}$ so that $\vdash A(\bar{n})$.

The first problem in applying this argument to higher order intuitionistic arithmetic (type theory) HAH is that it does not have enough terms. One thus considers a new theory T^+ defined so that whenever HAH $\vdash \exists x \forall y(y \varepsilon x \leftrightarrow A(y))$ (closed), where x and y are of the appropriate types, a new constant C_A of the same type as x is introduced together with the new axiom $\forall y(y \varepsilon C_A \leftrightarrow A(x))$.

One now faces a much deeper problem of defining slash for the atomic formulas of T^+ , so that the comprehension scheme:

$$\exists x \forall y(y \varepsilon x \leftrightarrow A(y))$$

(where x and y are of the appropriate types) is slashed: picking a term for x depends on $\mid A$, so $\mid (y \varepsilon x)$ would depend on $\mid A(y)$, leading into circularity (A is of arbitrary complexity) .

In 1971, Friedman [Fr1] (cf. also the exposition in [Tr2,§3.1.21], and in [LS]) solved this problem by considering a richer theory $T^{\#}$ conservative over T in which all comprehension constants are impredicatively indexed by sets of the suitable type (satisfying some additional technical requirements). These are treated as constants (of complexity 0).

For each formula A of $T^{\#}$, let A^- be the formula of T^+ obtained from A by erasing all set-indices of comprehension constants occuring in A. The theory $T^{\#}$ is then defined by requiring $T^{\#} \vdash A$ iff $T^+ \vdash A^-$.

One then defines

$$|(v \in C_{A,X}) \quad \text{iff} \quad v \in X ,$$

and then rest analogously to Kleene's definition. The soundness theorem then implies EP as follows: say $HAH \vdash \exists x A(x)$ (closed), then $|\exists x A(x)$, so in particular $T^{\#} \vdash A(C_{B,X})$ for some constant $C_{B,X}$. Thus $T^+ \vdash A(C_B)$, and so $HAH \vdash \exists x(A(x) \wedge \forall y(y \varepsilon x \leftrightarrow B(x)))$, done.

In 1978 Freyd gave a short, conceptual, category-theoretic argument establishing DP and EP for HAH and other higher-order theories, which does not rely on syntax. His proof establishes essentially the same semantics as the Friedman slash [SS].

Friedman and Myhill [M1] extended the slash to ZFI_R, establishing DP and EP. DP (and $EP_{\mathbf{N}}$) were shown for ZFI by Beeson [Be], using a different method. Friedman has recently proved that EP is false for ZFI (cf. section 5).

Myhill [M3] showed how to obtain DP (and $EP_{\mathbf{N}}$) for $ZFI_R + RDC$ by amending the Friedman slash to suit it for RDC :

$$\forall z(A(z) \wedge \forall x(A(x) \rightarrow \exists y(A(y) \wedge B(x,y))) \rightarrow$$
$$\rightarrow \exists \text{function } f(\text{dom}(f) = N \wedge f(0) = z \wedge \forall n \varepsilon N.(A(f(n)) \wedge B(f(n),f(n+1))))$$

Indeed, suppose the antecedent is slashed. Iterating, one obtains a sequence $t_0,t_1,\ldots,t_n,\ldots$ of terms so that $|A(t_i) \wedge B(t_i,t_{i+1})$ for each $i \in \mathbf{N}$. But now one lacks a term for this sequence, needed to slash the consequent. Thus one defines a new theory T^+ starting from ZFI_R by transfinite induction: at the successor stages, one introduces the comprehension constants C_A, and the choice constants at the limit stages (together with the appropriate new axioms specifying them). Again, one considers a richer theory $T^{\#}$, conservative over $ZFI_R + RDC$, and proves the soundness theorem. But the proof of EP given above for HAH breaks down at the last step, since one might have $T^+ \vdash A(t)$ for a choice constant t (instead of

$T^+ \vdash A(C_B))$, which cannot be reduced to a fact about $ZFI_R + RDC$. It does work for $EP_{\mathbf{N}}$, in which case the term needed is a numeral (or a corresponding set definable in ZFI_R).

Myhill [M3] in fact showed DP and $EP_{\mathbf{N}}$ for a set theory with RDC weaker than $ZFI_R + RDC$: more precisely, he considers only Δ_0-separation , and Exponentiation rather than Power Set. However, the methods of [M3] extend to $ZFI_R + RDC$ in a straightforward way.

Note that the difficulty arises already for HAS + RDC (HASC) . In 1982, Friedman introduced a new way of combining the slash with 1945 realizability that establishes EP for HASC , which was then further extended to stronger systems [FS2]. The key idea is as follows: introduce choice constants for recursive sequences only (obtained by considering the antecedent of RDC slashed). They are given essentially by recursive indices, so the problematic case in the attempted proof of EP suggested above goes through by an appeal to EP_N . The problem is, of course, that one then does not have enough choice terms: slashing the antecedent of RDC gives arbitrary sequences of terms. We need to assume that every such sequence is recursive: work in the metatheory HASC + CT! There are two problems with this: formalizing the slash might run against reflection principles, and on the other hand, defining T^+ by transfinite induction will not work, since in the intuitionistic metatheory, ordinals do not behave well. Thus we define slash for finite fragments T_0 only, and define T_0^+ by an inductive definition, like Kleene's $\mathcal{O}$.

Theorem 4.1. a) Let T be HAS, HAH , or intuitionistic Zermelo set theory, and let TC be T with Countable Choice or RDC added. Assume $TC \vdash \exists x A(x)$ (closed). Then for some formula $B(y)$ with exactly y free, $\exists x(\forall y(y \in x \leftrightarrow B(y)) \wedge A(x))$ is provable in TC .

b) Let TI be the schema of transfinite induction over all (true) primitive recursive well founded relations on ω . Let TI^- be the fragment of TI referring to such relations provably well founded in ZFI + RDC . Suppose $\exists x A(x)$ is a provable sentence in $ZFI_R + RDC$ ($ZFI_R + RDC + TI$, resp.). Then for some formula $B(y)$ with precisely y free, $\exists x(\forall y(y \in x \leftrightarrow B(y)) \quad A(x))$ is provable in $ZFI_R + RDC + TI^-$ ($ZFI_R + RDC + TI$, resp.).

<u>Sketch of proof</u>. a) Assume $TC \vdash \exists x A(x)$ (closed). Let TC_0 be a finite fragment of TC for which $TC_0 \vdash \exists x A(x)$. Working in TC + CT define TC_0^+ by an inductive definition like Kleene's $\mathcal{O}$, introducing choice constants for recursive sequences of constants only (observe that it is still a bounded system), and define the Friedman slash for $TC_0^{\#}$. The soundness theorem then gives

$$TC + CT \vdash (\exists \text{constant } \xi \text{ with } TC_0^+ \vdash A(\xi)) .$$

Provability in TC_0^+ is like being in $\mathcal{O}$, so let us write the above as:

$$TC + CT \vdash \exists n . n \in \mathcal{O} .$$

Applying 1945-realizability, one obtains:

$$TC \vdash \exists m . \exists n . \; m \;\underset{\sim}{r}\; (n \in \mathcal{O}) .$$

One can check that:

$$TC \vdash (m \;\underset{\sim}{r}\; n \in \mathcal{O}) \to n \in \mathcal{O} ,$$

so that:

$$TC \vdash \exists n . n \in \mathcal{O} ,$$

i.e.:

$$TC \vdash (\exists \text{constant } \xi \text{ with } TC_0^+ \vdash A(\xi)) .$$

Partial reflection principle allows us to give the semantical interpretation to new constants of TC_0^+, so:

$$TC \vdash \exists \text{term } t \text{ with } A(t) .$$

But now use $EP_{\mathbf{N}}$ for TC w.r.t. gödelnumber of the (comprehension) term t to obtain:

$$TC \vdash A(t) \text{ , for some } t \text{ ,}$$

more precisely:

$$TC \vdash \exists x(A(x) \wedge \forall y(y \varepsilon x \leftrightarrow B(y))) \text{ ,}$$

where $B(y)$ is a formula of TC whose only free variable is y. The proof of (b) differs slightly in the second half.

Remark. The Friedman slash can be easily modified to show DP , EP , $EP_{\mathbf{N}}$ for formulae with nonnumerical parameters. In particular, it then shows that if $\vdash \forall x(A(x) \vee B(x))$ (closed) , then $\vdash \forall xA(x)$ or $\vdash \forall xB(x)$.

5. Replacement and Collection in intuitionistic set theory.

Most of the models of ZFI mentioned in the introduction can in fact be considered as translations of ZFI into ZFI , but not as translations of ZFI_R to ZFI_R . In particular, this is true for the translations discussed in sections 1-3.

In 1983, Goodman realized that the use ofKripke models might establish the gap between Replacement and Collection. Shortly afterwards, Friedman sharpened this method and combined it with forcing in ZF to prove that ZFI does not have the Existence Property, in which it differs from ZFI_R (cf. section 4). These methods have in turn been strengthened in [FS4] to prove much stronger results of this kind. In particular, adding RDC and all classically true Σ_1 sentences to ZFI_R still does not yield Collection. Furthermore, Friedman showed that ZFI_R does not prove all π_2^0 theorems of ZFI . Here we only outline Friedman's basic construction and refer to [FS4] for details.

Theorem 5.1. Assume Con(ZFC) . Then there is a ZFI-provable sentence of the form

$$\exists v(\exists yB(y) \to \exists y\varepsilon v.B(y)) \text{ ,}$$

for which there is no formula $C(z)$ with exactly z free so that the sentence

$$\exists! zC(z) \wedge \exists v(C(v) \wedge (\exists yB(y) \rightarrow \exists y\varepsilon v.B(y)))$$

is provable in ZFI .

<u>Outline of the proof</u>. Let N be a countable model of ZFC + V = L , and let λ be the least ordinal in N greater than all definable ordinals in N . Collapse λ by forcing to make it countable. Let M be the resulting forcing extension. Let M^* be the structure defined as follows. Elements of M^* are all pairs $\langle x,y\rangle$, where $x,y \in M$. Let $\rho(\langle x,y\rangle,\langle u,v\rangle)$ iff $M \models \langle x,y\rangle \in u$. (M^*,ρ) satisfies all axioms of ZF except Extensionality and Power Set. Let $\equiv$ be the equivalence relation on M^* defined by $(\forall c)(\rho(c,a) \leftrightarrow \rho(c,b))$. We let $J: M^* \rightarrow N$ be a definable function such that

(i) $\rho(x,y)$ implies $J(x) \in J(y)$,

(ii) J maps each equivalence class $[x]$ with respect to $\equiv$ one-to-one onto $\{z\varepsilon N| \forall y(\rho(y,x) \rightarrow J(y) \in z\}$.

Notice that the range of J is as large as possible subject to (i). Also, such J is unique up to isomorphism, and furthermore, it is defined by transfinite recursion on the rank of elements of M^* .

We consider the following Kripke structure (the reader may consult Smoryński's contribution in [Tr1] about Kripke models in general). There are two moments, $1 < 2$. The objects at both moments are the elements of M^* . If $\phi(x_1,\ldots,x_n)$ is a formula of ZFI with exactly $x_1,\ldots,x_n$ free, we let

$$2 \Vdash \phi(x_1,\ldots,x_n) \text{ iff } N^* \models \phi(J(x_1),\ldots,J(x_n)) ,$$

with the ordinary abuse of notation. Kripke forcing at moment 1 is defined by

$1 \nVdash \bot$,

$1 \Vdash y \varepsilon x$ iff $\rho(y,x)$,

$1 \Vdash \phi \wedge \psi$ iff $1 \Vdash \phi$ and $1 \Vdash \psi$,

$1 \Vdash \phi \vee \psi$ iff $1 \Vdash \phi$ or $1 \Vdash \psi$,

$1 \Vdash \phi \to \psi$ iff $1 \Vdash \phi$ implies $1 \Vdash \psi$, and $N \models \phi \to \psi$,

$1 \Vdash \exists x \phi(x)$ iff $1 \Vdash \phi(x)$ for some $x \in M^*$,

$1 \Vdash \forall x \phi(x)$ iff $1 \Vdash \phi(x)$ and $N \models \phi(J(x))$, for every $x \in M^*$.

(As usual in Kripke models, one readily shows by induction on the complexity of $\phi(x_1,\ldots,x_n)$ that $1 \Vdash \phi(x_1,\ldots,x_n)$ implies $2 \Vdash \phi(x_1,\ldots,x_n)$. The proof then proceeds via several lemmas:

Lemma 5.2. All axioms of ZFI are Kripke forced at 1 . In particular, $x=y$ iff $1 \Vdash \forall z(z \varepsilon x \leftrightarrow z \varepsilon y)$.

Lemma 5.3. Let Dec be the sentence $\forall x \forall y(x \varepsilon y \vee x \not\varepsilon y)$. For any formula ϕ , let ϕ^* be the result of replacing every atomic subformula ψ of ϕ by $\psi \vee$ Dec . then $1 \Vdash \phi^*$ iff $M^* \models \phi$.

Lemma 5.4. Suppose $1 \Vdash \exists ! x \phi(x)$, where ϕ has exactly x free. Then for the unique $x \in M^*$ such that $1 \Vdash \phi(x)$, $J(x)$ is the unique solution to $\phi(x)$ in N . In particular, $rk(x)$ is bounded by some definable ordinal of N .

Lemma 5.5. Let $A(y)$ be a formula stating that "y is an uncountable ordinal", e.g. "y is transitive, linearly ordered by ε , and for some set x satisfying the Axiom of Infinity, there is no one-to-one onto map from x to y". Then there is no formula $C(z)$ with exactly z free so that ZFI proves the sentence

$$\exists ! z C(z) \wedge \exists v(C(v) \wedge (\exists y A^*(y) \to \exists y \varepsilon v . A^*(y))) .$$

The reader will note that by Separation, the schema of Collection is equivalent to the schema

$$\exists v \forall x \varepsilon u(\exists y \phi(x,y) \to \exists y \varepsilon v . \phi(x,y)) ,$$

of which the sentence

$$\exists v(\exists y A^*(y) \to \exists y \varepsilon v . A^*(y))$$

is an instance (with u a singleton). Thus, by Lemma 5.2, it is Kripke forced at 1. Because $M^* \models \exists y . A(y)$, Lemma 5.3 gives $1 \Vdash \exists y A^*(y)$ and thus $1 \Vdash \exists y \varepsilon v . A^*(y)$. By Lemma 5.3 again, $M^* \models A(y)$ for some $y \varepsilon M^*$ such that $\rho(y,v)$. Therefore $rk(y) < rk(v)$, so it is bounded by a definable ordinal $< \lambda$ in N (Lemma 5.3). But λ is countable in M, and y is an uncountable ordinal.

Remark. All we need is that $\lambda > rk(v)$, where v is definable by $C(v)$.

6. Weak set theories as foundations for analysis.

Errett Bishop's work on constructive analysis [Bi] has motivated several attempts to provide intuitionistic formal systems which would faithfuly reflect Bishop's approach. On the other hand, one may attempt to provide an alternative conceptual framework (and formal systems based on it), more analogous to the one for ordinary classical analysis, in which, nevertheless, Bishop's ideas can be expressed in a natural way. Myhill [M2], Friedman [Fr5], and Aczel [Ac] have proposed such systems based on weakenings of intuitionistic Zermelo set theory. Friedman [Fr5], in addition, has studied the logical strength of such (and several related) systems, and in particular the relationship between constructive analysis and ordinary analysis [Fr5, 6a,b]. It is possible to give classical systems of low logical strength (in particular: conservative over PA) which suffice for ordinary analysis (first attempts were in [Fr6,a,b]; the resulting work of Friedman and others in this area is discussed in [Si]). For constructive analysis, Myhill [M3] has argued against the inclusion of the Power Set, but used instead the existence of the set of all total functions from a given set x to a given set y (Exponentiation). Furthermore, not more than Δ_0-separation is really used in practice in [Bi], [BC], [Br] and other works on constructive analysis (and it would not be easy to defend the constructivity of stronger, impredicative versions of Separation).

In [Fr5], Friedman proposes a set-theoretic system **B** for foundations of constructive analysis, and shows it to be equiconsistent with first-order arithmetic. Myhill's system CST of [M3] is shown in [Fr5] to be equiconsistent with ε_0-ramified analysis.

The system **B** is based on Heyting's predicate calculus with equality. Its nonlogical symbols are binary predicate symbol ε (for membership), unary predicate symbol N ($N(x)$ for "x is a natural number"), unary predicate symbol S ($S(x)$ for "x is a set"), binary predicate symbol s ($s(x,y)$ for "y is a successor of x"), and the constant 0. Since **B** and other weak set theories we consider will have Extensionality, Pairing, Infinity, Successor, Union, and Δ_0- Separation, the auxiliary function symbols $\{x\}$ $\{x,y\}$ $\langle x,y\rangle$, $\mathrm{dom}(x)$, $\mathrm{rge}(x)$, $'$, $\mathbf{N}$ can be introduced together with the appropriate axioms. The axioms of **B** are as follows:

(A) Ontological axioms:

$$N(x) \vee S(x) ,\ N(x) \to \neg S(x) ,\ N(0) ,\ s(x,y) \to N(x) \wedge N(y) ,$$
$$x \varepsilon y \to S(y) .$$

(B) Extensionality:

$$S(x) \wedge S(y) \wedge \forall z(z\varepsilon x \leftrightarrow z\varepsilon y) \to x=y .$$

(C) Successor axioms:

$$s(x,y) \wedge s(x,z) \to y=z$$
$$s(y,x) \wedge s(z,x) \to y=z$$
$$\neg s(x,0) .$$

(D) Infinity: $\exists x \forall y(y\varepsilon x \leftrightarrow N(y))$

(E) Induction

(F) Pairing

(G) Union

(H) Exponentiation:

$$\exists x \forall y(y \varepsilon x \leftrightarrow y \text{ is a function} \wedge \mathrm{dom}(y) = u \wedge \mathrm{rge}(y) \subseteq v) .$$

(I) Limited Dependent Choice:

$$\forall x \varepsilon z . \exists y \varepsilon z . B(x,y) \rightarrow$$
$$\rightarrow \forall x \varepsilon z . \exists f\colon \ \mathbf{N} \rightarrow z.(f(0) = x \wedge \forall n \varepsilon \mathbf{N}. B(f(n),f(n+1))) ,$$

where $B(x,y)$ is a Δ_0- formula.

(J) Δ_0-Separation :

$$\exists x \forall y(y \varepsilon x \leftrightarrow y \varepsilon z \wedge A(y)) ,$$

where $A(y)$ is a Δ_0-formula , and x is not free in $A(y)$.

(K) Abstraction:

Let $A(y_1,\ldots,y_k,u)$ be a Δ_0-formula , x a set. Then $\{\{u \varepsilon x\colon A(y_1,\ldots,y_k,u)\}\colon \ y_1,\ldots,y_k \ \varepsilon \ x\}$ exists.

No Power Set is needed to show that the set of all compact subsets of a metric space exists, since they are the closures of totally bounded countable sets. For constructive measure theory, one considers [BC] which renders chapters 6,7 of [Bi] obsolete, dispensing with the Borel sets. Note that domains of measurable functions are not arbitrary subsets of a measure space - to give a measurable function f is to give a sequence of continuous functions converging to f in a suitable sense. Thus one does not really need Power Set to define a measure space, either.

We also consider set theories based on Heyting's predicate calculus with equality, with several of the following axioms:

(1) Ontological axioms
(2) Extensionality
(3) Successor axioms
(4) Infinity
(5) Induction: $0 \ \varepsilon \ x \qquad y \varepsilon a . y' \varepsilon a \rightarrow \mathbf{N} \qquad a$
(6) Pairing
(7) Union
(8) Exponentiation
(9) Bounded Dependent Choice:

$$\forall x \varepsilon z . \exists y \varepsilon z . B(x,y) \rightarrow$$

$\rightarrow \forall x \varepsilon z. \exists f: N \rightarrow z.(f(0) = x \wedge \forall n \varepsilon N.B(f(n),f(n+1)))$

(10) Δ_0-Separation

(11) Strong Collection:

$\forall x \varepsilon u. \exists y A(x,y) \rightarrow \exists v(\forall x \varepsilon u. \exists y \varepsilon v.A(x,y) \wedge \forall y \varepsilon v. \exists x \varepsilon u.A(x,y))$,

where v is not free in $A(x,y)$.

(12) Foundation:

$(x$ transitive $\wedge \forall y((y \varepsilon x \wedge y \subseteq z) \rightarrow y \varepsilon z)) \rightarrow x \subseteq z$.

(13) Induction Scheme:

$A(0) \wedge \forall x \varepsilon \mathbf{N}(A(x) \rightarrow A(x')) \rightarrow \forall x \varepsilon \mathbf{N}.A(x)$.

(14) RDC

(15) ε-induction

(16) Separation .

Note that the primitive recursive functions on $\mathbf{N}$ are definable in (1)-(12). Let τ_1 consist of (1)-(12), τ_2 consist of (1)-(14), τ_3 consist of (1)-(15), τ_4 consist of (1)-(16). (Note that τ_2 is essentially Myhill's CST of [M3], and that $\mathbf{B}$ is a subsystem of (1)-(11)). Let $\mathcal{S}$ be the intuitionistic system obtained from τ_1 by replacing Bounded DC with Countable Choice, and induction is restricted to Σ_1^0 or Π_1^0 sets.

Let $R(<\varepsilon_0)$ be the classical ε_0-ramified analysis, let $R^-(<\varepsilon_0)$ be $R(<\varepsilon_0)$ in Heyting's predicate calculus. Let $ID(\mathcal{O})$ be the classical first order theory of Kleene's $\mathcal{O}$ as an inductive definition, and let $ID^-(\mathcal{O})$ be $ID(\mathcal{O})$ in Heyting's predicate calculus (for a precise description, cf. [Fr5], pp. 20-21). Let PAS be the classical second-order arithmetic.

<u>Theorem 6.1</u> [Fr5,8]. The Π_2^0 sentences provable in $\mathcal{S}$, $\tau_1,\tau_2,\tau_3,\tau_4$ are, respectively, the same as those provable in primitive recursive arithmetic, HA , $R^-(<\varepsilon_0)$, $ID^-(\mathcal{O})$, and HAS .

<u>Theorem 6.2</u> [Fr5,7]. Every arithmetic sentence provable in $\tau_1,\tau_2,\tau_3,\tau_4$, respectively, is provable in HA , $R(<\varepsilon_0)$, $ID(\mathcal{O})$, and PAS.

<u>Remark</u>. Theorems 6.1 and 6.2 hold for $\mathbf{B}$ in place of τ_1 as well.

7. Partially intuitionistic set theories

Inspection of the usual proofs of most results of basic classical set theory (facts about ordinals, cardinals, the cumulative hierarchy, etc.) shows that Replacement, Separation, and the Law of Excluded Middle are applied only for Δ_0 -formulas, or $\Delta_0(\mathcal{P})$- formulas at worst ($\Delta_0(\mathcal{P})$-formula has only quantifiers of the form $\exists x \subseteq y$ or $\forall x \subseteq y$). On the other hand, notice that the Law of Excluded Middle holds in HA for Δ_0- formulas (of HA). These considerations led to the investigations of several fragments of ZFC in intuitionistic logic, which contain classical logic for important classes of formulas [Th], [Fr1,§6], [Fr3], [W1,2]. The methods and results of Friedman and others for ZFI described in the previous sections were modified and extended to such partially intuitionistic set theories in R. Wolf's Stanford dissertation [W1], directed by Friedman.

All fragments of ZFC considered here have Heyting's predicate calculus and the Law of Excluded Middle for Δ_0- formulas as the underlying logic, the usual forms of the Extensionality, Regularity, Pairing, Union, Power Set, Infinity, Δ_0- Separation, and Δ_0- collection. In addition, let K_1 have $\Delta_0(\mathcal{P})$- Replacement, "Every well-ordering is isomorphic to an ordinal", and ε -induction (in the presence of full Separation, Regularity and ε - induction would be equivalent, but in the present context, Regularity cannot even prove the existence of $R_{\omega+\omega}$, cf. [Ba]). Note that one can use $\Delta_0(\mathcal{P})$ -Collection, Π_1 -Replacement, or Π_1 -Collection instead of $\Delta_0(\mathcal{P})$- Replacement.

Let K_2 be K_1 in classical logic (i.e. essentially ZF with Replacement restricted to $\Sigma_1(\mathcal{P})$ -formulas). Let K_3 be K_1 with a strong form of Replacement which allows transfinite recursive definitions (for details, cf. [W2]). Let KL_i be $K_i + V = L$, KC_i be $K_i + AC$. Finally, we describe system $ZFC^{1/2}$ of [Fr1,§6] which is possibly stronger than K_3 ; it has Separation for decidable formulas (i.e. for which the law of Excluded Middle holds for each instance), Collection, Choice, and Transfinite Induction over well-founded decidable definable classes, besides the basic axioms given above. (Certainly $ZFC^{1/2}$ contains KC_1 + Collection.)

Friedman [Fr1,§6] and Tharp [Th] have independently given set-theoretic versions of Kleene's 1945-recursive realizability. It is used in [W1] to show the interpretability of KL_3 into KL_1 , $ZFC^{1/2} + V = L$ into KL_1 with

Transfinite Induction over well-founded decidable Δ_2 -classes; and KL_3 with set-theoretic modifications of CT and MP into KL_2. It is also used in [Fr3] to show that the non-classical theories described above do not prove the existence of a Hanf bound of the infinitary language $\mathcal{L}_{\omega_1\omega_1}$ (i.e. a cardinal κ so that for each formula ϕ of $\mathcal{L}_{\omega_1\omega_1}$, if ϕ has a model of power $\geq \kappa$, then ϕ has arbitrarily large models); although KC_2 proves the existence of Hanf bounds of the infinitary languages $\mathcal{L}_{\kappa\kappa}$ for all infinite cardinals κ.

Combining the negative interpretation with set-theoretic versions of Godel's Dialectica interpretation, it is shown in [Wl] that KL_2 interprets into KL_1. Thus all theories K_i, KL_i, KC_i are equiconsistent. Moreover, KL_i (KC_i, respectively) all prove the same Π_3 formulas, but not Σ_3 formulas. Although these systems do not have Disjunction Property, some modifications of it do hold.

References

[Ab] O. Aberth, Computable analysis, McGraw-Hill, New York, 1980.

[Ac] P. Aczel: The type-theoretic interpretation of constructive set theory: choice principles; in: The L.E.J. Brouwer Centenary Symposium, ed. by A. S. Troelstra and D. van Dalen, North-Holland, Amsterdam, 1982, pp. 1-40.

[Ba] J. Barwise: The Hanf number of second order logic, J. Symb. Logic 37 (1972), pp. 588-594.

[Be] M. Beeson: Continuity in intuitionistic set theories, in: Logic Colloquium 78 (ed. by M. Boffa, D. van Dalen, and K. McAloon), North Holland, Amsterdam, 1979, pp. 1-52.

[Bi] E. Bishop: Foundations of constructive analysis, McGraw-Hill, New York, 1967.

[BC] E. Bishop, H. Cheng: Constructive measure theory, Memoirs of the A.M.S. 116, 1972.

[Br] D. Bridges: Constructive functional analysis, Pitman, London-San Francisco-Melbourne, 1979.

[Bw] L.E.J. Brouwer: Cambridge Lectures on Intuitionism (ed. by D. van Dalen), Cambridge University Press, 1982.

[D] R. Diaconescu: Axiom of choice and complementation, Proc. A.M.S. 51 (1975), pp. 176-178.

[Fo1] M. P. Fourman: The logic of topoi; in: Handbook of Mathematical Logic (ed. by J. Barwise), North-Holland, Amsterdam, 1977, pp. 1054-1090.

[Fo2] M. P. Fourman: Sheaf models for set theory, J. Pure Appl. Algebra 19 (1980), pp. 91-101.

[FoS] M. P. Fourman, D. S. Scott: Sheaves and logic, in: Applications of Sheaves (ed. by M. P. Fourman, C. J. Mulvey and D. S. Scott), Springer LNM 753, Berlin, 1979, pp. 302-401.

[Fr1] H. Friedman: Some applications of Kleene's methods for intuitionistic systems, in: Cambridge Summer School in Mathematical Logic, Proceedings 1971 (ed. by A.R.D. Mathias and H. Rogers), Lecture Notes in Mathematics 337, Springer-Verlag, Berlin, 1973, pp. 113-170.

[Fr2] H. Friedman: The consistency of classical set theory relative to a set theory with intuitionistic logic, J. Symb. Logic 38 (1973), pp. 315-319.

[Fr3] H. Friedman: On existence proofs of Hanf numbers, J. Symb. Logic 39 (1974), pp. 318-324.

[Fr4] H. Friedman: The disjunction property implies the numerical existence property, Proc. Nat. Acad. Sci. U.S.A. 62 (1975), pp. 2877-2878.

[Fr5] H. Friedman: Set theoretic foundations for constructive analysis, Annals of Math., 105 (1977), pp. 1-28.

[Fr6a] H. Friedman: The arithmetic theory of sets and functions I, abstract, 1976.

[Fr6b] H. Friedman: The intuitionistic theory of sets and functions, abstract, 1976.

[Fr7] H. Friedman: Conservation of set theories without power set over HA , abstract, 1977.

[Fr8] H. Friedman: A system weaker than arithmetic for constructive analysis, abstract, 1978.

[Fr9] H. Friedman: Classically and intuitionistically provably recursive functions, in: Higher Set Theory (ed. by G.H. Müller and D.S. Scott), Lecture Notes in Mathemtics 669, Springer-Verlag, Berlin-Heidelberg-New York, 1978, pp. 21-27.

[FS1] H. Friedman, A. Ščedrov: Large sets in intuitionistic set theory, Annals Pure Appl. Logic 27 (1984), pp. 1-24.

[FS2] H. Friedman, A. Ščedrov: Set existence property for intuitionistic theories with dependent choice, Annals Pure Appl. Logic 25 (1983), pp. 129-140; Corrigendum, ibid., 26 (1984), p. 101.

[FS3] H. Friedman, A. Ščedrov: Arithmetic transfinite induction and recursive well-orderings, Advances in Math., to appear.

[FS4] H. Friedman, A. Ščedrov: The lack of definable witnesses and provably recursive functions intuitionistic set theories, Advances in Math., to appear.

[FS5] H. Friedman, A. Ščedrov: Intuitionistically provable recursive well-orderings, Annals Pure Appl. Logic, to appear.

[GM] N. D. Goodman, J. Myhill: Choice implies excluded middle, Z. Math.Logik Grundl. Math. 24 (1978), p. 461.

[Gr] R. J. Grayson: Heyting-valued models for intuitionistic set theory; in: Applications of Sheaves (ed. by M. P. Fourman, C. J. Mulvey, D. S. Scott), Springer LNM 753, Berlin, 1979, pp. 402-414.

[H1] S. Hayashi: On set theories in toposes; in: Springer LNM 891, Berlin, 1981, pp. 23-29.

[H2] S. Hayashi: A note on bar induction rule; in: The L.E.J. Brouwer Centenary Symposium (ed. by A. S. Troelstra and D. van Dalen), North-Holland, Amsterdam, 1982, pp. 149-163.

[Hy] J.M.E. Hyland: The effective topos, in: The L.E.J. Brouwer Centenary Symposium, ed. by A. S. Troelstra and D. van Dalen, North-Holland, Amsterdam, 1982, pp. 165-216.

[K] S. C. Kleene: Introduction to metamathematics, Van Nostrand, Princeton, NJ, 1950.

[Ko] A. Kolmogorov: Sur le principe "tertium non datur", Recueil. Math. de la Soc. Math. de Moscou 32 (1925), pp. 647-667. English translation: On the principle of excluded middle, in: From Frege to Gödel(ed. by J. van Heijenoort), Harvard Univ. Press, Cambridge, Mass., 1967, pp. 414-437.

[Kr1] G. Kreisel: Church's Thesis: a kind of reducibility axiom for constructive mathematics, in: Intuitionism and Proof Theory, Proceedings Buffalo 1968 (ed. by A. Kino, J. Myhill, R. E. Vesley),

North-Holland, Amsterdam, 1970, pp. 121-150.

[Kr2] G. Kreisel: Which number-theoretic problems can be solved in recursive progressions on π_1^1 -paths through $\mathcal{O}$, J. Symbolic Logic 37 (1972), pp. 311-324.

[LS] J. Lambek, P. J. Scott: Intuitionistic type theory and the free topos, J. Pure Appl. Algebra 19 (1980), pp. 215-257.

[L] D. Leivant: Intuitionistic formal systems, this volume.

[Mc] D. C. McCarty: Realizability and Recursive Mathemtics, Dissertation, Carnegie-Mellon University, 1984.

[Mu] P. S. Mulry: Generalized Banach-Mazur functionals in the topos of recursive sets, J. Pure Appl. Algebra 26 (1982), pp. 71-83.

[M1] J. Myhill: Some properties of intuitionistic Zermelo-Fraenkel set theory, in: Cambridge Summer School in Mathematical Logic, Proceedings 1971 (ed. by A.R.D. Mathias and H. Rogers), Lecture Notes in Mathematics 337, Springer-Verlag, Berlin, 1973, pp. 206-231.

[M2] J. Myhill: Embedding classical type theory in intuitionistic type theory, Proc. Symp. Pure Math. 13, part I (1974), pp. 267-270; 13, part II (1974), pp. 185-188.

[M3] J. Myhill: Constructive set theory, J. Symb. Logic 40 (1975), pp. 347-382.

[S1] A. Ščedrov: Consistency and independence results in intuitionistic set theory; in: Constructive Mathematics, Proceedings 1980 (ed. by F. Richman), Springer LNM 873, Berlin, 1981, pp. 54-86.

[S2] A. Ščedrov: The independence of the fan theorem in the presence of continuity principles; in: The L.E.J. Brouwer Centenary Symposium (ed. by A. S. Troelstra and D. van Dalen), North-Holland, Amsterdam, 1982, pp. 435-442.

[SS] A. Ščedrov, P. J. Scott: A note on the Friedman slash and Freyd covers, in: The L.E.J. Brouwer Centenary Symposium (ed. by A. S. Troelstra and D. van Dalen), North-Holland, Amsterdam, 1982, pp. 443-452.

[Sco] D. S. Scott: Relating theories of the λ -calculus; in: To H. B. Curry: Essays on combinatory logic, lambda calculus and formalism (ed. by J. P. Seldin and J. R. Hindley), Academic Press, New York, 1980, pp. 403-450.

[Si] S. G. Simpson: Subsystems of second order arithmetic, this volume.

[Th] L. Tharp: A quasi-intuitionistic set theory, J. Symb. Logic 36 (1971), pp. 456-460.

[Tr1] A. S. Troelstra: Metamathematical investigation of intuitionistic arithmetic and analysis, Lecture Notes in Mathematics 344, Springer-Verlag, Berlin, 1973.

[Tr2] A. S. Troelstra: Aspects of constructive mathematics, in: Handbook of mathematical Logic (ed. by J. Barwise), North-Holland, Amsterdam, 1977, pp. 973-1052.

[W1] R. Wolf: Formally intuitionistic set theories with bounded predicates decidable, Dissertation, Stanford University, 1974.

[W2] R. Wolf: A highly efficient "Transfinite recursive definitions" axiom for set theory, Notre Dame J. of Formal Logic 22 (1981), pp. 63-75.

HARVEY FRIEDMAN'S RESEARCH ON THE FOUNDATIONS
OF MATHEMATICS, L.A. Harrington et al. (editors)

ALGORITHMIC PROCEDURES, GENERALIZED TURING ALGORITHMS, AND ELEMENTARY RECURSION THEORY

J.C. SHEPHERDSON

Mathematics Department
University of Bristol
England

1. INTRODUCTION:

The theory of recursive or computable functions has been generalized in many ways:[1]

(1) To provide techniques for use in other parts of mathematics, particularly logic, set theory, and classical recursive function theory itself.

(2) To gain better understanding of advanced recursion theory, e.g. degree theory and hierarchy theory.

(3) To gain better understanding of the nature of computation.

The generalizations resulting from (1) and (2) have been very successful but have not helped with (3) very much because they usually involve generalizing the notion of finite, and the use of infinitely long computations. The present paper[43], though it has valuable spin-off to (1) and (2), is squarely aimed at (3). It is a major contribution aimed at answering the fundamental question,

Q 'What becomes of the concepts and results of elementary recursion theory if, instead of considering only computations on natural numbers, we consider computations on data objects from any relational structure?'

This has some relevance to real life programming where one performs e.g. arithmetic operations on real and complex numbers as well as on natural numbers. Admittedly in any implementation on a particular machine these continuous quantities are represented by finite discrete approximations, but for many purposes it is appropriate to make the idealization, which the abstract form of the programming language does, that one can perform these basic operations with perfect accuracy. Just before this paper Herman and Isard[13] proved two theorems about what could and could not be computed or decided by programs whose basic instructions were addition, multiplication, etc. for real and complex numbers. And Luckham and Park[19] building on an earlier idea of Ianov[15] introduced the idea of an abstract program scheme whose basic instructions, e.g. '$x := F(y)$', 'if $R(x,y)$ go to i else go to j', were thought of not as applying to a particular structure but as being applicable to any relational structure of the right type (i.e. in this case having a one argument function F, a two place predicate R). Their aim was to separate questions about the control structure of a program from the undecidable mathematical problems which immediately arose from the 'accidental' fact that the data operated on in the classical case were natural numbers. [Unfortunately it turned out (Luckham, Park, Paterson [20]) that the control structure or abstract program scheme could by itself be used to simulate the action of a Turing machine and that the unsolvability of the halting problem for Turing machines implied the unsolvability of the equivalence and optimization problems even for program schemes.]

In this paper Friedman independently attacks Q *ab initio*. He starts with a definition of computation in terms of programs. A formalised algorithmic procedure or *fap* is a finite list of instructions each of one of the three forms

$$y := F(x_1,\ldots,x_k)$$

if $R(x_1,\ldots,x_\ell)$ go to i, else go to j

stop.

If this program scheme is interpreted by taking a particular relational structure with domain D, and interpreting $x_1,x_2,\ldots,y,\ldots$ as variables ranging over D and all the F,R as functions and predicates over D then it becomes a program which computes a (partial) function over D.

In classical recursion theory one is computing over the natural numbers and they not only form the data objects but can be used for auxiliary arithmetic operations such as counting. In general one would expect that adding counters i.e. allowing variables, m, n, $n_1,\ldots$ ranging over the natural numbers, with instructions

$$m := 0$$

$$m := m+1$$

if $m_1 = m_2$ go to i else go to j

giving a formalized algorithmic procedure with counting (*fapc*), would increase the set of functions and relations which could be computed. Friedman shows that this is the case but singles out a large class of structures for which the two types of procedure are equivalent, (because one can construct surrogate natural numbers from the data elements). Is there anything else like this which would normally be considered part of computation, which one needs to add to *fapc* so as to be able to compute all functions and relations which are intuitively computable using the given basic functions and text predicates? Friedman shows that the answer is 'yes!' but to do this he must first give the appropriate notion of 'all computable functions'. He does this first in the form of generalized Turing algorithms (*gTa*), the natural generalization of a Turing machine. He then gives an equivalent definition with a much simpler logical structure. This is the notion of an effective definitional scheme (*eds*). This is essentially a definition by a recursively enumerable list of cases, where each case is given by a finite conjunction of atomic formulae and their negations. This elegant and powerful concept is much easier to work with than *gTa* and represents a considerable advance. Friedman shows that *eds* and hence *gTa* are in general more powerful than *fapc* but gives some conditions on structures sufficient to ensure that *fapc* are no more powerful than *fap*, and stronger conditions which ensure that *fap* are equivalent to *eds*. He also shows how to add to a structure so as to make it satisfy these conditions. It turns out to be equivalent to adding the equality relation as a basic test, i.e. allowing instructions of the form:

if $x = y$ go to i else go to j,

and allowing the use of push down stacks containing elements of D.

The second section of the paper is a lightning comprehensive survey of how the theorems of elementary recursion theory lift to arbitrary structures when the notion of recursive is taken to mean computable by a *gTa*. He divides 22 theorems of elementary recursion theory into 6 groups and shows e.g. that those in the first group lift to all structures, those in the second group lift to all structures with an equality relation, those in group 4 lift to all structures which are computably finitely generated. This throws new light on elementary recursion theory; as Friedman says, it isolates the computational meaning of the proofs in elementary recursion theory.

In a final brief section he suggests half a dozen directions for future work, which have by no means been exhausted yet.

This is a classic paper which shows Friedman's ability to go straight to the heart of the matter and pick out the most significant concepts. My own first reaction on reading it was one of chagrin because I had been thinking a little about these things myself and was disappointed to find that he had got there first. But this soon turned to admiration for the elegance and power with which, in a first paper in the field, he went so far. Instead, like most of us, of making timid forays near the edge, he strides boldly into a new field knocking down theorems left and right.

Unfortunately the paper did not at first receive the attention it deserves, particularly among theoretical computer scientists. This may have been partly because Friedman did not go out of his way to speak their language. In the final paragraph he says

> 'Computer scientists tell me that our algorithmic procedures without counting are similar to their program schemata, and have referred me to Paterson [32]. It would be interesting to see whether more of the work of computer scientists fits into the framework presented here. Perhaps the way in which the basic definitions of Section 1 are set out lead to the best ways of expressing basic distinctions familiar to computer scientists'.

But he does not spend time himself here looking into the precise relationship between the two ways of looking at things probably because he only heard of program schemata when he'd written this paper. However I think the neglect is mainly due to this paper having been published in the proceedings of a logic conference instead of in an international mathematical journal. Also, due to delay in publication, by the time this paper had appeared a parallel stream of development (of the material of most interest to computer scientists, i.e. section 1 of the paper) had appeared in computer science journals [see sections 4,5 below] and these naturally became the best known sources for computer scientists.

Friedman says 'In many cases only the intuitive idea of the proof is conveyed. The emphasis here is on presenting basic notions and ideas; this is not the place to look for delicate mathematical arguments.' But despite this - or perhaps because of it - this paper is still compulsory reading for anyone working in the field, not only for its fundamental results but also for its still potent spring of inspiration for future development.

2. SUMMARY OF SECTION 1:

Here and above we follow the computer science writers and simplify and recast Friedman's definitions very slightly so that they look more like conventional programming languages. A <u>formalised algorithmic procedure</u> or *fap* is a finite list of instructions each of one of the three[2] forms:

$y := F(x_1,\ldots,x_k)$

if $R(x_1,\ldots,x_\ell)$ go to i, else go to j

stop.

Here i,j are the labels or numbers of instructions, the understanding being, as usual, that you start on the first instruction and, after an instruction of the first type you proceed to the next instruction on the list. The F is a k-ary function or operation symbol and R an l-ary predicate or test symbol drawn from given finite sets[3] $\left\{F_1^{a_1},\ldots,F_r^{a_r}\right\}$, $\left\{R_1^{b_1},\ldots,R_m^{b_m}\right\}$ of function and predicate symbols of arities (number of arguments) $a_1,\ldots,a_r$, $b_1,\ldots,b_m$ respectively. This program scheme becomes a program if we take a particular structure $(D,f_1,\ldots,f_r,$ $r_1,\ldots,r_m)$ with some domain D and of the right similarity type, i.e. with each f_i a function $D^{a_i} \to D$, each r_i a predicate or relation $D^{b_i} \to \{\text{true, false}\}$, and replace each F_i by the corresponding f_i, each R_i by the corresponding r_i, and interpret the variables $x, x_1, \ldots, y, y_1, \ldots$ as ranging over D. If input variables or registers $x_1, \ldots, x_n$ and an output variable or register y are specified then this program computes a partial function $\phi: D^n \to D$; namely $\phi(d_1, \ldots, d_n)$ is defined and equal to d if and only if the program, when started with $x_1 = d_1, \ldots, x_n = d_n$ and all other variables with no values assigned (i.e. their registers empty), stops with $y = d$. If you want to compute relations instead of functions you can add boolean variables. This is often done but adds nothing to the power and can be avoided by agreeing that halting with output $y = d$ for some $d \in D$ means 'true', halting with y undefined (i.e. y register empty) means 'false', not halting means 'undefined'. In this way, when input variables $x_1, \ldots, x_n$ are specified each program computes a partial n-ary relation (i.e. whose values are true, false or undefined).

As described above the *fapc*[4] or formalized algorithmic procedure with counting has, in addition to the data variables $x, x_1, \ldots, y, y_1, \ldots$ also number variables $m, m_1, \ldots, n, n_1, \ldots$ and instructions

$$m := 0$$

$$m := m+1$$

$$\text{if } m_1 = m_2 \text{ go to } i \text{ else go to } j.$$

The generalized Turing algorithm (*gTa*) is like an ordinary Turing machine except that instead of the finite number of ordinary symbols (now called auxiliary symbols) it can also 'place' elements of the data structure D on the tape. It can tell whether the scanned square contains a data element or an auxiliary symbol and in the latter case can read it. When used to compute a function or relation of n arguments it starts with $x_1, \ldots, x_n$, the n elements of D which are the argument values, on n adjacent squares of the tape, the scanning head on the leftmost of these, and all other squares blank. In addition to the usual Turing machine moves it can switch the contents of the scanned square with that of its left or right neighbour, test the contents of the squares immediately to the right of the scanned square for satisfaction of a relation R, or compute a basic function F of these contents. What this means is that if $r_i^{b_i}$ is the corresponding basic relation of the data structure on which it is operating them it can test $r_i^{b_i}(y_{k_1}, \ldots, y_{k_{b_j}})$ where[5] $1 \le k_1, \ldots, k_{b_j} \le b_j$ and $y_1, \ldots, y_{b_j}$ are the contents of the b_j squares to the right of the scanned square (if one of the y_k is not an element of D then the result is undefined). Similarly if $f_i^{a_i}$ is one of the basic functions it can replace the content of the scanned square by $f_i^{a_i}(y_{k_1}, \ldots, y_{k_{a_j}})$, where[5] $1 \le k_1, \ldots, k_{a_j} \le a_j$. For computing a function the output value is taken to be the element of D on the scanned square (if when it halts it is scanning an element of D); for computing a relation, scanned square = blank means 'false', scanned square = non blank means 'true', doesn't halt means 'undefined'.

Finally an effective definitional scheme or *eds* is a recursively enumerable set of clauses

$$E_1 \ \& \ E_2 \ \& \ldots \& \ E_k \to t$$

where t is a term built up from the input data variables $x_1,\ldots,x_n$ and function symbols $F_1,\ldots,F_r$, and $E_1,\ldots,E_k$ are atomic formulae or their negations i.e. of the form $R_i(t_1,\ldots,t_{b_i})$ or $\neg R_i(t_1,\ldots,t_{b_i})$ where $t_1,\ldots,t_{b_i}$ are terms (in $x_1,\ldots,x_n$) and R_i is a basic relation symbol of arity b_i. The intended meaning is

'if E_1 & E_2 &...& E_k holds the value of the function defined by this procedure is to be t'.

In order that the output function should be single valued the condition is imposed that distinct clauses must have incompatible antecedents i.e. one of the conjuncts of one must be the negation of one of the conjuncts of the other. For defining or computing relations the t on the right hand side is replaced by 'true' or 'false'. When applied to a specific relational structure of the right type the function or relation defined or computed by such an *eds* is the one obtained by using the interpretation above, replacing each function and predicate symbol F_i, R_j by the corresponding function or predicate f_i, r_j of the structure. This simple notion strikes one immediately as capturing the general notion of computability by a finite procedure in a finite number of steps, for each computation uses a finite amount of information about the structure of exactly the kind expressed by E_1 & E_2 &...& E_k. It is reminiscent of some definitions of relative recursiveness in degree theory.

An easy argument (an effective version of the theorem that every open set in the Cantor space can be written as the union of pairwise disjoint basic open sets) establishes the useful lemma that the condition that distinct clauses in an *eds* must have incompatible antecedents may be replaced by the weaker condition that two elements whose antecedents are not incompatible must have the same consequent t.

Friedman's first main theorem, is

THEOREM 1.1 *eds are at least as strong as fap or gTa.*

SKETCH PROOF. The argument applies not only to *fap* and *gTa* but to any conceivable kind of computation and convinces us that in the *eds* he really has captured the most general notion of computability over abstract structures. In outline it goes like this. Given an *fap*, *gTa* or other algorithm $\mathscr{A}$ consider the set γ of clauses $C \rightarrow t$ such that:

C contains enough information about the input variables $x_1,\ldots,x_n$ for you to follow the action of $\mathscr{A}$ symbolically; and this action eventually stops and outputs a value equal to the value of the term t.

By the following the action symbolically, we mean with terms instead of their values, so that this can be carried out in ordinary recursion theory, or by an ordinary computer working with natural numbers or words on a finite alphabet as data. We have to show that this set γ of clauses is an *eds* and is equivalent to the given algorithm $\mathscr{A}$. To see that γ is recursively enumerable we obtain it by recursively enumerating all clauses and, as each clause C is enumerated trying to follow the action $\mathscr{A}$ would take on $x_1,\ldots,x_n$ if C were true. If $\mathscr{A}$ calls for information about the truth of atomic formulae not in C then we discard this clause and pass on to the next. If however we succeed in following the action through to a halt with output of a term t then we put $C \rightarrow t$ in γ. We have to do all this in a step by step manner, coming back to try and follow the computation for earlier C's one step further because it is possible that C might have enough information to follow the action of $\mathscr{A}$ completely, but this action might be to go on forever without halting. The other condition needed for γ to be an *eds*, that if two of its clauses $C_1 \to t_1$, $C_2 \to t_2$ have compatible antecedents C_1, C_2 they must have $t_1 = t_2$, is satisfied because if both C_1, C_2 are compatible and have enough information to follow the computation of $\mathscr{A}$ through to a halt then that information must be in their common part and their action must be the same. By definition of γ if γ gives a value for arguments $x_1,\ldots,x_n$ then $\mathscr{A}$ gives the same value. So it remains to show that on all arguments $x_1,\ldots,x_n$ on which $\mathscr{A}$ gives a value γ does too. This follows from the remark above, that if $\mathscr{A}$ gives a value it does so after a finite number of steps and the only way the values of $x_1,\ldots,x_n$ can influence the computation - since the only basic calculations allowed on D are computations of the f_i and testing the r_j - is via the truth or falsity of a finite number of atomic formulae, so there is some C which contains all this information, and the corresponding term $C \to t$ will therefore be in γ.

Note that this is a uniform result; γ is equivalent to $\mathscr{A}$ over all structures of the right type.

THEOREM 1.2 <u>*gTa* are at least as strong as *eds*.</u>

SKETCH PROOF. Given an *eds* γ we construct a *gTa* which enumerates the clauses of γ and, for each clause $C \to t$ replaces all occurrences of variables by their values and then, starting with subterms of lowest complexity, replaces all terms in C by their values, then evaluates all atomic formulae in C. If C evaluates to true it evaluates t and outputs it; otherwise it goes on to the next clause.

Here again the simulation of the *eds* by the *gTa* is uniform i.e. works over all structures of the right type.

THEOREM 1.3 <u>In general *eds* are stronger than *fapc*.</u>

SKETCH PROOF. In a given *fapc* you can only use a fixed finite number of variables ('registers', 'locations') but in general when evaluating terms you may need to have simultaneously available an unbounded number of subterms. For example let F be a binary and U singularly function symbol. Consider the array in Figure 1, where $U^2(x)$, etc. stands for $U(U(x))$ etc.

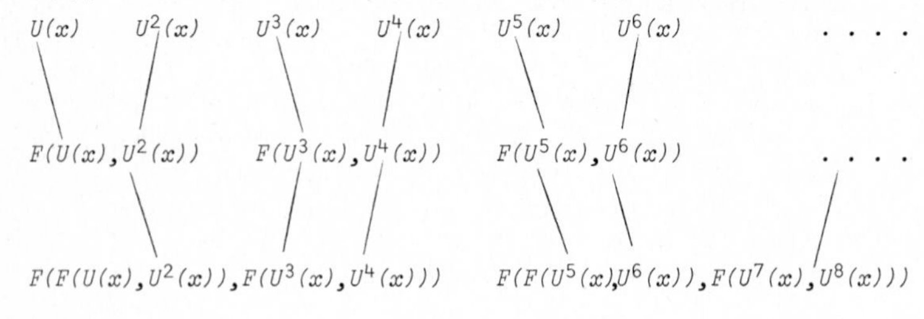

Figure 1. A 'wide' array of terms

To compute terms on level i needs at least i registers. This can easily be proved by induction or it can be seen intuitively using the pebble game[6] of Paterson and Hewett, who give a similar example[33] where you are allowed i pebbles and you try and cover the term you want by a pebble, the rules being that you can only move a pebble onto a term if all terms immediately above it are covered by pebbles. Let us now add a singulary relation symbol R and consider the algorithm $\mathscr{A}$ which on an argument x searches down the leftmost branch of the above tree for the first element (if any) satisfying R and, if it finds one outputs this as the value. By Theorem 1.1 there is an *eds* β equivalent to the algorithm.

Define a 'free' structure $\mathscr{F} = (D,u,f,r)$ of this type where D is the set of terms built up from the constant symbol c, with u,f,r defined by $u(t_1) = U(t_1)$, $f(t_1,t_2) = F(t_1,t_2)$, $r(y)$ true if and only if there is an i such that y is the ith term down the left hand branch of the above array based on $x = U^i(c)$. Then the function on $\mathscr{F}$ computed by β is not computable by any algorithm like *fap* or *fapc* which can only use a fixed finite number of registers each holding only one data element. [Friedman's proof of 1.3 is defective; lemma 1.3.1 on p373 fails; all the terms can be computed with 3 registers because there are common subterms.]

Here too the result is as strong as possible; we have a fixed structure M on which there is an *eds* not equivalent to any *fapc*. We can also throw in the equality relation as a basic relation without affecting the result.

THEOREM 1.4 <u>For a ω-rich structure (see below), *fap* are equivalent to *fapc*.</u>

SKETCH PROOF. An ω-rich structure is by definition one in which surrogates for the natural numbers and the counting instructions can be defined i.e. such that there exists an injection $g = \omega \to D$ and for some constant (i.e. zero-ary function) $c \in D$ and some f_j and binary R_e we have $g(n+1) = f_j(g(n),\dots,g(n))$ and, for all natural numbers n_1, n_2 we have $R_e(g(n_1),g(n_2)) \leftrightarrow (n_1 = n_2)$. There is then no need for counters; we simply use the data element $g(n)$ instead of the number n. Again the reduction is uniform.

THEOREM 1.5 <u>In general *fapc* are stronger than *fap*.</u>

SKETCH PROOF. Let M_n be the structure formed by the natural numbers cut off at n, with $n+1 = 0$ i.e. the structure $(\{0,1\dots,n\},f,=)$ where $f(i+1) = i+1$, if $i < n$, $= 0$ if $i = n$. Any *fap* of the right type (i.e. with one one-place function and one two-place relation) for computing a one-place function, and which has q instructions and uses ν variables, halts when applied to M_n with input 0, if and only if it halts in $\leq q(n+2)^{\nu}$ steps (otherwise it would repeat some total configuration and get into a loop). Since all recursive functions of natural numbers are computable on an *fapc* we can construct an *fapc* β which, when applied to M_n with input 0 first finds the value of n (by counting until $f^n(0) = 0$) then works out whether the nth *fap* would halt on M_n with input 0, and proceeds to do the opposite. This shows that there is no *fap* which is equivalent to the *fapc* β over all the structures M_1, M_2, ... You can give a fixed structure M (including = as a basic relation) on which no *fap* is equivalent to β by glueing all the M_k together, i.e. $M = (D,f,=)$ where $D = \{(n,m): n,m \in \omega$ and $0 \leq m \leq n\}$, $f((n,m)) = (n,m+1)$ if $n \neq m$, $= (n,0)$ if $n = m$.

THEOREM 1.6 <u>Over structures with = (as a basic relation) which are structural (see below) *fapc* are as strong as *eds*.</u>

SKETCH PROOF. A structure M is called <u>structural</u> if for each n there is some number p such that every term in $x_1,\dots,x_n$ over M is equal to a term which can be evaluated using not more than p registers. Most of the familiar algebraic structures are structural. For groups, semigroups or associative systems the associative law enables you to compute all terms using only one register (in addition to the n registers needed to hold the initial arguments); for (associative) rings the usual 'sum of products' form shows that two additional registers are enough; for fields three will do. Similarly for Lie rings, semilattices, lattices, boolean algebra. See Tucker[41].

This property of structurality, the ability to evaluate terms using only a fixed number of auxiliary registers is clearly an algebraic property but doesn't appear to be related to any well known algebraic properties. Perhaps it has not been studied precisely because most familiar algebraic systems are structural. Or is it the other way round? That the importance of associative and distributive laws is that they enable one to get a normal form for terms which allows their calculation with a small number of registers. Certainly one of the unpleasant things about dealing with non-associative systems is this lack of a simple normal form.

Since most of the familiar algebraic systems are structural and most of the infinite ones are ω-rich this theorem and Theorem 1.4 show that, despite the large theoretical gulf between *fap* and *eds* shown by Theorems 1.3 and 1.5, in practice, for algebraic systems of the usual kind they are equivalent, that the *fap* is then a general purpose computer: so Herman and Isard[13] were not restricting themselves when considering computations over fields (at least for fields of characteristic 0) by using only *fap*.

The proof of Theorem 1.6 is similar to that of Theorem 1.2 but simpler because you have random access to registers instead of the restriction to adjacent registers which, on the Turing machine, constantly force you to shift data and rough work in order to get access to other things. What it amounts to is verifying that the only difficulty in simulating an *eds* by an *fapc* is in evaluating terms. Since we have only required that each term is equal to one which can be evaluated using p registers, in order to evaluate a term you have to list its sequence of subterms, then enumerate all sequences of p-terms (terms computable with p-registers) until you find one in which the values of successive terms are built up in the same way. For example to evaluate $((x_1 o x_2) o (x_2 o x_1))$ for $x_1 = d_1$, $x_2 = d_2$ in an associative system you list a sequence for building up the given term, $((x_1 o x_2) o (x_2 o x_1))$:

$$x_1, x_2, (x_1 o x_2), (x_2 o x_1), ((x_1 o x_2) o (x_2 o x_1))$$

then you enumerate successively all sequences of five 1-terms

$$t_1(x_1,x_2), t_2(x_1,x_2), t_3(x_1,x_2), t_4(x_1,x_2), t_5(x_1,x_2)$$

until you find one which satisfies

$$\begin{aligned} t_1(d_1,d_2) &= d_1 \\ t_2(d_1,d_2) &= d_2 \\ t_3(d_1,d_2) &= t_1(d_1,d_2) o t_2(d_1,d_2) \\ t_4(d_1,d_2) &= t_2(d_1,d_2) o t_1(d_1,d_2) \\ t_5(d_1,d_2) &= t_3(d_1,d_2) o t_4(d_1\ d_2) \end{aligned}$$

e.g. the sequence

$$x_1, x_2, (x_1 o x_2), (x_2 o x_1), (((x_1 o x_2) o x_2) o x_1).$$

The value $((d_1 o d_2) o (d_2 o d_1))$ of the given term is the value of $t_5(d_1,d_2)$ for such a sequence. If q is the maximum of the arities $a_1,\ldots,a_r$ of the basic functions of the structure then to do this checking exercise you need, at any one time, the values of at most q p-terms, so $n+p(q+1)$ registers are sufficient (5 in the above example of terms in x_1, x_2 over an associative system).

DEFINITION. A formalized algorithmic procedure with stacks or *faps* is an *fap* augmented with any finite number of push-down stacks. Each of these can store a finite sequence of data elements on a last-in first-out basis. The allowable operations on a stack s are 'push' - add a data element to the 'top' of the stack and 'pop' - take the data element off the 'top' of the stack and put it somewhere else. These are expressed by the instructions:

$$push(s,x)$$

$$\text{if } s = \emptyset \text{ go to } i \text{ else } pop(s,x) \text{ and go to } j$$

where s is the name of a stack and x is a data variable.

If we use s to denote also the contents of the stack s, with the bottom at the left, top at the right, then

$$push(s,x) \text{ means } s := (s,x)$$

and,

$$\text{if } s = (s_1,y)$$

$$pop(s,x) \text{ means } x := y,\ s := s_1.$$

THEOREM 1.7 On structures with =,[7] *fap* with stacks are as strong as *eds*.

SKETCH PROOF. We have expressed the theorem like this for the sake of easier comparison later with the 'computer science' writers (see section 4 below). Friedman expressed the result in a model theoretic form which he would probably prefer and to be fair to the computer scientists it was they who introduced the notion of a stack. What Friedman actually said was than an *eds* operating on a structure M could be simulated by an *fap* operating on an augmented structure M^* obtained from M by adding to domain D all finite strings of elements of D and introducing new operations on this enlarged domain corresponding roughly to *push* and *pop*.

To prove Theorem 1.7 we use Theorem 1.6. Since stacks can obviously be used as counters we have only to show, in Friedman's terms that the augmented structure M^* is structural, in our terms, how to evaluate all terms using stacks. One way of doing this is to build up on a stack a sequence of subterms ending with the given term. You can do this by using a second stack to store the top section of the first when you want to peel it off to get at an earlier element and a counter to store a number coding the sequence giving the numbers starting from the top, of the earlier terms used, e.g.

$$x_1, x_2, (x_1 o x_2), (x_2 o x_1), (x_1 o (x_2 o x_1)), ((x_1 o (x_2 o x_1)) o x_1$$

$$(2,1) \quad (2,3) \quad (4,1) \quad (5,1)$$

In the last step, to put the last element on the stack, you get the pair (5,1) from the coded sequence of ordered pairs, pop the top element $(x_1 o (x_2 o x_1))$ off the stack into argument register 1, pop the next 3 elements onto the auxiliary stack, pop the 5th element x_1 into argument register 2, compute $((x_1 o (x_2 o x_1)) o x_1)$ in the result register as the product of the elements in argument registers 1 and 2, then push the elements back onto the first stack in the right order and finally push the new term on the top of it.

As formulated here there is no need for the assumption that equality is a basic relation. We used this in the proof of Theorem 1.6 to evaluate terms simply from knowing their value was equal to the value of a p-term. If we have stacks there is, as we have just seen, no difficulty in evaluating all terms.

THEOREM 1.8 *fapc are as strong as eds over any structure which has equality computable by an fapc and a weak pairing mechanism (see below) computable by fapc.*

SKETCH PROOF. A weak pairing mechanism on a structure with domain D is a pair (P,f) such that P is an injection $D^2 \to D$ and f a partial function $D \to D$ such that, for all $x,y \in D$ we have $f(P(x,y)) = y$. We rely on Theorem 1.6. Having equality computable by an *fapc* is just as good as having it as a basic relation. Although the structure itself may not be structural it is when P,f are adjoined. But P,f are obtained by *fapc* so the proof of Theorem 1.6 applies. Friedman shows that if $P(x,y) \neq y$ for all $x,y \in D$ then *fapc* may be replaced by *fap* throughout in Theorem 1.8 (but *fapc* can be allowed for the computation of f).

3. SUMMARY OF SECTION 2:

Friedman here takes up the question of how the results of elementary recursion theory generalize or, lift, to this new situation where the data are taken to be elements of a relational structure $(D,f_1,\ldots,f_r,r_1,\ldots,r_m)$ instead of natural numbers. He considers various definitions of recursive and recursively enumerable set which are equivalent in classical recursion theory and asks what, if any, additional assumptions on the structure are needed to preserve these equivalences. To cut out obviously singular cases, throughout this section he makes:

Assumption 1. All structures have infinite domains.

And to avoid considering some relatively uninteresting cases he makes

Assumption 2. All structures have no constants; but arbitrary constants can be used in computations.

Thus if we were considering *fap* we would simply allow an instruction $x: = d$ for each element d of D. In point of fact we want the most general notion of computable so we use *gTa* or equivalently (by Theorem 1.1, 1,2), *eds*. So for *gTa* we are allowed new instructions allowing any element d of D to be placed on a square of the tape; for *eds* we allow any finite number of elements of D to be used as well as variables $x_1,x_2,\ldots,x_n$ in building up terms.

Our generalization of partial recursive function is function computable by such a *gTa* or *eds*. The partial relations computable by such a *gTa* or *eds* are our generalization of partial recursively enumerable relation. The computational meaning of this is: given arguments we can test them for the relation. We may eventually terminate testing and give yes or no, or we may never terminate testing.

We shall omit proofs altogether in our discussion of this section.

THEOREM 2.1 *The following results lift to all structures.*

1(a) domains of partial recursive functions are the same as domains of partial recursively enumerable relations

(b) every domain of a monadic partial recursive function is the range of some monadic partial recursive function

(c) sets whose total characteristic relation is recursively enumerable are just the domains of partial recursive functions whose complement is the domain of some partial recursive function

(d) the domains of partial recursive functions are closed under pairwise intersection and union

(e) the partial recursive functions are closed under generalized composition

(f) the partial characteristic function of a partial recursively enumerable relation is partial recursive.

Condition I. Equality is a basic relation.

THEOREM 2.2 The following results lift to all structures satisfying condition I above.

2(a) the ranges of partial recursive functions are closed under pairwise intersection and union

(b) the ranges of total recursive functions are closed under pairwise intersections and unions

(c) the partial recursively enumerable relations are just those partial relations whose partial characteristic function is partial recursive

(d) the partial characteristic relation of a partial recursive function is partial recursively enumerable.

Condition II. There is a pairing mechanism and the set of relations is non-trivial.

Here a pairing mechanism is a triple (P,f,g) of gTa computable functions $P\colon D^2 \to D$, injective and total, $f\colon D \to D$, $g\colon D \to D$ such that $f(P(x,y)) = x, g(P(x,y)) = y$ for all $x,y \in D$. To say that the set of relations is non-trivial means that there is some gTa computable relation which is not always true or always false.

THEOREM 2.3 The following results lift to all structures satisfying Condition II above.

3(a) there is a partial recursive enumeration of the r-ary partial recursive functions

(b) there is a partial recursively enumerable enumeration of the k-ary partial recursively enumerable relations

(c) the domains of partial recursive functions are not closed under complementation.

Condition III. The structure is computably finitely generated.

This means that there are elements $d_1,\ldots,d_k \in D$ and total gTa computable functions $f_1,\ldots,f_\ell$ such that every element of D is expressible as a term involving only the f_i and d_j.

THEOREM 2.4 The following results lift to all structures satisfying Condition III.

4(a) the domains of partial recursive functions are closed under projection

(b) every partial recursively enumerable relation can be uniformized by a partial recursive function

(c) every partial function with partial recursively enumerable partial characteristic relation is partial recursive.

4(b) means that if $R(x_1,\ldots,x_k,y)$ is a k+1-ary gTa computable relation then there exists a gTa computable function $f(x_1,\ldots,x_k)$ such that $(\exists y)R(x_1,\ldots,x_k,y) \leftrightarrow (\exists y)(f(x_1,\ldots,x_k) = y)$.

THEOREM 2.5 The following theorem lifts to all structures satisfying conditions I and II:

5. The ranges of partial recursive functions are not closed under complementation.

THEOREM 2.6 The following theorems lift to all structures satisfying conditions I and III (and these imply condition II).

6(a) the range of every partial recursive function is the domain of some partial recursive function

(b) the ranges of total recursive functions are exactly the nonempty domains of monadic partial recursive functions

(c) the domain of every partial recursive function is the projection of some set with partial recursive total characteristic function

(d) there is a total recursive successor function which exhausts ω

(e) every infinite recursively enumerable set is the range of some one-one total recursive function.

Friedman conjectures that all these results are best possible i.e. no result numbered 3 lifts to all structures satisfying II and III; none numbered 3 lifts to all structures satisfying III; none numbered 4 lifts to all structures satisfying I and II; the one numbered 5 does not lift to all structures satisfying I nor to all structures satisfying II and III; the ones numbered 6 do not lift to all structures satisfying I and II nor to all structures satisfying II and III.

4. CHANDRA AND MANNA'S SURVEY OF THE POWER OF PROGRAMMING FEATURES:

We give now the comprehensive survey of the relative strength of *fap* with a variety of additional facilities to be found in the delightful paper of Chandra and Manna[5]. In addition to the counters and stacks described above they also consider:

Queues: like stacks only first-in first-out instead of last-in first out.

Arrays: an array A is a sequence of locations $A(0), A(1), A(2), \ldots$ each of which can hold one data value. The allowable operations are

$$x := A(c)$$

$$A(c) := x$$

where c is a counter (natural number variable). Arrays add nothing without counters so when we talk about programs with arrays we also include an arbitrary number of counters.

Recursive schemes: these are not just a feature to be added to *fap* but a separate method of defining functions. A recursive scheme for defining functions $G_1, \ldots, G_k$ of arities $c_1, \ldots, c_n$ is a set of n equations

$$G_1(\underline{x}_1) = \text{if } \alpha_1(\underline{x}_1) \text{ then } t_1(\underline{x}_1) \text{ else } t'_1(\underline{x}_1)$$

$$\ldots\ldots \qquad \ldots\ldots \qquad \ldots\ldots$$

$$G_n(\underline{x}_n) = \text{if } \alpha_n(\underline{x}_n) \text{ then } t_n(\underline{x}_n) \text{ else } t'_n(\underline{x}_n)$$

Here $\underline{x}_i$ denotes a vector $x_{i1}, \ldots, x_{ic_i}$ of c_i data variables, $t_i(\underline{x}_i), t'_i(\underline{x}_i)$ are terms in $\underline{x}_i$ (i.e. terms built up from the $\underline{x}_i$, the basic function symbols $F_1, \ldots, F_r$ of the structure and the functions $G_1, \ldots, G_n$ being defined) and $\alpha_i(\underline{x}_i)$ is $R(s_1, \ldots, s_j)$ where R is a basic predicate symbol and $s_1, \ldots, s_j$ are terms in $\underline{x}_i$. Recursive schemes for defining relations are defined similarly.

The relations between *fap* with various combinations of these features is neatly summed up in the Chandra-Manna diagram overleaf.

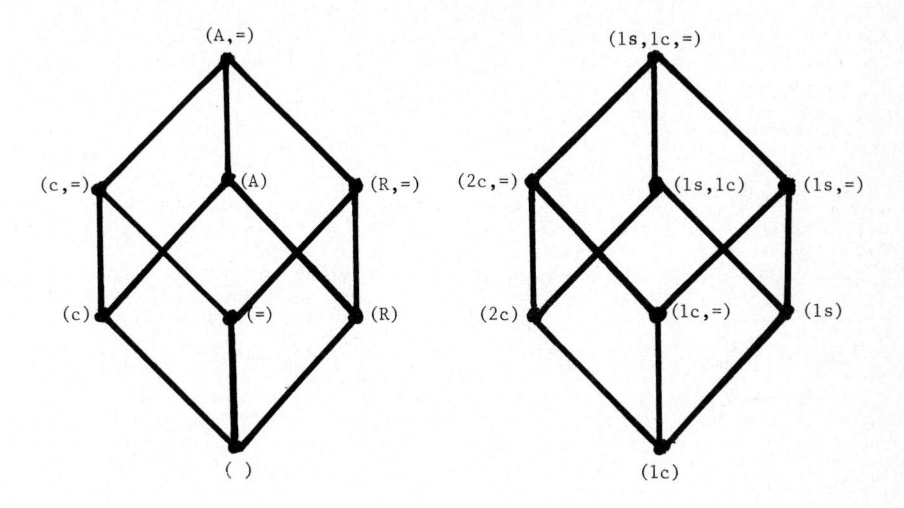

Figure 2. Chandra-Manna diagram

Here c,s,A,R stand for 'counter', 'stack', 'array', 'recursive', ( ) denotes the class of *fap*, (c,=) *fap* with = and any number of counters, (R) the class of recursive schemes, (1s,2c) the class of *fap* with one stack and two counters and so on. If there is an ascending path from a class C_2 to a class C_1 it means that C_1 is a strictly stronger class than C_2 i.e. there is a structure over which there is a function computable by a program in C_1 but not by any program in C_2, but for each program in C_2 there is a program in C_1 which is equivalent to it over all structures of the correct type. If two classes are not linked by an ascending path they are incomparable. The two cubes in the diagram are the same i.e. corresponding vertices are to be identified (A) = (1s,2c) i.e. *fap* with arrays (which, by definition includes any number of counters) are equivalent to *fap* with one stack and two counters. There are some further equalities:

$$(A) = (1s,1c) = (2s) = (1q) = (1A) = (s,a,A)$$

and similarly with the addition of =. This shows that (A,=) is a maximal class; in fact an arbitrary *fap* with = and any number of counters, stacks queues and arrays can be effectively translated into an equivalent scheme with equality and one array. By Theorems 1.1, 1.7 above we may confidently equate the strength of this maximal class to that of (*eds* with equality tests) = (all possible computations allowing equality tests) and similarly (A) = (*eds*) = (all possible computations without equality tests).

Observe that a queue is as strong as two stacks and that addition of further stacks or queues add no more strength. Similarly two counters are as strong as any number of counters but, curiously, one counter is no use at all by itself i.e. ( ) = (1c) (Plaisted[34]). But the data variables or registers do not behave like this; Chandra and Manna show that for all n, *fap* with n+1 data registers are stronger than *fap* with n data registers even for computing functions of one variable).

Chandra and Manna sum all this up succinctly:
"It is reasonable to ask what it is about the various features we have discussed that makes one class of schemas more powerful than another. An observation of the arguments involvement in proving the interrelationships shown in Figures 1 and 2 suggest three intuitive factors that determine the power of the various features.

(a) The amount of data space (x-axis of Figure 2 -- "add a stack and delete a counter"). Simple Algol-like schemas, and even those with counters and equality, have a fixed amount of data space. This limitation is shown by the fact that these schemas just cannot compute certain terms which are too large. The addition of a data variable to simple Algol-like schemas increases the power, as may be expected. Recursive schemas act as if they had an unbounded amount of data space available to them, as do schemas with stacks, queues or arrays.

(b) The control capability (y-axis of Figure 2 -- "add a counter"). The control capability of a schema signifies the ability of the schema to decide what to do next. Boolean variables and counters are examples of features that help in making such decisions. Boolean variables however add no inherent power, while two counters add as much control power as one might want. A pushdown stack provides, in addition to an unlimited amount of data space, some control capability because a stack can simulate a counter, but it does not have as much control capability as two counters. A queue, on the other hand, provides in addition to unlimited data space, as much control capability as two counters.

One can also consider other programming features that provide control capability. One such example is the boolean stack; a boolean stack is strictly more powerful than one counter but strictly less powerful than a pushdown stack or two counters. Two boolean stacks, however, are just as powerful as two counters (as is also one boolean queue) see also Green, Elspas and Levitt[12].

(c) The structure of terms (z-axis of Figure 2 -- "add equality"). In our discussion we observed that the addition of terms containing equality increases the power of schemas. This illustrates that if we enrich the structure of terms allowed we may increase the power of schemas. On the other hand, if we restrict the structure of terms, such as by limiting the depth of data terms, we may decrease the power."

The reader should be warned that the definitions above are not exactly the same as Chandra and Manna's e.g. they allow copying and permit boolean values to be stacked with the data elements, which amounts to allowing two markers (similarly in queues and arrays). They seem to me to be equivalent but this should be checked. Chandra and Manna refer to Chandra's thesis[3] for most of the proofs. It would be very useful to have more easily accessible a full collection of elegant proofs to complement their pretty survey of results.

From now on we shall adopt Chandra and Manna's succinct notation. For the sake of comparison with the notation used above we relabel the bottom square of the Chandra-Manna diagram with the notation used above. Since the right hand vertex has not been considered above we give it a bastard name *fapls*. We haven't included the top square since what the Chandra-Manna diagram makes clear is that in this sort of computation equality behaves like any other basic relation.

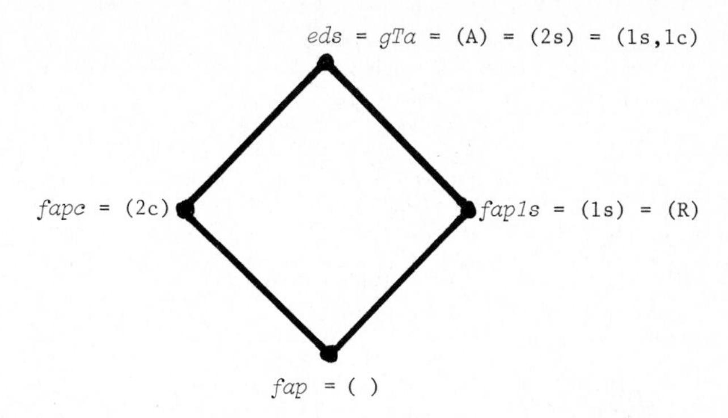

Figure 3. Chandra-Manna diagram with notation used above

5. HISTORICAL SURVEY OF PARALLEL WORK IN COMPUTER SCIENCE JOURNALS:

(a) strength of programming facilities

We have already mentioned that Luckham and Park[19] introduced program schemata, which are essentially the same as *fap*; but they were not widely known until Luckham, Park, Paterson[20]. Paterson and Hewitt[33] showed that recursive schemes were not reducible to *fapc* (i.e. (R) $\nleq$ (c) with the notation of the last section); Chandra and Manna[5] credit McCarthy[22] with an earlier proof that (R) $\geq$ ( ) and Constable and Gries[6] say that McCarthy[21] showed that recursive schemes are equivalent to recursive programs, i.e. to *fap* plus recursive schemes. Paterson and Hewitt's proof used a structure which was not structural i.e. in which there was no bound on the number of data registers need to compute terms. They defined a subclass of recursive schemes, the linear recursive schemes which could be programmed on *fap*. They also said
"We are currently investigating the relations between various augmented forms of program schema, such as program schemas with counters, push-down stores, or stacks. .... For example, program schemas with counters are not 'universal' in this theory, because the first proof of Theorem 1 shows that no such schema can be equivalent to the recursive schema V. However, we can show that under an appropriate definition of push-down program schemas, they are equivalent to the class of recursive schemas. .... We have a clear notion of effective schemes of computation involving uninterpreted functions, and should like to have a fairly natural augmentation of program schemas capable of representing any such effective computation. A good candidate for this supreme position in the hierarchy of schemas would seem to be program schemas with two push-down stores. Provided the schema has the ability to put special control constants in its stores and to subsequently recognize them, the "universality" of this model appears assured."

So they had proved (c) $\ngeq$ (R) and claimed (R) = (1s) (presumably) and (2s) = (all computations), at least if one is allowed to stack markers as well as data elements. de Bakker and Scott[1] are said (by Plaisted[34], Garland and Luckham [10] to have shown that certain recursion schemes were not equivalent to single variable schemes and considered the effect of adding counters to these schemes. Strong[38,39] also showed that (R) > ( ) and (R) $\nleq$ (c)[39] gave a definition (effective functional) equivalent to *eds*, proposed it as a candidate for universality and essentially showed that (1s,c) was equal to it. He took a slightly different but equivalent point of view to Friedman's, thinking not of a program operating on arbitrary data structures but of a program as representing a functional of the subroutines it calls.

Constable and Gries[6] is a major contribution. They discussed the use of arrays, stacks, counters, recursive schemes, labels (of statements) as values (of variables) and markers as values. They only use one kind of variable, so that use of counters means the ability to assign numbers as well as data objects to variables, and if stacks are used the ability to use markers as values means that they can be used as markers in the stack.

The instructions for the use of labels as values are

$$x := l$$

$$\text{go to } x$$

where l is the label of some statement of the program. For markers as values the instructions are, (essentially)

$$x := l$$

$$\text{if } x = m \text{ go to } i \text{ else go to } j$$

where m is a marker. Like Chandra and Manna they allow copying ($y := x$) but they have a weaker notion of stack which we shall denote by s-, which doesn't have a test for the bottom (emptiness) of a stack; an attempt to pop an empty stack now simply does nothing. Nor do they allow boolean variables, which Chandra and Manna could use as markers for stacks and arrays. They have also a weak notion of array which we shall denote by A- [Our (A-) is their P_A; our (A) their P_{Ae}], which has only weak counters. This means that only the operations $n := 0$, $n := n+1$ are allowed, not a test of $n_1 = n_2$ or $n = 0$. This is introduced to try and separate the effect of arrays from that of counters, the idea being that these operations on numbers are to be used only to access arrays and not for arithmetic computation. However with markers as well as arrays you can represent numbers as arrays of markers and simulate an equality test on numbers by the equality test on markers. Using L to denote use of labels, M the use of markers, the main results are:

$$(\ ) < (R) \leq (1s-)$$

$$(R) < (A-) = (A) = (1s-,c) = (eds)$$

$$(A) = (A,L) = (A,M) = (s-,M) = (2s-,1M) = (s-,L)$$

also that only one array is needed, i.e. A,A- can be replaced by 1A, 1A-.

All except one of the above inclusions $C_1 \leq C_2$ are effective and uniform, i.e. there is an algorithm which, given a program in C_1 produces a program in C_2 which is equivalent to it over all structures of the correct type. The exception is (A) ≤ (A-) which is uniform but provably non effective. The reason for this is that in order to replace a program in (A) by one in (A-) one needs to construct some data elements which can serve instead of markers (which can then be used as above to test equality of numbers). If one could find data elements or vectors of data elements which gave different truth values on some basic relation these could serve as markers. On the other hand if all predicates have constant truth values on all data elements encountered in the program then one doesn't need markers. The trouble is that to decide which of these is the case, and whether to go on looking for such candidates for markers requires knowledge of whether the program would halt on the various combinations of constant truth values for the basic relations and there is no algorithm for deciding this. They also showed the equivalence of (A) with Strong's effective functionals.

Brown, Gries and Szymanski[2] continued this work. They showed that (1s-) ≤ (R) completing the proof that (1s-) = (R) (also to be found in Chandra[3] and, possibly, Green, Elspas and Levitt[12]. They showed that the addition of global variables to recursion schemes (i.e. variables whose values are determined elsewhere in the program) did not increase their strength. They showed that, as one would expect, the addition of markers usually makes no difference e.g.

$$( ) = (M),\ (1s-) = (1s-,M) \text{ and, as above, } (A) = (A,M).$$

But it does for (2s-) because this is shown below to be non-universal whereas we have seen above that with the addition of one marker it becomes universal. They also showed that the use of multi-dimensional arrays was reducible to one dimensional ones. Their main interest was in the effect of adding a test for emptiness or bottom of a stack.

We shall denote by s_b this kind of stack; it's still not quite as strong as Chandra and Manna's stack because that has boolean values which can be used to get two or any number of markers. So

$$(s) = (s_b,M).$$

We also have

$$(s) = (s-,M)$$

because you can use a marker to test for the bottom of the stack. They showed

$$(c) < (2s_b)$$

$$(1s) = (1s-) = (1s_b) < (2s-) = (s-) < (2s_b)\ ?\ (3s_b) = (A),$$

where the ? means either ≤ or <; they conjectured <, though as they observe $(2s_b)$ is very close to universal power; a single 'chip' of which there is only one copy which it can place anywhere in its stacks and test for, makes it universal. So the only classes which have not been definitely identified with one of those in the Chandra-Manna diagram are (2s-) = (s-) and $(2s_b)$. The former is definitely not in; the latter will be iff it is universal.

Kfoury[16,17] gives an equivalent *eds*-like definition of the universal class of schemes. This is based on the idea of unwinding the loops of a program so that it becomes an infinite tree. His effective schemes are infinite binary trees each node of which, apart from the input node (which is labelled with the list of input variables) is either labelled with an assignment instruction $y := x$ or $x := F(x_1,\ldots,x_n)$ and has one successor, or is labelled with a test $R(x_1,\ldots,x_n)$ and has two branches, the left hand one labelled 'true', the right hand one labelled 'false', leading to successor nodes; or is a terminal node labelled with an output variable y (if it is computing a function; 'true' or 'false' if it is computing a relation). The schema must be effective i.e. the label of a node must be a recursive function of its number if the nodes are numbered from top down and left to right. He also introduces a smaller class than *fap*, the quasi-loop-free schemes where no go to instruction goes to a strictly earlier instruction.

Gordon[11] also introduces a slight variation of the *eds* in which the antecedents C_i are also allowed to contain disjunction signs and where the incompatibility condition is removed, uniqueness of the result being guaranteed instead by supposing that a definite recursive enumeration of the clauses $C_1 \to t_1$, $C_2 \to t_2, \ldots$ is given, and taking the result to be the value of the first t_n such that C_n holds. He shows it is equivalent to Moschovakis[28] notion of absolute prime computability and says that a similar characterization can be obtained for the search computable functions.

(b) Conditions on a structure for equivalence of programming facilities.

Kfoury[16,17] apparently independently discovered some of the results of section 1 of this paper. His definition corresponding to ω-rich (which provides surrogate numbers) is a little more general: the function f_j and relation R_e don't need to be basic functions of the structure; it is enough that they should be *fap* computable. He also shows that *fap* (indeed even quasi loop-free schemas) are universal (i.e. equivalent to *eds*) over any structure with the property that for each n there exists m such that every substructure generated by n elements has cardinality less than m. He shows that *fap* are universal over rings and fields of characteristic zero and that over a distributive lattice quasi loop-free schemes are universal. In [16] he gives some model theoretic criteria e.g. (1) With effective schema defined as above (equivalent to *eds*) and loop-free schema defined as a finite effective schema he shows that if $\mathcal{K}$ is a class of structures closed under ultraproducts then an effective schema is total over all structures in $\mathcal{K}$ iff it is equivalent over $\mathcal{K}$ to a loop-free schema. (2) If $\mathcal{A}$ is an infinite algebraic structure there exists a structure $\mathcal{B}$ of any given infinite cardinality less than that of $\mathcal{A}$ such that every effective scheme is total over $\mathcal{A}$ iff it is total over $\mathcal{B}$. (3) Totality of schemas is equivalent to loop-freeness over any structure $\mathcal{B}$ whose theory is $\aleph_0$-categorical if $\mathcal{A}$ is countable, or α-categorical for some infinite α if $\mathcal{A}$ is uncountable.

McKay[23] gives conditions on structures and classes of structures for the equivalence of any two of *fap*, *fapc*, *fap1s*, *eds*. He restricts himself to the case of structures with = as a basic relation. For such structures he proves the converse of Theorem 1.6 so proving that *fapc* are equivalent to *eds* (i.e. universal) over such a structure iff it is structural. Similarly *fap* ≡ *eds* over a structure iff (i) it is in a certain sense locally *fap*-enumerable and (ii) for each n there exists a bound m such that for all elements $d_1,\dots,d_n$ of D, if the substructure generated by $d_1,\dots,d_n$ is finite then it has $\le m$ elements. And the same with *fap* replaced by *fap1s*. For equivalence of *fap* to *fapc* and quasiloop free *fap* to *fap* and *fapc* he only gives different necessary conditions and sufficient conditions and for equivalence of *fap* and *fap1s* only sufficient conditions. Similar results are obtained for equivalence over classes of structures. It would be interesting to know whether these conditions for equivalence could be put in a more familiar algebraic form but we do not know how to do this even for the structurality condition. McKay also investigates the decidability of the equivalence problem for *fap* i.e. whether two arbitrarily given *fap* are equivalent over a given structure or class of structures. E.g. the recursive enumerability of the weak equivalence relation for *fap* over M is enumeration reducible to the axiomatisability of a specific part of the theory of M. He shows that both weak and strong equivalence of *fap* are decidable over boolean algebras or distributive lattices but these appear to be the only familiar algebraic structures for which this happens.

Tucker[41] also has some results on the equivalence of *fap* to *eds* (e.g. over the algebraic closure of a finite field Z_p *fap1s* < *fapc* = *eds*) and starts to examine to what extent familiar algebraic algorithms can be carried out by *fap* or *eds*. E.g. the membership question for finitely generalized ideals of polynomial rings is *eds*-computable. In [42] he exhibits families of locally finite abelian-groups and fields for whose theory of computation some basic theorems of degree theory and complexity theory can be proved.

(c) Serial and parallel procedures over partial structures. So far, we have been talking of computing partial functions and relations but we have assumed that the basic functions and relations of all structures are total. Paterson and Hewitt[33] and Strong[40] observed that over partial structures parallel procedures are stronger than serial ones. For example let (D_1, f_1, f_2) be a partial structure and define $f: D \to D$ by $f(x) = x$ if $f_1(x)\downarrow$ or $f_2(x)\downarrow$, undefined otherwise. (Here $f_1(x)\downarrow$ means '$f_1(x)$ is defined'). In a serial procedure one must choose one of $f_1(x)$, $f_2(x)$ to evaluate first and it is possible that this one is not defined but the other one is. In a parallel procedure one can simultaneously set in action two processing units one computing $f_1(x)$ and the other $f_2(x)$ and give out the value $f(x) = x$ if and when either of them reports back. Strong defined an effective functional which differs from an *eds* in allowing the conjuncts E_i of a clause E_1 & E_2 &...& $E_k \to t$ to be of the form $t_i\downarrow$ as well as atomic formulae and their negations. E.g. the function above is computed by the effective functional $\{f_1(x)\downarrow \to x, f_2(x)\downarrow \to x\}$. He put forward effective functionals as a candidate for universality over partial as well as total structures.

Constable and Gries[6] gave a corresponding program type definition which they called a multischeme. It is an extension of their universal class (A) of programs discussed above using arrays, and is based on the idea that an instruction $y := F(x_1,\ldots,x_n)$ first empties y then starts evaluation of $F(x_1,\ldots,x_n)$, which may take an unknown time and continues in parallel with the main program which goes on to the next line. A new instruction

if $y \neq \phi$ go to i else go to j

is added to test for completion of such evaluations. To deal with possibly partial predicates the old instruction type

if $R(x_1,\ldots,x_n)$ go to i else go to j

is replaced by

$$y := R(x_1,\ldots,x_n)$$

using their convention that only one type of variable is used, so that here y will hold the boolean value of $r(x_1,\ldots,x_n)$, if this exists, when the processor has evaluated it. This can be tested for by the above instruction 'if $y \neq \phi$ then go to i else go to j'. A multischeme is well defined iff any output value is independent of the times taken by the function and predicate processors. [It appears that they meant well defined on all structures; see below]. They show that well defined multischemes are equivalent to Strong's effective functionals. They also show that the earlier class (A) which is universal over total structures is equivalent over partial structures to Strong's deterministic effective functionals. The latter are effective functionals which can be recursively enumerated so that whenever they give a value, if C_o is the first clause which gives this value then in each earlier clause C all expressions up to the first one which is false are defined. In other words they are intended to correspond to functions which can be computed by serial procedures, and this is what Constable and Gries confirm.

This is elaborated in Shepherdson[37] where a very general notion of synchronous parallel procedures *spp*, is considered (and shown to be equivalent to asynchronous procedures). Over total structures they are shown to be equivalent to *eds*, to *cap* (countable algorithmic procedures:- infinite but recursive *fap*) and to *fapir* (*fap* with index registers; equivalent to arrays i.e. to the class (A)). For partial structures the *cap* and the *fapir* are strengthened by allowing computation of basic functions and predicates to continue in parallel with the main program like Constable and Gries do; these versions are called *capp* (countable algorithmic parallel procedure) and *fappir*. The *eds* is generalized as Strong did by allowing conjuncts $t\downarrow$, and then further to a *reds* (recursively enumerable definitional scheme) which is any r.e. set of clauses. All these kinds of parallel procedure may produce more than one result or a result which depends on the time taken by the various processors to return their values; when this is not the case they are said to be determinate. [This is not quite the same as Constable and Gries's notion 'well-defined' for they tacitly assume that the time taken to evaluate $f(d_1,\ldots,d_k)$ depends only on f and $d_1,\ldots,d_n$; an assumption which allows one to get information about equality of elements of D (*op.cit.* p500)]. It is shown for *reds*, *capp* and *fappir* that each is equivalent to one of the others over all structures on which it is determinate. Strong's effective functionals are equivalent to *reds* which are determinate on all structures. Thus for partial structures the Chandra Manna diagram must be extended as below

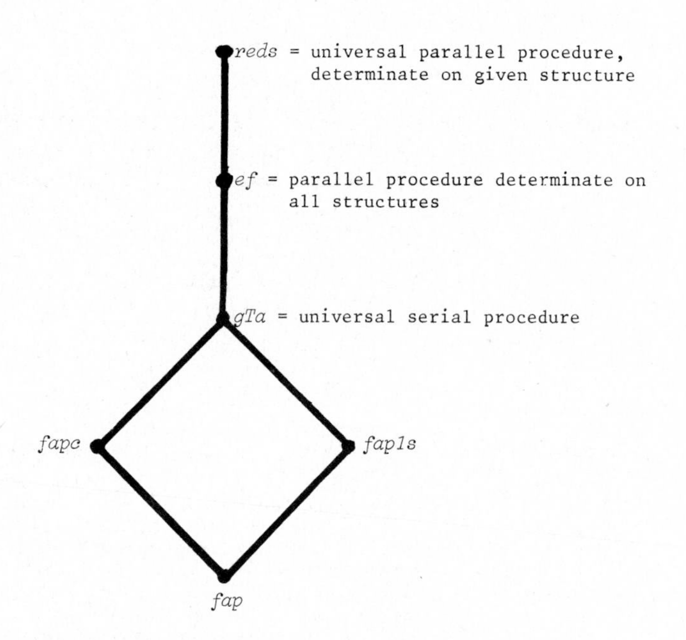

Figure 4. Chandra-Manna diagram for partial structures

Writing *ef* = effective functional, *MS* = multischeme, *esk* = effective scheme of Kfoury (see (b) above), *det* = determinate over all structures we can add the equivalences

$$reds = spp = Ms = capp = fappir$$

$$ef = det\ Ms = det\ capp = det\ fappir$$

$$gTa = cap = fapir = esk = treelike\ reds = deterministic\ ef = (A)$$

A *treelike reds* is a *reds* such that if $C_1 \to t_1$, $C_2 \to C_2$ are distinct clauses of it then there are conjuncts of C_1, C_2 which are negations of each other, and C_i, C_j agree up to these conjuncts. The *eds* no longer fit naturally into the picture since an *eds* need not 'act serially' on partial structures. [Consider the example

$\{R_1(x)\ \&\ \neg R_2(x) \rightarrow \text{true},\ R_1(x)\ \&\ \neg R_3(x) \rightarrow \text{true},\ R_3(x)\ \&\ \neg R_1(x) \rightarrow \text{true}\}]$.
Presumably all the other equivalences such as (A) = (1s,1c), (c) = (2c) still hold on partial structures but I haven't checked them all.

This difference between serial and parallel procedures which shows up on partial structures is familiar in recursion theory when one deals with degrees of partial functions i.e. defines the notion 'g is recursive in f'. The broadest notion 'g is partial recursive in f' corresponds to allowing any parallel procedure which is determinate on f. The strictest notion 'g is Turing-reducible to f' i.e. computable by an oracle Turing machine, amounts to permitting only serial procedures. The intermediate notion of Davis[7] or Rogers[36], 'there is a recursive operator Φ such that $g = \Phi(f)$' amounts to allowing only parallel procedures which are determinate on all partial functions.

6. QUESTIONS RAISED IN SECTION THREE:

This section raises several questions which still have not been answered completely:

(i) "We have indicated throughout the introduction that Turing's analysis could probably be significantly improved in relation to configurational computability; it seems, instead, to be an analysis of linear computability. The problem is to fill in the blank in: Turing's operations are to finite linear configurations as are ———— to arbitrary finite configurations. Perhaps one will first successfully deal with some restricted class of finite configurations."

The synchonronous parallel procedures of Shepherdson, may be a candidate for this but they lack the elegance and simplicity of Turing machines.

(ii) "Our effective definitional schemes provide informative equivalents to generalized Turing algorithms. It would be interesting to find similarly informative equivalents to formalized algorithmic procedures both with and without counting. Perhaps one can use suitably restricted effective definitional schemes for this purpose."

I don't know of any work on this, though it is easy to show that *fapc* are equivalent to *eds* for which there exists p such that all terms used are p-terms. For *fap* or *fap1s* some restriction on the way the terms of the *eds* are enumerated seems to be needed.

(iii) "It should be possible to elegantly generalize the usual schemata for defining partial recursive functionals via primitive recursion and μ-operator to obtain schemata for defining (equivalence type of) formalized algorithmic procedures with or without counting, and generalized Turing algorithms. This might bear on the relationship between our notions and the prime "computability" of Moschovakis[28]; there Moschovakis introduces prime "computability" not in terms of the action of devices, but in terms of schemata. We have not investigated the relation between our work and that of [28]."

Moldestad, Stoltenberg-Hansen and Tucker[25] show that *fap1s* correspond to inductive functions in the sense of Platek[35], (for a published account see Moldestad[24]) which are defined by schema. One of these schema is the least fixed point schema so this result is related to the earlier result that (1s) = (R). They show that *fap* correspond to directly inductive functions which are obtained by restricting the form of the terms used in the schemata. It would be interesting to know whether schemata closer to the usual Kleene ones could be found. As mentioned above Gordon[11] in an abstract claimed that *eds* correspond to absolute prime computability. This is taken up more fully in Moldestad, Stoltenberg-Hansen and Tucker[26], who also relate it to the notions of Fenstad [8,9], showing that *eds* correspond to the minimal computation theory with code set ω.

(iv) "We conjectured in Section 2 that our theorems of Section 2 were best possible. Some parts of this conjecture are trivial. Others seem to require genuine combinatorial arguments.

It is clear from glancing at the arguments in Section 2 that only certain basic facts are used about generalized Turing computability. It would be interesting to isolate these facts, and perhaps, obtain decision procedures. Of course the basic facts are dependent on which of the conditions I, II, III are placed on the structures M beforehand."

I don't know of any further work in this area.

We have seen that many other people were thinking about these concepts about the same time as Friedman but this paper stands out for the way in which it goes directly to the heart of the matter, for its clarity of presentation, the wealth of results it contains and its fertility as a source of inspiration for further work.

FOOTNOTES:

[1]For an excellent comprehensive summary see Kreisel[18]. Another contribution of Friedman's to this area is 'Axiomatic recursive function theory', discussed in section VD of this work.

[2]Friedman's list was longer and included copying i.e. $y := x$, but this is inessential if one is allowed to choose the output variable or register (by similar argument to Garland and Luckham[10]). Similarly for clearing registers and testing for emptiness; you can always keep track of whether a register is empty or not.

[3]For clarity we have interchanged Friedman's sub- and super-scripts.

[4]Friedman class these also *fap*, distinguishing the earlier *fap* as '*fap* without counting'.

[5]Friedman states these in the weaker form where $k = 1,\ldots,k_{b_j} = b_j$ but the form form given here appears to be necessary to yield universality (i.e. Theorem 1.2) unless the switching operations, which allow one to move data, are strengthened to allow copying (which Friedman doesn't allow here). This is the obverse of footnote 2; if you aren't allowed to use the same variable twice as an input you do need copying.

[6]A fuller discussion of the relevance of this part of Friedman's paper to pebble games and the logic of programs is found in Kfoury's contribution to this volume.

[7]The assumption is unnecessary: see end of proof.

REFERENCES:

[1] de Bakker, J.W. and Scott, Dana, A Theory of Programs, unpublished report, Aug. 1969.

[2] Brown, S., Gries, C., and Szymanski, T., Program schemes with pushdown stores', SIAM J. Comput. 1 (1972).

[3] Chandra, A., On the Properties and Applications of Program Schemas, Ph.D. Thesis, Comp. Sci., Stanford, Report No. CS-336, AI-188 (1973).

[4] Chandra, A. and Manna, Z., Program schemas with equality, Proc. 4th ACM Symposium on Theory of Computing (1972).

[5] Chandra, A. and Manna Z., On the power of programming features, Stanford AI Memo 185 (1973).

[6] Constable, R.L. and Gries, D., On classes of program schemata, SIAM J. Comput. 1 (1972) 66-118.

[7] Davis, M., Computability and Unsolvability (McGraw Hill, New York, 1958).

[8] Fenstad, J.E., Computation Theories: an axiomatic approach to recursion on general structures, pp 143-168 of G. Müller, A Oberschelp and K. Potthoff (eds.) Logic Conference, Kiel 1974 (Springer-Verlag, Berlin - Heidelberg - New York, 1975).

[9] Fenstad, J.E., General Recursion Theory. An axiomatic approach (Springer-Verlag, Berlin - Heidelberg - New York, 1980).

[10] Garland, S.J. and Luckham, D.C., Program schemes, recursion schemes and formal languages, J. Comp. Syst. Sci. 7 (1973) 119-160.

[11] Gordon, C.E., Finitistically computable functions and relations on an abstract (Abstract), J. Symbolic Logic 36 (1971) 704.

[12] Green, M.W., Elspas, B. and Levitt, K.N., Translation of recursive schemas into label-stack flowchart schemas, preliminary draft, Stanford Research Institute, Menlo Park, California (June 1971).

[13] Herman, G.T. and Isard, S.D., Computability over arbitrary fields, J. London Math. Soc. 2 (1970) 73-79.

[14] Hewitt, C., More comparative schematology, MIT Artificial Intelligence Memo 207, Project MAC (1970).

[15] Ianov, Iu.I., The logical schemes of algorithms, Problems of Cybernetics (USSR) 1 (1960) 82-140 (English version).

[16] Kfoury, D., Comparing Algebraic Structures up to Algorithmic Equivalence, in: Nivat (ed.) Automata, Languages and Programming (North-Holland, 1973).

[17] Kfoury, D., Translatability of schemas over restricted interpretations, J. Comp. Syst. Sci. (1974) 387-408.

[18] Kreisel, G., Some Reasons for Generalizing Recursion Theory, pp 134-198, in: Gandy and Yates (eds.) Logic Colloquium (North-Holland, 1971).

[19] Luckham, D. and Park, D., The undecidability of the equivalence problem for program schemata, Bolt, Beranek and Newman Inc. Report No. 1141, 1964.

[20] Luckham, D., Park, D. and Paterson, M., Formalized computer programs, J. Comp. Syst. Sci. 4 (1970) 220-249.

[21] McCarthy, J., Recursive functions of symbolic expressions and their computation by machine, Part I, Comm. ACM, 3 (1960) 184-195.

[22] McCarthy, J., Towards a mathematical science of computation, Proc. IFIP (1962) 21-34.

[23] McKay, R.L., The Equivalence of Program Schemes in Structures with Equality, Ph.D. Thesis, Bristol University (1976).

[24] Moldestad, J., Computations in Higher Types (Springer-Verlag, Berlin - Heidelberg - New York, 1977).

[25] Moldestad, J., Stoltenberg-Hansen, V. and Tucker, J.V., Finite algorithmic procedures and inductive definability, Math. Scand. 46 (1980a) 62-76.

[26] Moldestad, J., Stoltenberg-Hansen, V. and Tucker, J.V., Finite algorithmic procedures and computation theories, Math. Scand. 46 (1980b) 77-94.

[27] Moldestad, J. and Tucker, V., On the classification of computable functions in an abstract setting, Mathematical Centre Report, Amsterdam, 1979.

[28] Moschovakis, Y.N., Abstract first-order computability I, Trans. Amer. Math. Soc. 138 (1969) 427-464.

[29] Moschovakis, Y.N., Abstract first-order computability II, Trans. Amer. Math. Soc. 138 (1969) 465-504.

[30] Moschovakis, Y.N., Axioms for Computation Theories - first draft, pp119-225, in: R.O. Gandy and C.M.E. Yates (eds.) Logic Colloquium '69 (North-Holland, Amsterdam, 1971).

[31] Paterson, M., Equivalence Problems in a Model of Computation, Doctoral Dissertation, Cambridge (1967).

[32] Paterson, M., Program schemata, Machine Intelligence 3 (1968) 19-31.

[33] Paterson, M.S. and Hewitt, C.E., Comparative schematology, MIT Artificial Intelligence Memo 201 (1970); also in: Project MAC Conference on Concurrent Systems and Parallel Computation.

[34] Plaisted, D.A., Flowchart schemes with counters, Conference Record of the 4th Annual ACM Symposium on the Theory of Computing (1972).

[35] Platek, R.A., Foundations of Recursion Theory, Ph.D. Thesis, Stanford University, Stanford (1966).

[36] Rogers, H., Theory of Recursive Functions and Effective Computability (McGraw Hill, New York, 1967).

[37] Shepherdson, J.C., Computation over Abstract Structures: serial and parallel procedures and Friedman's effective definitional schemes, in: Logic Colloquium '73 (North-Holland) 445-513.

[38] Strong, H.R., Translating recursion equations into flow charts, 2nd Annual ACM Symposium on Theory of Computing (1970) (abstract only); J. Comp. Syst. Sci. 5 (1971) 254-286.

[39] Strong, H.R. Jr., Translating recursion equations into flowcharts, J. Comp. & Sys. Sci. 5 (1971) 254-285.

[40] Strong, H.R., High level languages of maximum power, Proc. IEEE Conference on Switching and Automata Theory (1971).

[41] Tucker, J.V., Computing in algebraic systems, Matematisk institutt, Universitetet i Oslo, Preprint Series No.12 (ISBN 82-553-0358-8), Oslo, 1980.

[42] Tucker, J.V., Sack's splitting theorem and Blum's theorem for computable functions on Abelian groups and fields, Universitet i Oslo (1982).

[43] Friedman, H., Algorithmic procedures, generalized Turing algorithms, and elementary recursion theory, in: Logic Colloquium '69 (North-Holland, Amsterdam, 1971) 361-389.

HARVEY FRIEDMAN'S RESEARCH ON THE FOUNDATIONS
OF MATHEMATICS, L.A. Harrington et al. (editors)

COMPUTATIONAL COMPLEXITY OF REAL FUNCTIONS

J.C. SHEPHERDSON

Mathematics Department
University of Bristol
England

1. INTRODUCTION:

The field of computational complexity has been very largely concerned with discrete problems. A notable exception is Smale's great and daunting paper[14] on Newton's method. This paper of Ko[15] and Friedman breaks new ground and gets well dug in to the application of complexity theory to real function theory. The aim is to present a natural definition of computational complexity of real functions and to study the relationship between complexity and analytical properties of real functions.

The first two sections define the notion of computable real function and establish basic continuity results. The third section contains the definition of computational complexity of recursive real numbers and functions and relates this to continuity, proving for example that a polynomial time computable function has a polynomially bounded modulus of continuity. The remaining sections examine some basic theorems in classical real analysis and study them from the viewpoint of computational complexity. E.g. a polynomial time computable real function may have a root of exponential or even higher time complexity; however if it is analytic its roots must have time complexity bounded by some polynomial. The polynomial time computable real numbers form a real closed field; the complex ones form an algebraically closed field. The standard example of a continuous nowhere differentiable function can be computed in polynomial time. The definite integral of a polynomial time computable function is polynomial space computable but is conjectured not to be polynomial time computable. The maximum of a polynomial time computable function is computable in polynomial time using some NP set as an oracle but is not known to be computable in polynomial time.

These last two results put these basic problems in real analysis into the long list of PSPACE and NP problems. The difficulty of the question "P = NP?", "P = PSPACE?" implies that it will be hard (if not impossible) to show that an exponential time algorithm for numerical integration or finding a maximum is the best possible.

This paper will undoubtedly lead to the growth of a new branch of complexity theory which will bring it in closer touch with practical numerical analysis.

2. SECTIONS 1,2. COMPUTABLE REAL FUNCTIONS:

A real number is considered as a sequence of dyadic rational numbers which converges to it.

DEFINITION. Let x be a real number, $0 \leq x \leq 1$. We say the function ϕ <u>binary converges</u> to x if $\phi(n)$ is of the form $m/2^n$ for integral m and

$$|\phi(n) - x| \leq 2^{-n} \text{ for all } n \in N.$$

A real number x, $0 \leq x \leq 1$ is <u>recursive</u> if there is a recursive function ϕ which binary converges to it.

Computable real functions are to be defined not only on the computable real numbers so a natural approach is via oracle Turing machines (OTM). These are Turing machines (TM) with an extra query-tape and a special query state. When the machine enters the query state the oracle (a function ϕ) replaces the string w on the query tape by $\phi(w)$and restarts the machine in a new state which must be specified by the TM program before entering the query state. Although we assume the oracle can provide the required information in one step, the TM must take $|\phi(w)|$ (length of the string $\phi(w)$) steps to read the output $\phi(w)$ from the query tape. Oracle machines which have a set instead of a function as an oracle are defined similarly, the answer to the question 'does the string on the query tape belong to the oracle set or not' being given by restarting the machine in one of two new special states, the yes state or the no state.

If M is an OTM with ϕ, a function, as an oracle and n as an input then we write $M^{\phi}(n)$ for the output and $M^{\phi}(n)\uparrow$ if M does not halt on input n with oracle ϕ. We denote M^{ϕ} the partial function whose value at n is $M^{\phi}(n)$.

For simplicity only functions with domains contained in [0,1] are considered.

DEFINITION. Let $S \subseteq [0,1]$ and $\bar{S} = [0,1] - S$. A function f: $S \to R$ is said to be <u>partial recursive</u> if there is a function-oracle Turing machine M such that for all x and ϕ, if ϕ binary converges to x, then if x is in S,M^{ϕ} binary converges to $f(x)$, if x is in $\bar{S}$ then M^{ϕ} is the everywhere undefined function.

A function is called <u>total recursive</u> if it is partial recursive on [0,1].

For total functions this is equivalent to the standard definitions of Grzegorczyk[2] and Lacombe[9]. It would be interesting to know whether the equivalent version of Pour-El[11,12] in terms of a recursively enumerable sequence of approximating polynomials would simplify any of the proofs. [Partial recursive functions of real variables do not appear to have been considered very much. For them it is clearly appropriate to replace, as the present authors do, the usual general recursive functional by a recursive functional (i.e. not to require M^{ϕ} to be total for all total ϕ); but at first sight it seems a little strong to require M^{ϕ} to be everywhere undefined when ϕ binary converges to a number not in the domain. The definition is very similar to the notion of weakly computable of Shepherdson[13]. Both have the desirable effect (which some of the definitions do not have) of making $1/x$ computable on [0,1]. Actually nearly all the results of this paper concern only total functions because there are difficulties (see §3, §8 below) with computational complexity for partial functions)].

The alternative treatment of Banach, Mazur, (see Grzegorczyk[1,2], Klaua[6]) where computable real functions are required to be computable only on the computable real numbers leads to pathological results and is clearly not the appropriate treatment here. Ko[7] apparently contains a full discussion of a variety of notions of computability of real numbers and functions.

It is shown that the domains of partial recursive real functions are precisely the <u>recursively open</u> sets i.e. the union of a recursively enumerable set of open intervals with rational endpoints. The known result that recursive real functions are continuous is given a converse: if f is continuous on [0,1] then there exists a set-oracle E of integers such that f is recursive in E. The familiar fact that recursive functions on [0,1] have recursive modulus of uniform continuity is proved. From this follows easily that the definite integral of a recursive function over [0,1] is a recursive real number.

3. COMPUTATIONAL COMPLEXITY OF REAL FUNCTIONS:

A <u>recursive real number</u> is said to have <u>time complexity</u> $\leq T$ (where T is a function) if there is a Turing machine M computing a function ϕ which binary converges to x

such that the time complexity function T_M of M is bounded above by T. Here $T_M(n)$ may be defined as the maximum number of moves M takes to compute $\phi(n)$ when n is input in unary form, i.e. as a string of length n. Intuitively a recursive real number has time complexity $\leq T$ if there is an effective, uniform method, which in $T(n)$ steps can get n significant bits of x. Similarly for space complexity where one counts the number of cells used by the Turing machine. The class PR, of polynomial time computable real numbers, consists of those recursive real numbers with time complexity bounded by a polynomial. All rational numbers, algebraic numbers, e and π are in PR. Similarly PSPACER is the class of polynomial space computable real numbers.

A total recursive function is said to have time complexity $\leq T$ if there is an oracle Turing machine M which computes f such that for each ϕ which binary converges to a number in [0,1], the number of steps taken by M with oracle ϕ is bounded above by T. [It is not appropriate to extend this definition to partial recursive functions because it is shown that unless such a function is uniformly continuous there is no recursive upper bound on its computation time]. The class PF of polynomial time computable real functions consists of those recursive real functions with time complexity bounded by a polynomial. Almost all commonly used functions e.g. e^x, $\sin x$, are polynomial time computable. If a function f is polynomial time computable then there is an oracle Turing machine M and a polynomial P such that for all x in [0,1], in order to get n significant bits of $f(x)$, M does not need any information except the first $p(n)$ bits of x. This is because M does not have time to read more than $p(n)$ bits from the oracle for x. That is to say f has a polynomially bounded modulus (of continuity), a polynomial p such that for all x,y in [0,1] and $n \in N$,

$$|x-y| \leq 2^{-p(n)} \rightarrow |f(x)-f(y)| \leq 2^{-n}.$$

There is a converse to this: if f is continuous on [0,1] and has a polynomially bounded modulus there exists an oracle set E such that f is polynomial time computable in E. Finally it is shown that a function is polynomial time computable if and only if there is a sequence of piecewise linear functions (with vertices at dyadic rationals) which in a sense uniformly converges to it at a polynomially bounded speed.

4. ROOTS:

A root of a recursive real function must be recursive but a root of a polynomial time computable function need not be polynomial time computable. It is shown there exists a polynomial time computable strictly increasing function with $f(0) < 0 < f(1)$ but $f^{-1}(0)$ not polynomial time computable. Indeed $f^{-1}(0)$ can be made to be any recursive real number in [0,1].

Despite the fact that the roots of some functions are hard to compute, many numerical methods of finding roots exist. This is explained by the result that if f is a strictly increasing analytic polynomial time computable function then f^{-1} is also polynomial time computable. This is true more generally, for every injective polynomial time computable f, such that f^{-1} has a polynomially bounded modulus on the range of f.

An easy consequence is that all roots of an analytic polynomial time computable function are polynomial time computable and hence that the polynomial time computable real numbers form a real closed field (and the complex ones form an algebraically closed field). It is shown that the above proof breaks down if analytic is replaced by $C^\infty[0,1]$ and it is conjectured that there is a polynomial time computable function in $C^\infty[0,1]$ which has a root which is not polynomial time computable.

Based on some fast root-finding algorithms for polynomials (e.g. Henrici and Gargantini[4]) they conjecture that there exists a multi-oracle TM which factors complex polynomials in polynomial time. The input to such a machine consists of two numbers k, the degree of the polynomial to be factored, and n the number of significant bits required in the answer. If the first input is k there are $2k+2$ function oracles available giving the real and imaginary parts of the complex numbers $a_o,\ldots,a_k$ which are the coefficients of the polynomial. The machine outputs $2k+2$ n-bit dyadic rationals viewed as representing approximations to within 2^{-n} to the factorisation numbers $b_o,\ldots,b_k$ given by $a_k z^k + \ldots + a_1 z + a_o = b_o(z-b_1)\ldots(z-b_k)$. It is not clear how they intend to deal with the difficulty mentioned above in connection with the complexity of partial functions, which occurs here when $a_k = 0$. E.g. if $k=1$, $b_1 = -{}^{a_o}/_{a_1}$ and if a_1 is sufficiently near zero it may take an arbitrarily long time to find the first digit of ${}^1/a_1$. Perhaps they intended to restrict to monic polynomials.

5. DERIVATIVES:

A polynomial time computable function need not be differentiable; indeed a standard example of an everywhere continuous nowhere differentiable function is shown to be polynomial time computable. It is not known whether if the derivative does exist it must be polynomial time computable, but it is shown that, if the derivative is continuous, it is polynomial time computable iff it has a polynomially bounded modulus. It follows that if an analytic function is polynomial time computable, then every coefficient of its power series is polynomial time computable. However to use the power series to compute the function in polynomial time you would also want the whole sequence of coefficients to be, in an obvious sense, polynomial time computable. Unfortunately the complexity of the sequence of coefficients is not known precisely, although from the Cauchy integral formula and results about integrals in the next section it follows that this sequence is polynomial space computable.

6. INTEGRALS:

Since all polynomial time computable functions are continuous, they are also Riemann integrable. In addition the integral function $I(x) = \int_o^x f$ of a polynomial time computable function f has a polynomially bounded modulus. But the exact complexity of integrals of polynomial time computable functions is not known. It is shown that if f is polynomial time computable then $\int_o^1 f$ is polynomial space computable and it is conjectured that if P $\neq$ PSPACE there exists a polynomial time computable function f such that $\int_o^1 f$ is not polynomial time computable. [It can be shown that P = PSPACE $\rightarrow$ PR = PSPACER].

It is shown that the complexity of a uniform method of integration is greater than polynomial time. This means that the functional $I(f) = \int_o^1 f$ defined on polynomial time computable f, is not polynomial time computable. This means there is no two-oracle Turing machine M, who computation time is bounded by a polynomial, which will compute a function ψ which binary converges to $I(f)$ when the second oracle is plugged in to a function f whose modulus is bounded by a polynomial and the first oracle is plugged in to a modulus m for f. The first oracle here answers questions like 'If we want n significant digits of $f(x)$ to be output from the second oracle f how many significant digits of x do we need to supply?' The second oracle f is a function oracle which answers questions like 'If d is an $m(n)$ bit dyadic number such that $|d-x| \leq 2^{-m(n)}$ please give me an

n-bit dyadic number e such that $|e-f(x)| \leq 2^{-n}$'.

7. MAXIMUM VALUES:

The maximum value of a recursive real function is known (e.g. Lacombe[9] to be a recursive real number). Unfortunately the situation in regard to polynomial time computability is similar to that for the integral. It is shown that if f is polynomial time computable than max f is computable in polynomial time using some NP set as an oracle. And the functional max f is not polynomial time computable. It is not known whether P $\neq$ NP would imply the existence of a polynomial time computable f with max f not polynomial time computable.

8. CONCLUSION:

Step functions are a useful tool in analysis and, if they have recursive jump points and we ignore the jump points they are partial recursive. But with the above definition of time complexity there is no polynomial time bound within which one can distinguish x from one of the step points x_o even if x_o is some easy number like $^1/2$. However for most numbers x we can quickly get an answer to the question whether $x < {}^1/2$ or not, and only for a small portion of numbers in [0,1] do we need a large amount of time to answer the question. So it is suggested it might be worth studying a notion of polynomial time approximability which is defined precisely and is roughly equivalent to saying that $f(x)$ can be computed in polynomial time for most x's in the domain of f. An extension of the theory of recursive measure to the study of 'polynomial time measurable sets' is also contemplated. It is to be hoped this switch from worst case to average time complexity would not lead to the enormous increase in difficulty which it does in other branches of complexity theory.

Since "P = NP?" and "P = PSPACE?" are known to be very hard problems the authors say it would be interesting at least to try and relate the integration and maximum value problems to known NP or PSPACE problems. Ko[8] has shown that real numbers cannot be NP-complete or PSPACE-complete unless P = NP or P = PSPACE respectively, so these problems seem to form a new class of NP problems which are structurally different from most NP combinatorial problems and well worth studying.

Many more open problems which have not been touched are mentioned, e.g. the complexity of a sequence of real numbers, of differential equations and the relation between computational complexity and other analytical properties of a real function. The authors believe that the study of these problems will make contributions to numerical analysis and operations research as well as to complexity theory.

There is no doubt that this paper provides a bridge between combinatorial complexity theory and analysis which looks set to creating a new branch of complexity theory for problems about continuous quantities.

REFERENCES:

[1] Grzegorczyk, A., Computable functionals, Fund. Math. 42 (1955) 168-202.

[2] Grzegorczyk, A., On the definitions of computable real continuous functions, Fund. Math. 44 (1957) 61-71.

[3] Grzegorczyk, A., Some approaches to constructive analysis, in: Heyting, A. (ed.), Constructivity in Mathematics (North-Holland, Amsterdam, 1959).

[4] Henrici, P. and Gargantini, I., Uniformly convergent algorithms for the simultaneous approximation of all zeros of a polynomial, in: Dejon, B and Henrici, P. (eds.), Constructive Aspects of the Fundamental Theorem of Algebra (Wiley-Interscience, New York, 1969).

[5] Herman, G.T. and Isard, S.D., Computability over arbitrary fields, J. London Math. Soc. 2 (1970) 73-79.

[6] Klaua, D., Prazisierung des Berechenbarkeitsbegriffes in der analysis mit Hilfe rationaler funktionale, Zeitschr. f. math. Logik und Grundlagen der Math. 5 (1969) 33-96.

[7] Ko, K., Computational Complexity of Real Functions and Polynomial Time Approximations, Ph.D. Thesis, The Ohio State University, Columbus (1979).

[8] Ko, K., The maximum value problem and NP real numbers, J. Comput. Sci. 24 (1985), 15-35.

[9] Lacombe, D., Extension de la notion de fonction récursive aux fonctions d'une ou plusieurs variables réeles, and other notes, Comptes Rendus 240 (1955) 2478-2480; 241 (1955) 13-14, 151-153, 1250-1252.

[10] Lacombe, D., Les ensembles récursivement ouverts ou fermès et leurs applications à l'analyse récursive, and other notes, Comptes Rendus 244 (1957) 838-840, 996-997; 255 (1957) 1040-1043

[11] Pour-El, M.B., Abstract computability and its relation to the general purpose analog computer, Trans. Amer. Math. Soc. 199 (1974) 1-28.

[12] Pour-El, M.B. and Caldwell, J.C., On a simple definition of computable function of a real variable with applications to functions of a complex variable, Zeitschr. f. math. Logik und Grundlagen der Math. 21 (1975) 1-19.

[13] Shepherdson, J.C., On the definition of computable function of a real variable, Zeitschr. f. math. Logik und Grundlagen der Math. 22 (1976) 391-402.

[14] Smale, S., The fundamental theorem of algebra and complexity theory, Bull. (New Series) Amer. Math. Soc., 4 (1981) 1-35.

[15] Friedman, H., and Ko, K., Computational complexity of real functions, J. Theoretical Computer Science, 20 (1982) 323-352.

[16] Ko, K., On the Definitions of some Complexity Classes of Real Numbers, Mathematical Systems Theory, to appear.

[17] Ko, K., Some Negative Results on the Computational Complexity of Total Variation and Differentiation, Information and Control, to appear.

[18] Ko, K., Approximation to Measurable Sets and Measurable Functions, J. Assoc. Computing Machinery, submitted.

Note added December 1984

Friedman has obtained further results on some of the open questions mentioned above in a recent paper

The Computational Complexity of Maximization and Integration, *Advances in Mathematics*, Vol.53, No.1, July 1984, 80-98.

He shows that if the maximum of a polynomial time computable real function is polynomial time computable then P = NP for sets of unary strings. He also proves that two apparently stronger forms of maximization principle are equivalent to the full P = NP. These are

Let g: $[0,1]\times[0,1] \rightarrow R$ be polynomial time computable. Then the function h: $[0,1] \rightarrow R$ given by $h(x) = \max_{y\in[0,1]} g(x,y)$ is polynomial time computable.

Let g: $[0,1] \rightarrow R$ be polynomial time computable. Then the function h: $[0,1] \rightarrow R$ given by $h(x) = \max_{y\leq x} g(y)$ is polynomial time computable.

Similarly he shows the polynomial time computability of $\int_0^1 g(x)dx$ for g polynomial time computable is equivalent to a certain well known discrete computational complexity problem, and he does the same for the stronger hypotheses that $\int_0^1 g(x,y)dy$ and $\int_0^x g(y)dy$ are polynomial time computable.

He also gives a different, but equivalent, treatment of the computational complexity of real functions which doesn't rely on oracle functions but on the polynomial time computability of approximations.

HARVEY FRIEDMAN'S RESEARCH ON THE FOUNDATIONS
OF MATHEMATICS, L.A. Harrington et al. (editors)

THE PEBBLE GAME AND LOGICS OF PROGRAMS

A.J. Kfoury
Department of Computer Science
Boston University
Boston, Massachusetts, U.S.A. 02215

1. Introduction

This paper is a computer scientist's commentary on just a small part of Friedman's article entitled "Algorithmic procedures, generalized Turing algorithms, and elementary recursion theory."

The pebble game is a basic technique in theoretical computer science. It has been particularly successful in studies that compare time and space requirements of computations. Thus, by far the most numerous applications of the pebble game have been in the area of time-space tradeoffs. In recent years, however, there have also been important applications in the areas of comparative schematology, logics of programs, code generation and optimization, among others.

Harvey Friedman did not write an article on the pebble game as such, nor did he use the expression himself. He was nevertheless one of its originators in his article mentioned above.

The two other people who independently invented the pebble game, approximately a year later, are Michael Paterson and Carl Hewitt [1970]. (Friedman's article appeared in the proceedings of the Logic Colloquium held at Manchester in August 1969; Paterson's and Hewitt's original report was dated November 1970.) Until a few years ago, computer scientists generally acknowledged Paterson's and Hewitt's contribution but not Friedman's. Thus, for example, in survey articles on the pebble game by a foremost expert, credit is given to Paterson and Hewitt while Friedman is not mentioned (Pippenger [1980,1982]).

Without dwelling too much on this history, there are at least two reasons for the fact that Friedman's early contribution was ignored until recent years. First, Paterson and Hewitt talked to computer scientists and worked among them, while Friedman primarily addressed himself to an audience of logicians (and his article was thus published in the proceedings of a Logic Colloquium which are seldom looked at by computer scientists). Second, Friedman buried his version of what is now called pebbling (or the pebble game) in a long and highly technical investigation. In fact, Friedman stressed other ideas in the article, such as the concept of an "effective definitional scheme" which also proved to be important in logics of programs many years later (Tiuryn [1981a,1981b]).

There was more than one (missed) opportunity in the 1970's to give Friedman's article the full attention it deserved. Friedman's ideas and results were reorganized, and compared with similar work by computer scientists, in a comprehensive study by Shepherdson [1975], which unfortunately seemed to escape the attention of computer scientists because it appeared in the proceedings of another Logic Colloquium! Another paper with a definite logical flavor, by another logician (Abramson [1978]), was also based on some of Friedman's ideas, in particular what we now call the pebble game, and some of the work on program schemes at the time. Abramson developed techniques that were later used in logics of programs, but he considered a problem ("interpolation questions for program schemes") which was then of limited interest. As a result, Abramson's paper seemed to find no audience among computer scientists at the time, even though it appeared in a computer science journal.

It is perhaps in the 1980's that computer scientists have given Friedman his due. Effective definitional schemes, which we now sometimes call "Friedman schemes", are a very convenient tool in logics of programs (see for example the articles by Tiuryn [1981a,1981b], Meyer and Tiuryn [1982], Urzyczyn [1981a], Kfoury [1983], Crasemann and Langmaack [1983]). Once we started using "Friedman schemes", we also discoveed in the same article that Friedman knew how to apply the pebble game to compare the expressive powers of programming formalisms, which we have subsequently used in our logical analysis of programs (Kfoury and Urzyczyn [1985], Tiuryn [1981c], Urzyczyn [1981b]).

For the many connections between Friedman's ideas, in particular effective definitional schemes, and the earlier theory of program schemes in the 1970's (extended by various logics of programs in the 1980's), the reader is referred to Shepherdson's contribution to this volume. For a history of the pebble game and its generalizations in relation to time-space tradeoff results, the reader is referred to Pippenger [1980,1982]. *

2. Friedman's Original Formulation of the Pebble Game

Friedman's original formulation was somewhat different from Paterson's and Hewitt's original formulation–and neither one was expressed in terms of pebbles and pebbling. (The words "pebbles" and "pebbling" were first used by Hopcroft, Paul, and Valiant [1975].) Friedman's formulation appeared more algebraic, while Paterson's and Hewitt's was in terms of graphs, which made the latter closer to the current formulation of the pebble game. Both were devised however to prove essentially the same result, namely, that programs with unbounded memory are more powerful than programs with bounded memory.

In what follows we identify definitions and results taken from Friedman's article. We have reorganized the material and introduced a few simplifications here and there, which do not change the essence of Friedman's analysis.

2.1 Definition (Friedman Definition 1.8). A vector of $n \geq 1$ variables is denoted $x = (x_1, \ldots, x_n)$. A finite set of function symbols, each of arity ≥ 0, is denoted $\mathcal{F} = \{F_1, \ldots, F_r\}$. ($\mathcal{F}$ will later correspond to the similarity type of a structure without, for the sake of simplicity, underlying relations.) The pair $(x, \mathcal{F})$ is called a *definitional type* and denoted α. The α-terms are defined inductively by:

* Of course, Friedman's contributions to theoretical computer science are not limited to things related to effective definitional schemes and the pebble game. There is in particular his remarkable work with Ker-I Ko on the computational complexity of real functions. More recently, he was invited to be on the Advisory Board of *Information and Control*, a leading journal in theoretical computer science.

(1) each variable $x_i, 1 \leq i \leq n$ is an α term.

(2) if $t_1, \ldots, t_j$ are α-terms and $F \in \mathcal{F}$ is of arity $j \geq 0$, then $F(t_1, \ldots, t_j)$ is an α-term.

The α-terms are the terms built up from the variables x and the function symbols $\mathcal{F}$.

2.2 Definition (Friedman, Definition 1.20 and 1.23). For every $p > 0$, the $p - \alpha$-terms are defined as a subset of the α-terms according to the following. A $p - \alpha$-sequence is a sequence of finite length $q > 0$, denoted $C_1, C_2, \ldots, C_q$ such that

(1) each C_i is a vector of length p whose entries are in $\{\alpha - \text{terms}\} \cup \{*\}$,

(2) every entry of C_1 is $*$,

(3) every C_{i+1} differs from C_i in exactly one place - and, in this place, the entry of C_i and that of C_{i+1} are not both α-terms (i.e., one of the two is a $*$),

(4) if the place where C_{i+1} differs from C_i is assigned an α-term of the form $F(t_1, \ldots, t_k)$ for some $k \geq 1$, then each t_j is an entry of C_i.

A $p - \alpha$-term is an α-term which is an entry of some C_i in some $p - \alpha$-sequence.

Thus, a $p - \alpha$-term t is an α-term that can be built up from x and $\mathcal{F}$ such that, at any intermediary point, at most p subterms of t are "remembered". The fact that the $p - \alpha$-terms for a fixed p are generally a proper subset of the α-terms is established by means of the pebble game.

The only thing missing from Definition 2.2 to put it in the current nomenclature of the pebble game, is the association of terms with their representations as *dag's* (directed acyclic graphs). With this association the preceding definition becomes the definition of finite dags whose "space requirement" (defined below) is p. For example, if $\alpha = ((x), \{F, G\})$ where F is unary and G binary, then the α-term

$$t = G[G[x, F(x)], G[F(F(x)), F(F(F(x)))]]$$

is represented by the following labelled dag.

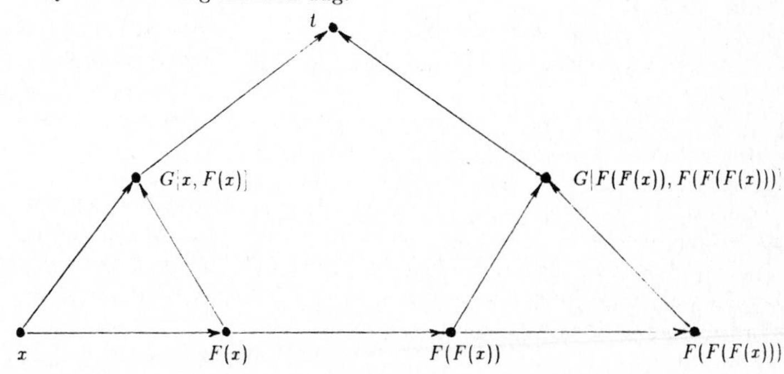

A node in a dag with in-degree = 0 (i.e., without predecessor nodes) is called an *input*; a node with out-degree = 0 is called an *output*. Clearly, an arbitrary finite dag with exactly one output is the dag of an α-term for some α, where node labels are omitted and the order of edges into a node is ignored.

2.3 Definition. The *pebble game* is a one-person game played on a finite dag $\mathfrak{D}$. At any point in the game some nodes of $\mathfrak{D}$ will have pebbles on them (one pebble per node), while the remaining nodes will not. A *configuration* is a subset of the nodes comprising just those nodes that have pebbles on them. A *move* in the game consists in placing a pebble on a node, or removing a pebble from a node, according to the following rules:

(1) if all the immediate predecessors of a node have pebbles on them, then a pebble may be placed on that node,

(2) a pebble may be removed from any node.

A legal move is represented by an ordered pair of configurations, the second of which follows from the first according to (1) or (2). Since an input has no immediate predecessors, a pebble may be placed on any input. A *calculation* is a sequence of configurations $C_1, C_2, \ldots, C_q$, such that

(1) $C_1 = \emptyset$,

(2) each pair (C_i, C_{i+1}) is a legal move.

a calculation is *complete* when every output of $\mathfrak{D}$ appears in some configuration C_i, i.e., has a pebble on it at some point. Intuitively, placing a pebble on a node corresponds to performing an operation and storing the result in a memory location, while removing a pebble from a node corresponds to freeing a memory location so that it can be re-used to store another value.*

There are two measures associated with a complete calculation on a dag $\mathfrak{D}$: *time* (number of moves) and *space* (maximum number of nodes in any configuration, i.e., maximum number of pebbles on the dag at any point). We can talk accordingly about the time and space requirements of $\mathfrak{D}$. For the dag of the term t (see figure above), the time requirement is 7 (using 7 pebbles) and the space requirements 4 (making at least 10 moves).

* What we have given here is one version of the pebble game, which corresponds to Definition 2.2 above. There is another version which allows the removal and placement of a pebble in a single move. In particular, the latter allows the additional rule that if all the immediate predecessors of a node have pebbles on them, then one of these pebbles may be moved onto that node. This rule mimics more directly assignment instructions in high-level programming languages, which may store their result in one of the same memory locations that hold their arguments. (Think, for example, of the instruction $X := X + Y$.)

There are also several generalizations of the pebble game, in either one of the two versions above, which have been introduced for various applications in computer science. Such are, for example, the "black- and-white pebble game", the "pebble game with auxiliary pushdown stores", the "edging game", and the "parallel pebble game" (Pippenger [1981,1982], Kozen [1984], and Savage and Vitter [1984], respectively.)

The connection between Definition 2.2 and the pebble game is now plain. α-term t is a $p-\alpha$-term iff the space requirement of the dag representing t is at most p. Put differently, a $p-\alpha$-term is an α-term whose evaluation requires at most p memory locations (Friedman, Lemma 1.6.1).

2.4 Lemma (Friedman, Lemma 1.3.1). Let $\mathfrak{F}$ be any set of function symbols containing at least one symbol of arity ≥ 2. Then for every p there is an α-term t which is not a $p-\alpha$-term.

Proof. The idea here is to construct a family of terms $\{t_n\}$ such that, (1) the dag representing t_n is a full binary tree of arbitrary height n, and (2) there are no repeating subterms in t_n, i.e. any two nodes in the dag representing t_n are labelled with distinct subterms. Since a full binary tree of height $n \geq 1$ has space requirement $n+2$ (easily proved by induction on n), the result follows.†

With no loss of generality, let $\alpha = ((x), \{G\})$ where G is a binary function symbol. First, define α-terms $\{u_n\}$ inductively, by setting $u_1 = x$ and $u_{n+1} = G(u_n, u_n)$. Second, define the α-terms $t_1(u_1, u_2)$, $t_2(u_1, u_2, u_3, u_4), \ldots, t_n(u_1, \ldots, u_{2^n}), \ldots$ inductively, by setting

$$t_1(u_1, u_2) = G(u_1, u_2),$$

$$t_{n+1}(u_1, \ldots, u_{2 \cdot 2^n}) = G(t_n(u_1, \ldots, u_{2^n}),\ t_n(u_{2^n+1}, \ldots, u_{2 \cdot 2^n})).$$

The dag representing t_n is obtained from a full binary tree by replacing the leaves with $u_1, \ldots, u_{2^n}$, and therefore contains no repeating subterm. ∎

The lemma fails for the case when all symbols in $\mathfrak{F}$ are of arity ≤ 1. In this case, the dag representing an α-term has a single node or is a linear chain, so that the space requirement is 1 or 2.

An effective definitional scheme (*eds* for short) of type $\alpha = (\mathbf{x}, \mathfrak{F})$ is a recursively enumerable sequence of clauses, each of the form

$$C \longrightarrow t$$

where t is an α-term and C is a finite conjunction $E_1 \& E_2 \& \ldots \& E_k$, where each E_j is $u = v$ or $u \neq v$ for some α-terms u, v. (Friedman, Definitions 1.10, 1.13).

The interpretation of an eds $(C_i \longrightarrow t_i | i \in \omega)$ in a structure $\mathfrak{A}$, with universe A and similarity type $\mathfrak{F}$, defines a partial function $f : A^n \longrightarrow A$ whose value at $\mathbf{a} \in A^n$ is

$$f(a) = \begin{cases} t_i^{\mathfrak{A}}(a), & \text{where } i \in \omega \text{ is the smallest index for which } \mathfrak{A} \models C_i[\mathbf{a}], \\ \mathit{undefined}, & \text{otherwise,} \end{cases}$$

where $t_i^{\mathfrak{A}}$ is the interpretation of t_i in $\mathfrak{A}$.

It turns out that eds's are another way of looking at certain programming languages. For example, for every eds there is an equivalent flowchart scheme augmented with a

† The proof in Friedman's article has a bug: he constructs a family of terms $\{t_n\}$ which satisfies (1), but not (2). As a result, the dag representing t_n is no longer a full binary tree when nodes labelled with identical subterms are merged. After this merging of nodes, the dag representing t_n has space requirement ≤ 4 (independent of n).

push-down stack and counters and vice-versa. (Two program schemes are equivalent if they define the same partial function over every structure in which they are interpreted.) Further connections between eds's and programming languages are discussed in Shepherdson's article in this volume.

Flowchart schemes are an example of a programming language which does not allow its individual members to each access more than a bounded number of memory locations. "Flowchart scheme" (or, equivalently, "**while**-program scheme") is the name given by computer scientists to what Friedman calls a "formalized algorithmic procedure" (*fap* for short). Augmenting a flowchart scheme wih a push-down stack is one way of allowing it access to unbounded storage space. Counters, on the other hand, do not add to the storage space accessible by a flowchart scheme; counters can hold integers, but not elements from the universe of whatever structure the scheme is interpreted in.

The following theorem is one way of stating that programs that can access unbounded storage space are more powerful than those that cannot.

2.5 Theorem (Friedman, Theorem 1.3). There is a flowchart scheme, augmented with a push-down stack and counters, which is not equivalent to any flowchart scheme augmented with counters only.

(In Friedman's terminology, there is an eds which is not equivalent to any fap with counters.)

Proof: Let $\alpha = ((x, y), \{G\})$ where G is a binary function symbol. Denote by $(t_i(x) | i \in \omega)$ the set of α-terms not containing y, recursively enumerated in some fixed order. The desired eds S of type α is

$$(y = t_i(x) \longrightarrow x | i \in \omega).$$

A structure $\mathfrak{A}$, over which S defines a partial function not programmable in the absence of unbounded storage space, is the following. The similarity type of $\mathfrak{A}$ is $\{G\}$, and for every $a, b, c, d \in \mathbf{A}$:

$$G^{\mathfrak{A}}(a, a) \neq a \text{ and}$$

$$(a, b) \neq (c, d) \text{ implies } G^{\mathfrak{A}}(a, b) \neq G^{\mathfrak{A}}(c, d).$$

These two conditions guarantee the existence of an infinite chain $a_1, a_2, a_3, \ldots$ such that $G^{\mathfrak{A}}(a_i, a_i) = a_{i+1}$ and any 2^n elements in this chain are the base of a full binary tree of height n. The argument of Lemma 2.4 shows that to generate all the elements from any $a \in \mathbf{A}$, an unbounded number of memory locations is required. The partial function $f : A \times A \longrightarrow A$ defined by S when interpreted in $\mathfrak{A}$ is:

$$f(a, b) = \begin{cases} a, & \text{if } b \text{ is generated by } a, \\ undefined, & \text{otherwise.} \end{cases}$$

No program scheme restricted to bounded storage space, in particular no flowchart scheme with counters, will compute f when interpreted in $\mathfrak{A}$-because no such program scheme can check whether b is generated by a, for all $a, b \in \mathbf{A}$. ∎

The proof technique of Theorem 2.5 is general. It can be used again to compare any two programming formalisms, one that allows each of its programs unbounded storage space and one that does not-and as long as their type contains at least one function symbol of arity ≥ 2. Thus, Paterson and Hewitt [1970] used the technique to show that flowchart schemes with recursive calls are more powerful than flowchart schemes. (Recursive calls can be simulated by push-down stacks, and vice-versa, they are therefore a way of accessing unbounded storage space.)

In the case when all function symbols are of arity ≤ 1, the pebble game or any of its generalizations cannot be used to establish such results. Access to unbounded storage may or may not add to the power of programs in this case. Here are some of the results when all function symbols are of arity ≤ 1.

(1) Every flowchart scheme augmented with a push-down stack and counters is equivalent to a flowchart scheme augmented with counters only (an immediate consequence of Friedman, Theorem 1.6).

(2) There is a flowchart scheme augmented with a push-down stack which is not equivalent to any flowchart scheme, provided there are at least two function symbols of arity = 1 (this follows from the main result in Berman, Halpern, and Tiuryn [1982], or in Urzyczyn [1983], or in Stolboushkin and Taitslin [1983], and Section 2 in Kfoury and Urzyczyn [1985]).

We do not know whether (2) still holds when there is exactly one function symbol which is of arity = 1; in fact if it does, it will settle a fundamental open problem of complexity theory, namely, DSPACE $(n) \neq \cup$ DTIME (c^n); and if (2) holds when there is exactly one function symbol of arity = 1 and one or more relation symbols of positive arity, it will settle another open problem of complexity theory, DSPACE (log n) $\neq$ PTIME (Tiuryn and Urzyczyn [1983].

3. Some Applications of the Pebble Game to Logics of Programs

A logic of programs is a formal language for reasoning about program properties. As such, it combines an "assertion formalism" and a "programming formalism" in a single language. There are many different logics of programs, depending on the possible choices for the assertion part, the programming part, and the way the two are combined.

For the sake of uniformity, our presentation is based on a specific logic of programs, first-order dynamic logic (Harel [1979]). The results below remain true for other logics programs, such as algorithmic logic (Engeler [1975], Salwicki [1970]) and logic of effective definitions (Tiuryn [1981a,1981b]). Our presentation is rather cursory; all missing details can be found in the references, along with relevant discussions and generalizations.

Our version of first-order dynamic logic fixes the assertion part and the way it is combined with the programming part of the logic, but allows the latter to vary. If $\mathcal{P}$ is a programming language, i.e., a class of program schemes satisfying some closure properties, let us denote by $DL(\mathcal{P})$ the corresponding first-order dynamic logic.

3.1 The Syntax of Dynamic Logic. Let $\mathcal{P}$ be a specific, but otherwise arbitrary programming formalism. Briefly, $DL(\mathcal{P})$ is conventional first-order logic to which we add a construct $\langle S \rangle \phi$ with the meaning that there is an execution of program $S \in \mathcal{P}$ after which the assertion ϕ holds. Hence, the symbols of $DL(\mathcal{P})$ are the usual symbols of first-order logic (including function symbols, predicate symbols, and equality) in addition to the pair of symbols $\langle\ \rangle$ (pronounced "diamond") and the symbols of the programming formalism

$\mathcal{P}$. There is another pair of symbols $[\,]$ (pronounced "box") which are strictly speaking not in the language, but which serve to abbreviate formulas. $[\,]$ is an abbreviation for $\neg\langle\rangle\neg$ just as $\forall$ is an abbreviation for $\neg\exists\neg$.

Terms and atomic formulas are formed exactly as in first-order logic. Formulas are also formed exactly as in first-order logic except that $\langle S\rangle$ may be used in place of $\exists x$, where S is any program in $\mathcal{P}$. The formulas of $DL(\mathcal{P})$ may thus be defined as follows:

(1) any atomic formula is a formula;

(2) for any formulas ϕ and ψ, $\neg\phi$ and $(\phi \vee \psi)$ are formulas;

(3) for any formula ϕ and variable $x, \exists x\phi$ is a formula;

(4) for any formula ϕ and program $S \in \mathcal{P}$, $\langle S\rangle\phi$ is a formula.

What we have called programs in this definition are strictly speaking program schemes, because they contain uninterpreted function and predicate symbols.

3.2 Adjustment. One possible choice for $\mathcal{P}$ is the class of eds's. But now, to properly define the semantics of $DL(\mathcal{P})$, we have to specify for every eds the variable (or variables) from which output values are retrieved. To keep notational details to a minimum, and with no loss of generality, the input variables $\mathbf{x}$ of an eds of type $\alpha = (\mathbf{x}, \mathcal{F})$ will also be its output variables; that is, clauses will henceforth be of the form $C(\mathbf{x}) \longrightarrow \mathbf{x} := tt(\mathbf{x})$ where $tt(\mathbf{x})$ is a vector of n α-terms–or, more simply, we shall write $C(\mathbf{x}) \rightarrow tt(\mathbf{x})$ with our new convention implicit in the notation.

The same adjustment is made for other classes of program schemes: their output variables will coincide with their input variables.

3.3 The Semantics of Dynamic Logic. Let $\mathcal{A}$ be a structure (in the appropriate similarity type), and $\sigma : V \longrightarrow A$ a valuation from the set V of all variables into the universe A of $\mathcal{A}$. Let ϕ be a formula of $DL(\mathcal{P})$. The definition of the relation $(\mathcal{A}, \sigma) \models \phi$ follows the inductive definition of $DL(\mathcal{P})$. For cases (1),(2), and (3), in 3.1, the definition of $\models$ is precisely the same as in first-order logic. For case (4) in 3.1:

if $\phi \in DL(\mathcal{P})$ and $S \in \mathcal{P}$, then $(\mathcal{A}, \sigma) \models \langle S\rangle\phi$ iff there is a finite computation of S over $\mathcal{A}$ which transforms σ into σ' and $(\mathcal{A}, \sigma') \models \phi$.

Note that, because $\mathcal{P}$ may allow non-deterministic computations, we say "there is a finite computation of S over $\mathcal{A}$ which transforms σ into σ' " rather than "the computation of S over $\mathcal{A}$ is finite and transforms σ into σ'". From this case it follows that $(\mathcal{A}, \sigma) \models [S]\phi$ iff for every finite computation of S over $\mathcal{A}$, if σ is transformed into σ' then $(\mathcal{A}, \sigma') \models \phi$.

As expected, the complexity and expressive power of $DL(\mathcal{P})$ varies greatly depending on the choice of the underlying programming language. At one extreme, if $\mathcal{P}$ is the trivial language of loop-free programs (for example, the fragment of PASCAL which does not use the mechanisms **while-do**, **repeat-until**, and recursive calls), then $DL(\mathcal{P})$ collapses into conventional first-order logic. This is because, in this case, every program in $\mathcal{P}$ can be uniformly represented† by a quantifier-free first-order formula.

The simplest, non-trivial choice for $\mathcal{P}$ is the language of **while**-programs or flowchart schemes, which allow unbounded iterations (loops). Such programs cannot be uniformly represented by first-order formulas. We can get still more powerful programming formalisms by adding to this $\mathcal{P}$ such features as: non-determinism, recursive calls (with and without parameters), counters, and others.

Sometimes the partial functions and relations computed by a programming formalism over a structure $\mathcal{A}$ are first-order definable in $\mathcal{A}$, even in the presence of loops and other programming mechanisms. This can also happen over a class of structures. For example, over the standard model of arithmetic $\mathcal{N}$, any programmable function (i.e., partial recursive function) is first-order definable in $\mathcal{N}$. This is of course because first-order formulas over $\mathcal{N}$ can code and decode finite sequences of natural numbers. In this case too, whatever can be expressed by first-order dynamic logic is already expressed by conventional first-order logic.

There are situations where the partial function computed by a program over a structure $\mathcal{A}$ is first-order definable in $\mathcal{A}$ -not because first-order formulas can code and decode finite sequences of elements, but because there is a uniform bound on the lengths of the terminating computations of that program over $\mathcal{A}$. In such a case, we say that the program unwinds in $\mathcal{A}$.

The unwind property was first stated in Engeler [1975], and derives its name from aspects of flowchart schemes to which it was first applied. It has a very succinct formulation in terms of Friedman's effective definitional schemes.

3.4 Definition. Let $\mathcal{A}$ be a structure of similarity type $\mathcal{F}$. An eds $(C_i \to tt_i | i \in \omega)$ of type $\alpha = (\mathbf{x}, \mathcal{F})$ unwinds in $\mathcal{A}$ if there is an integer n such that

$$\mathcal{A} \models (C_o \vee \cdots \vee C_n) \longleftrightarrow \bigvee_{i \in \omega} C_i \ ,$$

in which case the partial function computed by the eds over $\mathcal{A}$ is first-order definable in $\mathcal{A}$ by

$$(C_o(\mathbf{x}) \wedge \mathbf{y} = tt_o(\mathbf{x})) \vee \ldots \vee (C_n(\mathbf{x}) \wedge \mathbf{y} = tt_n(\mathbf{x})).$$

This definition directly applies to any program scheme that can be translated into an eds, in particular any flowchart scheme augmented with a push-down stack and counters. It is easily extended to more general program schemes, in particular to non-deterministic ones (Kfoury [1985]).

If an eds unwinds in a structure $\mathcal{A}$ then it unwinds in every structure $\mathcal{B}$ elementary equivalent to $\mathcal{A}$, because (using the notation of 3.4)

$$\mathcal{A} \models C_i \to C_o \vee \ldots \vee C_n \text{ iff } \mathcal{B} \models C_i \to C_o \vee \ldots \vee C_n$$

for all $i > n$ so that also

$$\mathcal{B} \models (C_o \vee \ldots \vee C_n) \longleftrightarrow \bigvee_{i \in \omega} C_i.$$

† "Uniformly represented" in the sense that the same formula represent the functions or relations computed by that program over all structures.

Again, this fact can be extended to more general program schemes, including non-deterministic ones.

If every program scheme in a programming formalism $\mathcal{P}$ unwinds in a structure $\mathcal{A}$, we say that $\mathcal{A}$ has the unwind property for $\mathcal{P}$.

3.5 Lemma. Let $\mathcal{P}$ be a programming formalism. If structure $\mathcal{A}$ has the unwind property for $\mathcal{P}$, then for every $\phi \in DL(\mathcal{P})$ there is a first-order formula ψ such that Th $(\mathcal{A}) \models \phi \leftrightarrow \psi$.

Proof. This is by induction on the syntax of $DL(\mathcal{P})$, using the fact (implied by the unwind property) that every formula $\langle S\rangle\theta$, where θ does not contain the symbols $\langle\ \rangle$, is Th $(\mathcal{A})$-equivalent to a formula θ' not containing $\langle\ \rangle$. This fact is especially transparent when S is translated into an eds $(C_i \rightarrow tt_i | i \in \omega)$: the desired formula θ' is

$$\exists \mathbf{y}\{[(C_o \wedge \mathbf{y} = tt_o) \vee \ldots \vee (C_n \wedge \mathbf{y} = tt_n)] \wedge \theta(\mathbf{y})\} \quad \blacksquare$$

Consider now the structure $\mathcal{M} = (\omega, \ =, g, 0)$ where $g : \omega \times \omega \rightarrow \omega$ is defined by

$$g(m,n) = \begin{cases} n+1, & \text{if } m = [\frac{n}{2}], \\ 0, & \text{otherwise.} \end{cases}$$

In 3.6, 3.7, and 3.8, we assume that all program schemes are in the similarity type of $\mathcal{M}$, containing one binary function symbol and one constant symbol.

3.6 Lemma. Let $\mathcal{P}$ be a programming formalism which restricts each of its program schemes to bounded storage space. $\mathcal{M}$ has the unwind property for $\mathcal{P}$.

Proof. This is an application of the pebble game. The idea is to show that with a finite supply of p memory locations, there is a bounded number (depending on p and n only) of elements that can be generated from any finite set of n natural numbers using the function g. From this, it follows that a program scheme must unwind in $\mathcal{M}$ if it has access to only p memory locations (think again in terms of eds's–it is easier). The details are in Kfoury [1983]. ∎

3.7 Lemma. Let $\mathcal{P}$ be the formalism of flowchart schemes each augmented with a push-down stack and counters. There is a formula $\phi \in DL(\mathcal{P})$ such that for no first-order formula ψ is it the case that Th $(\mathcal{M}) \models \phi \leftrightarrow \psi$.

Proof. We can take $\mathcal{P}$ to be the class of eds's. Let $\alpha = ((x), \{G\})$, where G is a binary function symbol, which will be interpreted as the function g of $\mathcal{M}$. Let $(t_i(x) | i \in \omega)$ be an r.e. sequence of all α-terms. It is easy to see that the set $\{t_i^{\mathcal{M}}(0) | i \in \omega\}$ exhausts the entire universe of $\mathcal{M}$, so that the eds $S = (x = t_i(0) \rightarrow x | i \in \omega)$ converges over $\mathcal{M}$ for every input value. This also means that over a non-standard model $\mathcal{M}'$ of Th $(\mathcal{M})$, the eds S does not converge for every input value. Hence, $\mathcal{M} \models \forall x(\langle S\rangle \text{true})$ but $\mathcal{M}' \not\models \forall x(\langle S\rangle \text{ true})$. The desired formula ϕ of $DL(\mathcal{P})$ is $\forall x(\langle S\rangle \text{ true})$. ∎

Theorem 3.8 was first established by Tiuryn [1981c] by a different method.* Tiuryn used a rather involved back-and-forth construction (a variant of Ehrenfeucht-Fraisse games) together with the pebble game.

* A third proof (in Russian) is in Erimbetov [1981].

3.8 Theorem. Let $\mathcal{P}$ be the formalism of flowchart schemes each augmented with a push-down stack and counters, and $\mathcal{P}'$ that of flowchart schemes augmented with counters. Then $DL(\mathcal{P})$ is more expressive than $DL(\mathcal{P}')$.

Proof. By 3.7 there is a formula of $DL(\mathcal{P})$ which can distinguish between two distinct models of Th $(\mathcal{M})$. By 3.5 and 3.6, no formula of $DL(\mathcal{P}')$ can. ∎

The preceding result is a general one, and can be proved again by the same technique for other programming formalisms $\mathcal{P}$ and $\mathcal{P}'$ such that: $\mathcal{P} \supset \mathcal{P}'$ and $\mathcal{P}$ allows access to unbounded storage space but $\mathcal{P}'$ does not (Tiuryn [1981c]). The theorem holds as well in the non-deterministic case. Our specific choice for $\mathcal{P}$ and $\mathcal{P}'$ was in order to directly use Friedman's effective definitional schemes (identified with the formalism of flowchart schemes each equipped with a push-down stack and counters), as well as to make explicit the sense in which Friedman's earlier result (Theorem 2.5 above) is extended by Theorem 3.8. The first is implied by the second, but not the other way around.

The comments at the end of Section 2 again apply here. When all function symbols are of arity ≤ 1, Theorem 3.8 cannot be proved by techniques based on the pebble game. If $\mathcal{P}$ is the formalism of flowchart schemes each augmented with a push-down stack, and $\mathcal{P}'$ the formalism of flowchart schemes, then $DL(\mathcal{P})$ is more expressive than $DL(\mathcal{P}')$, provided there are at least two function symbols of arity $= 1$ (see the references mentioned at the end of Section 2). When there is exactly one function symbol which is of arity $= 1$, we do not know whether the result holds.

4. Acknowledgements

Thanks are due to Steven Homer, Jerzy Tiuryn, P. Urzyczyn, and J.C. Shepherdson for various corrections and comments on an earlier draft of the paper. I am especially grateful to Steve Homer who took charge of putting the paper in its final form. Tom Orowan was responsible for typesetting the paper in TEX.

5. References

[1] Abramson, F.G. [1978], "Interpolation theorems for program schemata," *Information and Control*, **36**, No. **2**, 217-233.

[2] Berman, P., J.Y. Halpern, and J. Tiuryn [1982], "On the power of non-determinism in dynamic logic," *Proceedings of the 9-th ICALP, Lecture Notes in Computer Science*, **140**, Springer-Verlag, 48-60.

[3] Crasemann, C. and H. Langmaack [1983], "Characterization of acceptable by Algol-like programming languages," *Proceedings of Workshop on Logics of Programs, Carnegie-Mellon Univ.*, edited by E. Clarke and D. Kozen (*Lecture Notes in Computer Science*, **164**, Springer-Verlag).

[4] Engeler, E. [1975], "Algorithmic logic," in *Mathematical Center Tract* No. **63**, edited by de Bakker, Amsterdam.

[5] Erimbetov, M.M. [1981], "On the expressive power of programming logics," *Proceedings of the Conference; Research in Theoretical Programming*, Alma-Alta 1981, 49-68 (in Russian).

[6] Friedman, H. [1970], "Algorithmic procedures, generalized Turing algorithms, and elementary recursion theory," *Logic Colloquium '69*, ed. by R.O. Gandy and C.M.E. Yates, North-Holland, 361-389.

[7] Harel, D. [1979], First-Order Dynamic Logic *Lecture Notes in Computer Science*, **68**, Springer-Verlag.

[8] Hopcroft, J., W. Paul and L. Valiant [1975], "On time versus space," *16th IEEE Symp. on Foundations of Computer Science*, 57-64. Journal version of the same article appeared in *J. of the ACM* **24**, (1977), 332-337.

[9] Kfoury, A.J. [1983], "Definability by programs in first- order structures," *Theoretical Computer Science, Fundamental Studies*, **25**, 1-66.

[10] Kfoury, A.J. [1985], "Definability by deterministic and non-deterministic programs (with new proofs of old and new results in first-order dynamic logic)," to appear in *Information and Control*.

[11] Kfoury, A.J. and P. Urzyczyn [1985], "Necessary and sufficient conditions for the universality of programming formalisms," to appear in *Acta Informatica*.

[12] Kozen, D. [1984], "Pebblings, edgings, and equational logic," *16th ACM Symp. on Theory of Computing*, 428-435.

[13] Meyer, A.R. and J. Tiuryn [1982], "A note on equivalences among logics of programs," *Logics of Programs Workshop IBM Yorktown Heights*, Proceedings edited by D. Kozen (*Lecture Notes in Computer Science* **131**, Springer-Verlag).

[14] Paterson, M. and C. Hewitt [1970], "Comparative schematology," *MIT A.I. Lab Technical Memo No. 201*, (also in Proceedings of Project MAC Conference on Concurrent Systems and Parallel Computation, 1970.)

[15] Pippenger, N. [1980], "Pebbling," *Fifth Symposium on Mathematical Foundations of Computer Science*, IBM Japan.

[16] Pippenger, N. [1981], "Pebbling with an auxiliary pushdown," *J. Computer and System Sciences* **23**, 151-165.

[17] Pippenger, N. [1982], "Advances in Pebbling," *Proceedings of the 9-th ICALP, Lecture Notes in Computer Science*, **140**, Springer-Verlag.

[18] Salwicki, A. [1970], "Formalized algorithmic languages," *Bull. Acad. Pol. Sci. Ser. Sci. Math. Astr. Phys.* **18** (5).

[19] Savage, J.E. and J.C. Vitter [1984], "Parallelism in space- time tradeoffs," *Proceedings of International Workshop on VLSI: Algorithms and Architecture*, Amalfi, Italy.

[20] Shepherdson, J.C. [1975], "Computation over abstract structures: serial and parallel procedures and Friedman's effective definitional schemes," in *Logic Colloquium, '73*, North-Holland, 445-513.

[21] Stolboushkin, A.P. and M.A. Taitslin [1983], "Deterministic dynamic logic is strictly weaker than dynamic logic," *Information and Control*, April 1983.

[22] Tiuryn, J. [1981a], "Logic of effective definitions," *Fundamenta Informaticae, Annales Societatis Mathematicae Polonae*, IV.3, 629-659.

[23] Tiuryn, J. [1981b], "Survey of the logic of effective definitions," *Logic of Programs Workshop, ETH Zurich*, proceedings edited by E. Engeler (Lecture Notes in Computer Science, **125**, Springer-Verlag).

[24] Tiuryn, J. [1981c], "Unbounded memory adds to the power of logics of programs," *Proceedings of 22nd Annual Symposium on Foundations of Computer Science*, 335-339 (full paper to appear in *Information and Control*).

[25] Tiuryn, J. and P. Urzyczyn [1983], "Some relationships between logics of programs and complexity theory," *Proceedings of 24th Annual Symposium of Foundations of Computer Science*.

[26] Urzyczyn, P. [1981a], "Algorithmic triviality of abstract structures," *Fundamenta Informaticae, Annales Societatis Mathematicae Polonae, IV.4*, 819-849.

[27] Urzyczyn, P. [1981b], "The unwind property in certain algebras," *Information and Control*, **50**, No. **2**, 91-109.

[28] Urzyczyn, P. [1983], "Non-trivial definability by flowchart programs," *Information and Control*, **58**, No. 1-3.

HARVEY FRIEDMAN'S RESEARCH ON THE FOUNDATIONS
OF MATHEMATICS, L.A. Harrington et al. (editors)

EQUALITY BETWEEN FUNCTIONALS REVISITED

R. STATMAN

Department of Mathematics
Carnegie-Mellon University
Pittsburgh, Pennsylvania 15213
U.S.A.

In this note we shall try to place Friedman's remarkable little paper ([3]) in the context of what we know today about the model theory of the typed λ-calculus. In order to do this, it is appropriate to survey recent work in this area, in so far as it touches on issues raised by Friedman. We don't intend a general survey; much will be left out. In particular, we shall omit discussion of unification, fragments of L.C.F., and polymorphic types, which are areas of interest to the author, and the specialist will surely find other omissions. However, we will try to show the reader how Friedman's paper lays the foundation for the general model theory of the typed λ-calculus.

The plan of this note is the following. We shall begin by a brief, informal introduction to the subject. The reader is then advised to read Friedman's paper. It is, after all, quite accessible and easy to read. We shall then proceed to consider the three major issues touched on in "Equality between functionals" as we view them today. These issues are

(1) completeness theorems,

(2) the solvability of higher type functional equations, and

(3) logical relations.

THE TYPED λ-CALCULUS

Let A be a nonempty set. Then we can form the function spaces $A^A, A^{(A^A)}, (A^A)^A, \ldots$ etc. Each function space will be assigned an expression (type) describing how it is constructed. 0 is assigned to A, $0 \to 0$ to A^A, $(0 \to 0) \to 0$ to $A^{(A^A)}$, $0 \to (0 \to 0)$ to $(A^A)^A$ etc. Each type τ can be written uniquely in the form $\tau(1) \to (\ldots(\tau(t) \to 0)\ldots)$. More generally a frame (for Friedman "pre structure") $\mathfrak{A}$ is a map from types to non-empty sets $\tau \to \mathfrak{A}^\tau$ such that

$$\mathfrak{A}^{\sigma\to\tau} \subseteq \mathfrak{A}^{\tau\,\mathfrak{A}^\sigma}.$$

When $\mathfrak{A}^{\sigma\to\tau}$ is always $\mathfrak{A}^{\tau\,\mathfrak{A}^\sigma}$, is the full type structure $\mathcal{S}_{\mathcal{A}}$

for $\mathcal{N} = |\mathfrak{A}^0|$.

Functions of several arguments can be included by "Currying". For example, if $\Phi \in A^{A\times A}$ for $a \in A$ define Φ_a by $\Phi_a x = \Phi(a,x)$ and $\hat{\Phi}$ by $\hat{\Phi}a = \Phi_a$ so $\hat{\Phi} \in (A^A)^A$.

Let $\mathfrak{A}^*$ result from $\mathfrak{A}$ by adjoining infinitely many indeterminates to each $\mathfrak{A}^\tau$ just as one does in algebra. $\mathfrak{A}$ is called a model (for Friedman "structure") if for each Φ in $\mathfrak{A}^*$ and each indeterminate x there exists a Ψ in $\mathfrak{A}^*$ which does not depend on x (as a function depends on its arguments) such that $\Psi x = \Phi$. In other words each Φ can be made an explicit function of its parameters. Such a Ψ is denoted $\lambda x\, \Phi$. We say that models are frames closed under "explicit definition".

Type theory is a language for describing models. Its equational part is the typed λ-calculus. Terms are built up from some set Σ of constants of various types, and variables $x^\sigma, y^\tau, z^\rho, \ldots$ of types $\sigma, \tau, \rho, \ldots$ by means of the application symbol $(\ )$ and abstraction symbols $\lambda x^\sigma, \lambda y^\tau, \lambda z^\rho, \ldots$. Terms have the obvious meaning in models $\mathfrak{A}$, where $(\ )$ is the union overall σ and τ of the maps $(\Phi,\Psi) \to \Phi\Psi$ in $\mathfrak{A}^{\tau\,\mathfrak{A}^{\sigma\to\tau}\times\mathfrak{A}^\sigma}$, and λx^σ is the union overall τ of the maps $\Phi \to \lambda x\Phi$ in $\mathfrak{A}^{*\sigma-\tau\,\mathfrak{A}^{*\tau}}$. The resulting set of terms is denoted $\Lambda(\Sigma)$ with the set of terms of type τ denoted $\Lambda(\Sigma)^\tau$. $\overline{\Lambda}(\Sigma)$ denotes the set of closed terms, with the set of closed terms of type τ denoted $\overline{\Lambda}(\Sigma)^\tau$. We reserve the letters M and N to denote member of $\overline{\Lambda}$, where $\Lambda = \Lambda(\emptyset)$.

We shall make the following notational conventions. We delete type superscripts when no confusion can result. We also delete application symbols associating to the left. Finally we write $\lambda x_1 \ldots x_n$ for $\lambda x_1 \lambda x_2 \ldots \lambda x_n$. Thus, for example, $\lambda x^{0\to(0\to0)}\lambda yz\ xz(yz)$ abbreviates the term $\lambda x^{0\to(0\to0)}\lambda y^{0\to0}\lambda z^0((x^{0\to(0\to0)}z^0)(y^{0\to0}z^0))$.

The language of type theory is built up from equations between terms by the usual propositional operations and typed quantifiers.

Now the equations

$$\lambda xX\ Y = [Y/x]X \qquad (\beta),$$

where $[Y/x]X$ is the result of substituting Y for x in X, and

$$\lambda x(Xx) = X \qquad (\eta),$$

for x not free in X, are clearly valid in all models. Let $\underset{\beta\eta}{=}$ be the congruence relation on terms generated by (β) and (η). Are there any other valid equations? Friedman's completeness theorems answer this question.

COMPLETENESS THEOREMS

It now seems fair to say that, with the 1-section theorem below, we have a satisfactory understanding of the completeness phenomenon. This was not the case in 1970. Friedman's completeness theorems stand as the first contributions to the model theory of the subject. Although his techniques have been superceeded in this area, they make a direct contribution to the theory of logical relations which will be discussed later. In addition, Friedman's emphasis on the universal algebraic aspects of completeness motivated the author to look for something like the 1-section theorem.

If $\mathcal{C}$ is a class of models we say $\mathcal{C}$ is complete if $M \underset{\beta\eta}{=} N \iff \forall \mathfrak{A} \in \mathcal{C}\ \mathfrak{A} \models M = N$. $\underset{\beta\eta}{=}$ is in some sense an equational theory (although, extensionality is required for models). Thus one expects that something like free models should exist. Assume that $\overline{\Lambda}(\Sigma)^{0} \neq \emptyset$. For $T_1, T_2 \in \overline{\Lambda}(\Sigma)^{\tau}$ define

$$T_1 \sim T_2 \iff \forall U_1 \in \overline{\Lambda}(\Sigma)^{\tau(1)} \ldots \forall U_t \in \overline{\Lambda}(\Sigma)^{\tau(t)}$$
$$T_1 U_1 \ldots U_t \underset{\beta\eta}{=} T_2 U_1 \ldots U_t .$$

If $\sim$ is a congruence relation then we can build a model $\mathfrak{M}_\Sigma$, consisting of $\sim$ congruence classes, satisfying

$$\mathfrak{M}_\Sigma \models T_1 = T_2 \iff T_1 \sim T_2 .$$

Friedman's first completeness theorem is just this. If for all τ Σ^{τ} is infinite then $\mathfrak{M}_\Sigma$ exists and $\mathfrak{M}_\Sigma \models T_1 = T_2 \iff T_1 \underset{\beta\eta}{=} T_2$. This generalizes to the

<u>Existence of Free Models</u> ([15])

> $\mathfrak{M}_\Sigma$ always exists. Moreover, only five local structures (see [2] pg. 496) are possible.

Friedman's first theorem has a short of converse which resembles Hilbert-Post completeness.

<u>Consistency Theorem</u> ([14])

> $\underset{\beta\eta}{=}$ is the maximal non-trivial congruence on closed terms satisfying typical ambiguity.

Here typical ambiguity just means that 0 can be replaced by any other type.

Friedman's <u>second completeness theorem</u> concerns the full type structure $\mathcal{S}_{\mathcal{U}}$ over a ground domain of size $\mathcal{U}$. In short, if $\mathcal{U} \geq \aleph_0$ then $\{\mathcal{S}_{\mathcal{U}}\}$ is complete. He also pointed out that for no finite m is $\{\mathcal{S}_m\}$ complete. In [13] we refined this to the

<u>Finite Model Property</u>

> For each M there exists an m, recursive in M, such that for all N
>
> $$\mathcal{S}_m \models M = N \iff M \underset{\beta\eta}{=} N .$$

Thus the typed λ-calculus exhibits a rather strong separation property with respect to its hereditarily finite full models.

If $\mathfrak{A}$ is a model the 1-section of $\mathfrak{A}$ is $\mathfrak{A}^0 \cup \mathfrak{A}^{0\to 0} \cup \mathfrak{A}^{0\to(0\to 0)} \cup \ldots$ (i.e. the first-order part of $\mathfrak{A}$). The next theorem says, somewhat surprisingly, that completeness depends only on 1-sections and not higher types. Let $\mathcal{F}$ be the free term algebra on a single constant and a single binary function symbol.

<u>1-Section Theorem ([13])</u>

> $\mathcal{C}$ is complete if and only if $\mathcal{F}$ can be embedded in the 1-section of some countable direct product of members of $\mathcal{C}$.

One consequence which Friedman observed of his second completeness theorem is that equality of λ definable functionals (i.e. those denoted by members of $\overline{\Lambda}$) in $\mathcal{S}_{\mu}$ is decidable. Unfortunately, when $\mu > 1$, there is no efficient decision procedure.

<u>Complexity of $\beta\eta$ Conversion ([12] and [13])</u>

> Any elementary recursive set of closed terms closed under $\underset{\beta\eta}{=}$ contains none or all of the terms of any given type.

SOLVING FUNCTIONAL EQUATIONS AT HIGHER TYPES

Much of the literature on the typed λ-calculus prior to 1970 concerned functional equations

$$E \equiv E(\vec{y},\vec{x}) \equiv M\,\vec{y}\,\vec{x} = N\,\vec{y}\,\vec{x}$$

which, given parameters $\vec{y}$, we wish to solve for $\vec{x}$. Roughly speaking the literature sorts itself into three topics; constructive solvability (e.g. Kleene [8], Kreisel [9], Gödel [4], Scott [11]), solvability in all models i.e. unification (e.g. Andrews [1], Gould [5], Guard [6]), and solvability of special classes of equations (e.g. Scott [11]). Friedman first recognized the significance of the <u>axiom of choice</u> (A.C.) in this area.

$$\text{(A.C.)} \qquad \forall x\, \exists y\, \mathcal{f}(x,y) \to \exists z\, \forall x\, \mathcal{f}(x,zx)$$

Functional equations are always closed under conjunction ([3],[13]). Under A.C. they are closed under negation. Let's call a sentence of type theory which asserts that a parameterless functional equation has a solution an $\exists$E sentence. Friedman observed that there exist a (polynomial time) computable map $\mathcal{f} \mapsto \mathcal{f}^+$ from sentences to $\exists$E sentences such that A.C. $\vdash \mathcal{f} \longleftrightarrow \mathcal{f}^+$. In [16] we refined this as follows. Let Δ be the sentence that asserts that the definition by cases functional (IF____THEN____ELSE____) of lowest type exists.

<u>Normal Form Theorem part 1</u>

> There exists a polynomial time computable map $\mathcal{f} \mapsto \mathcal{f}^*$ from sentences to $\exists$E sentences s.t.
>
> $$\text{A.C.} \vdash \mathcal{f} \longleftrightarrow \mathcal{f}^*$$
> $$\Delta \vdash \mathcal{f}^* \longrightarrow \mathcal{f}$$

Models of Δ are just models of the $\lambda\delta$ calculus where

$$\delta xyuv = u \quad \text{if} \quad x = y$$
$$= v \quad \text{if} \quad x \neq y.$$

The normal form theorem extends to theories.

Normal Form Theorem part 2

There exists a polynomial time computable set of $\exists E$ sentences $A.C.^{\exists E}$ such that if $\mathcal{T}$ is any theory at least as strong as A.C.

$$\mathcal{T} \text{ is equivalent to } \mathcal{T}^* + A.C.^{\exists E} + \Delta$$
$$\mathcal{T} \vdash \varphi \Longleftrightarrow \mathcal{T}^* + A.C.^{\exists E} \vdash \varphi^*$$

One consequence of Friedman's observation is that if a functional equation is not solvable in a model of A.C. then it is not solvable in any extension (in the obvious sense) of that model. This corollary motivated the author to search for a result which equates the unsolvability of E in every extension of $\mathfrak{A}$ with the solvability of some other $\tilde{E}$ in , for general models $\mathfrak{A}$. Given E as above define $\tilde{E}_n \equiv \tilde{E}_n(z_0 z_1 \vec{y}, \vec{u})$ for $z_o, z_1 \in 0$ to be

$$\lambda\vec{x}\ z_o = \lambda\vec{x}\ u_1\vec{x}\ (M\vec{y}\ \vec{x})(N\vec{y}\ \vec{x}) \ \wedge$$

$$\lambda\vec{x}\ u_1\vec{x}\ (N\vec{y}\ \vec{x})(M\vec{y}\ \vec{x}) = \lambda\vec{x}\ u_2\vec{x}\ (M\vec{y}\ \vec{x})(N\vec{y}\ \vec{x}) \ \wedge$$

$$\vdots$$

$$\lambda\vec{x}\ u_n\vec{x}\ (N\vec{y}\ \vec{x})(M\vec{y}\ \vec{x}) = \lambda\vec{x}\ z_1$$

No Counterexample Theorem ([18])

$E(\vec{\Phi},\vec{x})$ is solvable in an extension of $\mathfrak{A}$ if and only if for each n and $a,b \in \mathfrak{A}^*$ s.t. $a \neq b$ $\tilde{E}_n(ab\vec{\Phi},\vec{u})$ is not solvable in $\mathfrak{A}$.

Friedman used his normal form theorem to conclude that the problem of determining if a funtional equation has a solution in all models of A.C. is undecidable (Σ_1^0 complete). It also follows from his normal form theorem that the problem of determining if a functional equation has a solution in some (general) model is Π_1^0 complete.

LOGICAL RELATIONS

Examples of logical relations occur throughout the literature on the typed λ-calculus. These are hereditarily defined classes of functionals of which Howard's hereditarily majorizable functionals ([7]) is typical. Friedman's notion of partial homomorphism, central to his second completeness proof, is easily cast in the form of a logical relation on the product of models. In [10] Plotkin introduced a general notion of logical relation in a model theoretic context. However, perhaps the most striking use of relations of this type is Tait's proof ([20]) of the normalization theorem, and his proof is purely syntactic. In addition, as Friedman pointed out, his notion makes perfectly good sense for frames ("pre structures") where λ-terms can fail to denote. Our theory of logical relations is designed to cover each of these cases. The form of words "logical relation" is particularly appropriate, since the principal properties of such relations are closure under existential quantifications, infinite conjunctions, and in the context of models, infinite disjunctions, when these are suitably defined.

Let Σ_i for $1 \leq i \leq n$ be sets of constants. A logical relation $\mathcal{R}$ is a map $\tau \to \mathcal{R}_\tau \subseteq \Lambda(\Sigma_1)^\tau \times \ldots \times \Lambda(\Sigma_n)^\tau$ satisfying

$$\mathcal{R}_{\sigma\to\tau}(X_1,\ldots,X_n) \iff \forall Y_1\ldots Y_n \mathcal{R}_\sigma(Y_1,\ldots,Y_n) \longrightarrow \mathcal{R}_\tau(X_1Y_1,\ldots,X_nY_n).$$

Such an $\mathcal{R}$ is called admissible if it is closed under coordinatewise head expansions ([2] pg. 169) at type 0. It is called normal if it is closed under coordinatewise $=_{\beta\eta}$ at type 0.

Let θ_i range over substitutions such that $\forall x\ \theta_i x \in \Lambda(\Sigma_i)$. Define

$$\mathcal{R}^*(X_1,\ldots,X_n) \iff \forall\theta_1\ldots\theta_n\ \forall x \mathcal{R}(\theta_1 x,\ldots,\theta_n x) \longrightarrow \mathcal{R}(\theta_1 X_1,\ldots,\theta_n X_n)$$

and

$$\exists\mathcal{R}(X_2,\ldots,X_n) \iff \exists X_1 \mathcal{R}^*(X_1,\ldots,X_n).$$

If $\mathcal{S}$ is a set of logical relations define $\wedge\mathcal{S} = \bigcap_{\mathcal{R}\in\mathcal{S}} \mathcal{R}^*$, and $\vee\mathcal{S}$ = the applicative closure of $\bigcup_{\mathcal{R}\in\mathcal{S}} \mathcal{R}^*$.

Closure Theorem ([17])

> Suppose $\mathcal{R}$ is admissible and so is each member of $\mathcal{S}$. Then $\mathcal{R}^*$, $\exists\mathcal{R}$ and $\wedge\mathcal{S}$ are logical, admissible and fixed by $*$. Moreover, if each member of $\mathcal{S}$ is normal then $\vee\mathcal{S}$ is logical, normal and fixed by $*$.

Fundamental Theorem of Logical Relations ([17])

> If $\mathcal{R}$ is admissible and $X_1,\ldots,X_n \in \Lambda$ then $X_1 =_{\beta\eta} \ldots =_{\beta\eta} X_n \Longrightarrow \mathcal{R}^*(X_1,\ldots,X_n)$.

The standard syntactic results about the typed λ-calculus such as the Church-Rosser property, normalization and strong normalization, the standardization theorem and η postponement follow quite easily from these theorems. Moreover, the above definitions carry over to frames in the obvious way. We have:

Characterization Theorem ([17])

1. Suppose $\mathfrak{A}$ is a frame. Then $\mathfrak{A}$ is a model if and only if for each frame $\mathfrak{B}$ and logical $\mathcal{R}$ on $\mathfrak{A}\times\mathfrak{B}$ $\exists\mathcal{R}$ is logical.
2. Suppose $\mathfrak{A}$ is a model. Then $\Phi \in \mathfrak{A}$ is λ-definable if and only if for each $\mathfrak{B} \supseteq \mathfrak{A}$ and each logical $\mathcal{R}$ on $\mathfrak{B}$, $\mathcal{R}(\Phi)$.

A similar characterization of λ-definability related to 2. is the

Definability Theorem ([14])

> In any model the λ-definable functionals are precisely those which are stable solutions of λ-free systems of functional equations.

Here stable solutions are roughly those which are unique and remain so under "perturbations" of the model by partial homomorphisms.

The outstanding open problem in the model theory of the typed λ-calculus is to find an effective characterization of λ-definability for the $\mathcal{P}_m$.

λ-Definability Conjecture

> λ-definability in the $\mathcal{P}_m$ is decidable.

For some consequence of this conjecture we refer the reader to [13].

REFERENCES

[1] Andrews, P. B., Resolution in type theory, J. Symbolic Logic 36 (1977) 414-432.

[2] Barendregt, H. P., The Lambda Calculus (North Holland, 1981).

[3] Friedman, H., Equality between functionals, Springer Lecture Notes in Math 453, edited by R. Parikh (Springer Verlag, 1975) 22-37.

[4] Gödel, K., Über eine bisher noch nicht benutzte, Erweiterung des finiten Standpunktes Dialectica 12 (1958) 280-287.

[5] Gould, W. E., A matching procedure for ω-order logic, Dissertation, Princeton Univ. (1966) Princeton, N.J.

[6] Guard, J. R. et al., Semiautomated mathematics, J. of the Assoc. for Comp. Machinery 16 (1969) 257-267.

[7] Howard, W. A., Hereditarily majorizable functionals of finite type, Springer Lecture Notes in Math. 344, editor A. S. Troelstra (Springer-Verlag, 1973) 454-461.

[8] Kleene, S. C., Recursive functionals and quantifiers of finite types I, Trans. Amer. Math. Soc. 91 (1969) 1-52.

[9] Kreisel, G., Interpretation of analysis by means of constructive functionals of finite type, Constructivity in Mathematics, edited by A. Heyting (North Holland, 1959) 101-128.

[10] Plotkin, G. D., Lambda definability in the full type hierarchy, Combinatory Logic, Lambda Calculus, and Formalism, edited by J. P. Seldin and J. R. Hindley (Academic Press, 1980) 363-373.

[11] Scott, D. S., A type theoretical alternative to ISWIM, CUCH, OWHY, unpublished manuscript, Oxford Univ. (1969).

[12] Statman, R., The typed λ calculus is not elementary recursive, Theo. Comp. Sci. 9 (1979) 73-81.

[13] Statman, R., Completeness, invariance and λ-definability, J. Symbolic Logic 47 (1980) 17-26.

[14] Statman, R., λ-definable functionals and βη conversion, Archiv for Math. Logik und Grund. 22 (1982) 1-6.

[15] Statman, R., On the existence of closed terms in the typed λ calculus III, manuscript.

[16] Statman, R., Embeddings, homomorphisms, and λ-definability, manuscript.

[17] Statman, R., Logical relations in the typed λ-calculus, to appear in Information and Control.

[18] Statman, R., Solving functional equations at higher types, to appear in the Notre Dame Journal of Formal Logic.

[19] Statman, R., A model of Scott's theory L.C.F. in Gödel's theory T based on a notion of rate of convergence, manuscript.

[20] Tait, W. W., Intensional interpretation of functionals of finite type, J. Symbolic Logic 32 (1967) 198-212.

HARVEY FRIEDMAN'S RESEARCH ON THE FOUNDATIONS
OF MATHEMATICS, L.A. Harrington et al. (editors)

Mathematical Aspects of Recursive Function Theory

Robert E. Byerly
Department of Mathematics
Texas Tech University
Lubbock, Texas 79409

In mainstream recursion theory, the notions of computation and computability play a central role. There are several instances in mathematics in which recursion theoretic concepts arise naturally from concepts apparently having nothing to do with computability, e.g., the Diophantine representability of recursively enumerable sets. Friedman has discovered a number of characterizations of recursion theoretic and logical notions in terms of concepts from mainstream mathematics.

In a different direction, it has often been observed that certain structural properties of recursive functions and their gödel numberings, which again apparently have nothing to do with the notion of a computation, play a central role. The enumeration theorem and s-m-n theorems are the most noticeable examples of this. This observation suggests two areas of study. First, one could study axiomatizations of recursion theory and its generalizations in which these structural properties, rather than the notion of a computation, occupy center stage. Second, one could look at the structures (models) naturally occurring in recursion theory in terms of traditional mathematical and logical issues such as definability theory, isomorphisms of structures, and automorphisms of a structure. Friedman has made substantial contributions to both of these areas.

In section 1 we will look at axiomatic recursion theory and related issues connected with the observations made in the preceeding paragraph. Friedman in [5] investigated the model theory of an axiomatization of recursion theory due to Strong [18] based on earlier work by Wagner. The emphasis was on two traditional metamathematical issues: relative categoricity and minimality. It turns out that many of the properties of the models of this axiomatization, as well as the existence of certain objects in these models, follow from the first order axioms of Strong, sometimes in addition to a few special assumptions. We will also discuss some related work arising from some of Friedman's ideas.

In section 2, we will look at two characterizations of recursion theoretic notions, due to Friedman, given in terms of some simple ideas from two parts of elementary mathematics: namely linear algebra and analysis.

SECTION 1

In examinining statements of theorems and proofs from ordinary recursive function theory, we observe that certain fundamental concepts play a central role. An analysis of some of these concepts due to Strong [18], based on earlier work by Wagner [19], led to the BRFT (basic recursive function theory) axioms to be described below. Before giving the axiomatization, we will first examine some of the considerations that give rise to the axiomatization.

The partial recursive functions were defined to formalize the notion of functions effectively computable by finite algorithms. A partial recursive function f is defined by some algorithm, or program, in some formal system. An ideal computing agent, not limited by bounded resources of time and memory, executes the algorithm on natural number inputs $x_1,\ldots,x_n$. The computation may or may not terminate in a finite amount of time. If it does terminate, we may assume it produces an output which we will denote by $f(x_1,\ldots,x_n)$. If the value of f is defined for all possible inputs we say that f is _total_, otherwise f is _partial_. A basic theorem from elementary recursion theory shows that these two cases cannot in general be distinguished by an effective analysis of the algorithm.

A crucial observation about the algorithms or programs for partial recursive functions is that they can be coded by natural numbers in such a way that an algorithm can be reconstructed from its code by an effective mechanical procedure, and vice versa. We will therefore often identify the algorithm and its code and speak of a natural number as an algorithm or program. (A possibly better alternative would be to formulate the partial recursive functions on the set of expressions in a suitable formal language rather than on the natural numbers. Such a language should be rich enough to formalize the algorithms for partial recursive functions as well as the data which serve as inputs for the functions. We will, however, follow the more traditional approach.)

For a particular formalism, we use the notation $\{e\}_n(x_1,\ldots,x_n)$ to denote the result (if any) of applying the program e for some n-ary function to inputs $x_1,\ldots,x_n$. When dealing with two formalisms at the same time, we may use square brackets [and] for the second formalism to distinguish the two.

It is not too surprising that in the natural formalisms for the partial recursive functions, an n-ary partial recursive _universal function_ exists for each n. I. e., for each n, there is a partial recursive function f of n+1 arguments satisfying $f(e,x_1,\ldots,x_n) \simeq \{e\}_n(x_1,\ldots,x_n)$. (The symbol $\simeq$ means that either both sides of an expression are undefined, or both sides are defined and have equal values.) Intuitively, f simulates the action of program e on inputs $x_1,\ldots,x_n$. We will use the notation ϕ_n to denote the n-ary universal function for a particular formalism. In dealing with two formalisms at once, we may use ψ_n for the second.

The Enumeration Theorem, which asserts the existence of n-ary universal functions for each n, plays a fundamental role in basic recursive function theory. Almost as important is a certain structural property of universal functions: Kleene's s-m-n theorem, which states that there is a computable procedure for incorporating data into a program. More formally, for each m, n > 0, there is a total recursive function s^m_n satisfying

$$\{e\}_{m+n}(x_1,\ldots,x_m,y_1,\ldots,y_n) \simeq$$
$$\{s^m_n(e,x_1,\ldots,x_m)\}_n(y_1,\ldots,y_n)$$

for all $x_1,\ldots,x_m,y_1,\ldots,y_n$. The data $x_1,\ldots,x_m$ have thus been incorporated into the program e. Some of the data $x_1,\ldots,x_m$ may code programs themselves and so be interpreted by e. In this case, they will be incorporated into e as subroutines.

An important consequence of the s-m-n theorem is the recursion theorem, which states that for each partial recursive function f of n+1 variables, there is a program e such that $\{e\}_n(x_1,\ldots,x_n) \simeq f(e,x_1,\ldots,x_n)$. Intuitively, e is a program that can in the course of a computation compute a copy of itself and use that copy in any specified way. For example, by applying the recursion theorem to a function f satisfying $f(e,x) \simeq \{e\}_1(x) + 1$, we get a program e that is undefined for all inputs (since $\{e\}_1(x) \simeq \{e\}_1(x)+1$ for all x.) For another example, we can prove the statement made earlier about there being no algorithm for determining of a program whether or not it computes a total function by the following reductio ad absurdum: Suppose there were such a program. Call it d. The recursion theorem can be used to construct a program e exhibiting the following adolescent behavior. Given any input x, e first computes a copy of itself and checks, using d, to see whether e should be total. If the answer is yes, e procedes to give the lie to program d by going into an infinite loop and failing to converge. Otherwise, $\{e\}(x) = 0$.

Before presenting the definition of a BRFT, we need some notation. If D is a set and F is a set of partial functions from D into D, then F_n denotes the set of all partial functions in F of n arguments. $\lambda x_1,\ldots,x_n R(x_1,\ldots,x_n)$, where R is an expression, denotes the function f defined by $f(x_1,\ldots,x_n) \simeq$ the value of the expression $R(x_1,\ldots,x_n)$.

1.1 Definition. A BRFT is a structure $(D, F, \phi_1, \phi_2, \ldots)$ satisfying:
(1) D is an infinite set and F is a collection of partial functions on D.
(2) ϕ_n is in F_{n+1} and enumerates F_n. I.e., every function in F_n is of the form $\lambda x_1,\ldots,x_n(\phi(e,x_1,\ldots,x_n))$.
(3) F is closed under generalized composition. I.e. , whenever g is an n-ary function in F, and $f_1,\ldots,f_n$ are m-ary functions of F, then $\lambda x_1,\ldots,x_m(g(f_1(x_1,\ldots,x_m),\ldots,f_n(x_1,\ldots,x_m)))$ is in F.
(4) Each U^m_n, $1 \le n \le m$, is in F, where $U^m_n(x_1,\ldots,x_m) = x_n$, for $x_1,\ldots,x_m$ are in F.
(5) Each C^m_x, for x in D, is in F, where $C^m_x(x_1,\ldots,x_m) = x$.
(6) The function $\lambda abcx(b$, if $x = a$; c, if $x \neq a)$ is in F.
(7) There are functions s^m_n in F_{m+1} satisfying

$$\lambda y_1,\ldots,y_n(\phi_{m+n}(x,x_1,\ldots,x_m,y_1,\ldots,y_n)) \simeq \lambda y_1,\ldots,y_n(\phi_n(s^m_n(x_1,\ldots,x_m),y_1,\ldots,y_n)),$$

for all $y_1,\ldots,y_n$ in D.

The BRFT axioms are sufficiently powerful to prove most of the basic theorems of elementary recursion theory, in particular, nearly all those whose statements and proofs do not explicitly involve the notion of the length of a computation. The recursion theorem, as well as the two consequences of it mentioned above, are true in all BRFT's.

To increase readability, we use the notation $\{e\}_n$ for $\lambda x_1,\ldots,x_n\phi_n(e,x_1,\ldots,x_n)$. If we are considering two BRFT's simultaneously, we use the notation [,], instead of {, } for the second.

In a sense, all <u>total</u> recursive functions can be represented in any BRFT [19]. That is, if $(D, F, \phi_1, \phi_2, \ldots)$ is a BRFT, then there is a sequence of elements $a_0, a_1, \ldots$ in D for which the following is true. For every total recursive function g, there is a function g' in F such that $g'(a_n) = a_{g(n)}$ for all n.

1.2 Definition. An ω-BRFT is a BRFT $(\omega, F, \phi_1, \ldots)$, where ω is the set of all natural numbers (i.e., non-negative whole numbers), such that the successor function $S = \lambda x(x+1)$ is in F.

1.3 Theorem (Wagner and Strong [18]). If $(\omega, F, \phi_1, \ldots)$ is an ω-BRFT, then F contains all the partial recursive functions.

Friedman [5; theorem 2.2] proved a generalization of this which informally says that the construction principles for constructing the partial recursive functions in a particular formalism can be represented in any ω-BRFT. One can in fact prove that in any ω-BRFT, a program e for any partial recursive function f can be explicitly described in terms of programs for the basic functions in the definition of BRFT, a program for the successor function in the ω-BRFT in question, 0, and a program for f in any standard formalism.

A standard formalism, then, gives rise to an ω-BRFT (ω, PR, $\phi_1,\ldots$) which is minimal in the sense that PR, which is the set of all partial recursive functions, is a subset of the set of partial functions associated with any other ω-BRFT. Alternatively, the partial recursive functions form the unique set of functions contained in all ω-BRFT's.

A strengthening of the minimality theorem yields the following result. If g is Turing reducible to the total function f, then g is in the set of partial functions of any ω-BRFT containing f. (g is Turing reducible to the total function f if, intuitively, there is a program that can compute any value of g if implemented on a machine which has access to an oracle for f. (I.e., a "black box" that will output f(x) when supplied with x, for any x.)) This result holds for partial functions f if we use the definition for Turing reducibility given in [17]. (The definition is the same, if we use the convention that the computation of a value of g is undefined if the program asks the oracle for f for a value of f(x) that is undefined.) The author [1] has also formulated and proved a similar minimality theorem for the notion of weak Turing reducibility as defined in [17] for ω-BRFT's satisfying a certain closure condition. No similar result is known for Rogers' definition of enumeration reducibility [15; section 9.7].

We remark that for any ω-BRFT (D, F, $\phi_1,\ldots$), F is the set of partial functions Turing reducible to ϕ_1.

We now turn to relative categoricity results. It can be shown that certain BRFT's are in a sense uniquely determined by their sets of functions.

1.4 Definition. (a) Two BRFT'S (D, F, $\phi_1,\ldots$) and (D, G, $\psi_1,\ldots$) are n-Rogers isomorphic if there is a 1-1 function h from D onto D satisfying $\{e\}_n = [h(e)]_n$ for all e in D. (b) Two BRFT's (D, F, $\phi_1,\ldots$) and (E, G, $\psi_1,\ldots$) are n-isomorphic if there is a 1-1 function h from D onto E satisfying $h(\{e\}_n(x_1,\ldots,x_n)) \simeq [h(e)]_n(h(x_1),\ldots,h(x_n))$ for all e, $x_1,\ldots,x_n$ in D.

1.5 Theorem. Any two ω-BRFT's (ω, F, $\phi_1,\ldots$) and (ω, F, $\psi_1,\ldots$) (observe that the sets of partial functions are the same) are both n-Rogers-isomorphic and n-isomorphic for all n.

The fact that the two BRFT's are n-Rogers-isomorphic can be proved as in the Rogers isomorphism theorem [16]. The fact that they are n-isomorphic is, from the logician's point of view, somewhat more interesting, as it is more closely akin to the usual model-theoretic definition of isomorphism, and implies that the structures are elementarily equivalent with respect to a fairly powerful language [5; theorem 1.5]. The second part of the theorem was first proved by Blum [15; theorem X(b)] for the special case where F is exactly the set of partial recursive functions. Friedman's theorem [5; theorem 2.5] is the general case.

This theorem tells us, among other things, that there is just one ω-BRFT up to isomorphism whose set of functions is precisely the set of partial recursive functions. We call such a BRFT a <u>standard</u> <u>ω-BRFT</u>. This suggests the following potentially useful proof technique. If we want to prove that some property holds of all acceptable enumerations of the partial recursive functions (f is an acceptable enumeration if f is ϕ_1 for some standard ω-BRFT.), all we need do is prove it for some acceptable enumeration, possibly one custom tailored to make the proof simple, and to prove that the property is preserved under isomorphism. As far as the author knows, this idea has been exploited in an essential way only in [3; section 3].

Let $(\omega, PR, \phi_1, \ldots)$ be a standard ω-BRFT, and consider the structure $\Omega = (\omega, p)$, where p is a ternary predicate on ω defined by p(a,b,c) iff $\phi_1(a,b) = c$. By theorem 1.5 Ω is uniquely determined in a categorical sense, i.e., up to isomorphism. By contrast, many of the objects constructed in all but the most elementary parts of recursion theory appear to vary considerably in their recursion theoretic properties according to the specific "programming system" used for defining the partial recursive functions, as well as such things as the specific method for defining the length of a computation, the specific enumeration of the programs, etc. (Cf. [10].) Since results of Wagner and Strong imply that many statements of recursion theoretic interest can be stated in the language of Ω, it is reasonable to consider sets and functions defined in Ω as being somehow more "natural" than those defined in terms of recursion theoretic objects that are not uniquely determined.

1.6 Theorem ([2;theorem 3.1]) A subset A of ω is definable without parameters in Ω iff it is definable in arithmetic and invariant with respect to automorphisms of Ω.

1.7 Theorem ([2;theorem 2.1]) A recursively enumerable subset A of ω that is invariant with respect to automorphisms of Ω is necessarily 1-complete.

The cited proof of 1.6 uses combinatory logic, but the result can also be proved using Matijasevic's theorem. The proof of 1.7 uses modifications of a "back-and-forth" technique first used by Friedman to get structural information about Ω.

If we regard the invariant recursively enumerable sets as being more "natural" than other non-recursive recursively enumerable sets, it may or may not be of interest to note that the "natural" recursively enumerable sets in our sense form a proper subset of Hartmanis' [9] natural 1-complete recursively enumerable sets, at least for programming systems having polynomial time computable s-m-n functions and effective composition functions. (To Hartmanis, a

1-complete set is natural if every r.e. set is 1-reducible to it via a function f such that f and f^{-1} are polynomial-time computable.)

The recursively enumerable sets invariant with respect to automorphisms of Ω include many of the most important natural recursively enumerable sets from elementary recursion theory, e.g., the set of all programs for non-empty functions, or the set $K = \{e: \{e\}_1(e)$ is defined$\}$. It would therefore seem to be of interest to obtain a decent characterization of them. A decent characterization is not known (see [2] for a discussion of their properties), but there is a fairly nice characterization of a special subclass of them.

To obtain this characterization, consider the structure (ω, E), where dEe iff $\{e\}_1(d)$ is defined (equivalently, iff d is in the eth recursively enumerable set in some standard enumeration.) Here, $\{e\}_1$ refers to the standard ω-BRFT. A modification of Blum's argument shows that the structure (ω, E) is unique up to isomorphism. A <u>basic</u> formula is a finite formula built up from atomic formulas $s = t$, $s \neq t$, and sEt by taking conjunctions only, where s and t are either the variable x or one of an infinite list of constant symbols $c_1, \ldots$. A gödel number e is called <u>reflective</u> if it is an index of ω, i.e., nEe for all n. An <u>interpretation</u> is a one-to-one map assigning a reflective gödel number to each constant $c_1, \ldots$. The truth value of a formula for a given interpretation and an assignment of a gödel number to the variable x is defined in the obvious way.

1.8 Theorem. ([2; theorem 4.2]) A recursively enumerable set A is invariant with respect to automorphisms of (ω, E) iff for some recursively enumerable set Σ of basic formulas and every interpretation I, n is in A iff for some $\sigma(x)$ in Σ, $\sigma(n)$ is true for the interpretation I.

The results about the structures (ω, E) discussed above suggest a study of an axiomatic theory of recursively enumerable sets analogous to the BRFT axioms. Indeed, Friedman in the very last paragraph of [5] suggests that such a theory would be more suitable for formalizing recursion theoretic properties of admissible sets than the BRFT axioms. In [8], Grzegorczyk gives such an axiomatic theory (which we will not describe) of recursively enumerable sets. Ideally, such a theory should contain as primitives only the universal predicates E_n, where the intuitive interpretation of $E_n(x_1,\ldots,x_n,y)$ is that the n-tuple $(x_1,\ldots,x_n)$ is in the yth recursively enumerable set of n-tuples. Grzegorczyk's formulation contains some additional functions as primitives, including a pairing mechanism and some shifting functions. For the purposes of the present discussion, Grzegorczyk's formulation can be made acceptable by using an observation of Rogers' [16]. E_1 is in some respects analogous to the membership relation ε of a rather weak set theory in which, among other things, foundation and extensionality fail. This set theory is nevertheless strong enough to formalize modifications of the usual set theoretic tricks for coding n-tuples and functions. Grzegorczyk's theory can be modified by adding suitable "set theoretic" axioms involving E_1, after which Grzegorczyk's axioms involving the additional primitives can be formulated in a suitably encoded form. After these admittedly rather _ad hoc_ changes have been made, analogs of 1.3, 1.5, 1.6, and 1.7 can be formulated and proved.

We briefly mention another result from [5]. Friedman's theorem 2.6 characterizes the sets Y of functions of several arguments on ω that can occur as the set of total elements of F, where $(\omega, F, \phi_1, \ldots)$ is an ω-BRFT. Y is such a set if and only if Y is a countable set of total functions closed under join and Turing reducibility (i.e., f is in Y whenever f is Turing reducible to some finite sequence of members of Y). 2.6 also shows that this is equivalent to Y's being the set of all total functions on ω weak Turing reducible to some partial function g.

The models of the BRFT axioms we have considered so far are based on notions of computability for which the computations are finite. Recursion theorists have considered various generalizations of ordinary recursive function theory in which computations may be infinite. Among these are the theory of the Π^1_1-functions, recursion theory on admissible ordinals, recursion in higher-type objects, and various set recursion theories.

The axiomatic approach we have discussed applies to many of these theories as well. Friedman in [5] discusses two cases: the theory of the Δ^1_1-functions (or partial hyperarithmetic functions) and admissible ordinals. The author in [1; chapter 3] looks at a number of other cases as well from the point of view of minimality and relative categoricity.

The functions Δ^1_1-in-h (or hyperarithmetic-in-h),where h is a total function, are all partial functions f on ω definable in both Π^1_1 and Σ^1_1 forms. f isΠ^1_1-in-h if it is definable by

$$f(a) = b \text{ iff } (\forall g \in \omega^{\omega})A(h,g,a,b),$$

where A is a formula of arithmetic (i.e., containing only natural number quantifiers, and only numerical terms built up from arithmetic operations and any function variables that may be present.) f is Σ^1_1-in-h if it is definable by a similar formula, but which the initial $\forall$ replaced by $\exists$. A partial function f is Δ^1_1 or hyperarithmetic if it is Δ^1_1 in some recursive function. There are several more complicated but more enlightening definitions which characterize the functions hyperarithmetic-in-h as those computable by an algorithm whose convergent computations (which may involve infinite searches) may have length bounded by some α, where α is a (possibly infinite) ordinal recursive in h (cf. [5; definition 3.2]). (An ordinal is recursive in h if it is the order type of a relation on ω^2 that is recursive in h.)

1.9 Definition. An ω-E-BRFT is an ω-BRFT $(\omega, F, \phi_1, \ldots)$ in which the functional $E:\omega^\omega \to \omega$, where $E(f) = 0$ iff 0 is in the range of f, is representable; i.e., there is a g in F such that $g(e) \simeq E(\{e\}_1)$, if $\{e\}_1$ is total; undefined, otherwise.

1.10 Theorem. If h is a total function and g is a partial function, then g is contained in every ω-E-BRFT containing h iff g is hyperarithmetic-in-h.

In particular, g is hyperarithmetic if and only if it is contained in every ω-E-BRFT. As a minimality theorem, theorem 1.10 is rather different from our previous minimality theorems since there is no ω-E-BRFT which is minimal in the same sense as the standard ω-BRFT. It is known that the set of partial functions partial recursive in the functional E is the set of partial Π^1_1 functions. (The notion of "partial recursive in a functional" is defined in [12].) It would be reasonable to expect that an ω-E-BRFT would contain all the partial functions recursive in E, i.e., all the Π^1_1 functions. As it is known that there are _partial_ Π^1_1 functions that are not hyperarithmetic (although all total ones are) theorem 1.10 implies that not every ω-E-BRFT contains all the partial Π^1_1 functions. The reason is that the omitting types theorem for ω-logic implies that for every set of integers that is Π^1_1 definable but not Δ^1_1 definable there is an ω-model for second order arithmetic in which it is not present. On the other hand, it can be shown that the set of partial functions Π^1_1 definable in a given ω-model of second order arithmetic forms the set of functions for an ω-E-BRFT.

The author in [1] has formulated the notion of an ω-G-BRFT, where $G:\omega^\omega \to \omega$, is a functional, and shown that an analog to 1.10 always holds. However, there are ω-G-BRFT's for which the stronger minimality theorem alluded to in the last paragraph holds, e.g., the functional which is the characteristic function for the set of all functions on ω coding a well-founded relation.

We briefly mention some more generalizations of these results. Friedman in [5] formulates the notion of an α-BRFT, where α is an admissible ordinal. (The author in [1] gives an improved version.) A decent minimality theorem holds for this notion, but a decent relative categoricity theorem analogous to 1.5 holds if and only if α is non-projectible. The author in [1] defines the notion of a GRFT (generalized recursive function theory) which models such generalizations of recursion theory as recursion in higher type objects, recursive operator theory, and enumeration operator theory. Minimality and categoricity theorems are proved for these notions.

Section 2

In this section we shall briefly describe some work of Friedman showing how some concepts of recursion theory arise naturally in linear algebra and analysis.

Some of Friedman's initial results in [6] are recursive unsolvability results. These can be regarded as generalizations of a result of Patterson [13], who showed the recursive unsolvability of the mortality problem, i.e., the problem of deciding of a given finite set of 3-by-3 matrices with integer coefficients whether or not the zero matrix can be expressed as a finite product of members of the set. Patterson's technique was based on an earlier result of Post [14]. Friedman's apparently more powerful technique was based on the solution to Hilbert's tenth problem. The author does not know whether Patterson's proof can be extended to prove Friedman's results.

We use R and Z to denote the real numbers and integers respectively. Recall that an affine transformation $T:R^n \to R^m$ is a linear transformation plus a constant.

2.1 Theorem. Let A be a subset of Z^n. A is recursively enumerable if and only if there is an integer k and a finite set of affine transformations $T_1,\ldots,T_n$ with integral coefficients from various R^i to R^j such that A = {x in R^n: the zero vector in R^k can be obtained by successive applications of the T_i to x}.

Recall that a set of the form {x: Tx = u}, for some u in R^n and some linear transformation $T:R^n \to R^n$, is called an <u>affine subspace</u> of R^n, and is called <u>integral</u> if T and u can be chosen to have integral coefficients. A set K of integral affine subspaces is called recursively enumerable if the set of all integral (T, u) such that {x:Tx = u} is in K is recursively enumerable. The following generalizes theorem 2.1.

2.2 Theorem. Let A be a subset of R^n. A is the union of an r.e. set of integral affine subspaces if and only if there is an integer k and a finite set of integral affine transformations $T_1,\ldots,T_n$ such that A = {x in R^n: the zero vector in R^k can be obtained by successive applications of the T_i to x}.

This theorem generalizes 2.1, since a single point in Z^n can be viewed as an integral affine subspace. Note the difference in the hypothesis of the two theorems. Theorem 2.1 requires that A be a subset of Z^n, whereas 2.2 allows A to be any subset of R^n.

We now turn to analysis. The results to be described next are a by-product of Friedman's work on the logical strength of mathematical statements [4].

Roughly speaking, Friedman considers theorems and theories of mainstream mathematics (i.e., mathematics not depending on abstract set theory) regarded in themselves as formal systems. It turns out that these theories in general are equivalent in logical strength to systems that have already occurred naturally in logical investigations.

In the course of these investigations, Friedman considered the smallest (countable) collections of real numbers and functions on Euclidean spaces that are closed under certain construction principles occurring naturally in analysis. It turned out that many classes of functions and reals that had previously been studied by recursion theorists and set theorists could be characterized as the functions and reals of such a collection. Conversely, the collections of reals and functions given by construction principles of analysis can usually be characterized in a natural way by logical and recursion theoretic notions. Furthermore, the Cantor ordinals of closed domains of functions in such a collection are countable ordinals which have arisen naturally in logical investigations.

We shall not give all the results in [7] as it would be necessary to give expositions of some advanced topics in logic and recursion theory to do so, but will present some samples (as well as definitions of the necessary notions) sufficient to give the flavor of the work. We first give some definitions. These definitions are given in two parts: first, logical and recursion-theoretic concepts, and secondly, analytic concepts.

2.3 Definition (logical and recursion-theoretic concepts).

(a) A nested sequence of rational intervals is a sequence $\{i_n: n \in \omega\}$, where each $i_n = (a_n/b_n, c_n/d_n)$, where a_n, b_n, c_n, and d_n are natural numbers, and where each i_{n+1} is contained in i_n. $\{i_n\}$ is said to be <u>recursive</u> if the functions taking n to a_n, b_n, c_n, and d_n respectively are recursive functions. A <u>recursive real number</u> is a real number which is the unique element of the intersection of the members of some recursive nested sequence of rational intervals $\{i_n\}$, (equivalently, if it has a recursive decimal expansion.) The concepts of <u>arithmetic real number</u> and <u>hyperarithmetic real number</u> are defined identically, <u>mutatis mutandis</u>. (An arithmetic function $f:\omega \to \omega$ is one satisfying $f(n) = m$ if and only if $A(n, m)$, where A is a formula of arithmetic. For the definition of arithmetic formula, as well as the definition of hyperarithmetic function, see the paragraph preceding 1.9.)

(b) <u>Church-Kleene</u> ω_1 is the smallest non-recursive ordinal. (For the definition of recursive ordinal, see the paragraph preceding 1.9.)

2.4. Definition (concepts from analysis). Throughout, K is a set of partially defined functions $f:R^n \to R$. Note that the domain of an element of K is often a proper subset of R^n.

(a) A real x is named in K if x = f(0) for some f in K. A subset A of R^n is named in K if it is the domain of some f in K. A partial function $f:R^n \to R^m$ is named in K if $f = (f_1,\ldots,f_m)$, where each f_i is in K. If f is n-ary, f_m denotes the function $f_m(x_1,\ldots,x_n) = f(m,x_1,\ldots,x_n)$, where defined; undefined otherwise. A function f with domain NxA is said to represent a sequence of functions on a subset A of R^m if $f_n:A \to R$ for all n in ω.

(b) We say that K is basic if (1) it contains the functions $\pi_n^k(x_1,\ldots,x_n) = x_k$, the function U(x) = 1, the function Z(x) which is x if $x \leqq 0$ and is undefined for positive x, the unary functions - and 1/( ), and the binary functions + and x, and (2) it is closed under generalized composition and finite sums and products. By the latter we mean that whenever f represents a sequence of functions defined on the subset A of R^n, then the n+1-ary functions g and h defined on NxA are in K, where $g(m,x_1,\ldots,x_n) = f(1,x_1,\ldots,x_n) + \ldots + f(m,x_1,\ldots,x_n)$ and $h(m,x_1,\ldots,x_n) = f(1,x_1,\ldots,x_n) \times \ldots \times f(m,x_1,\ldots,x_n)$. (Friedman subdivides the concept of basic further, but this will not be necessary for our purposes.)

(c) K is said to be closed under pointwise convergence if whenever f represents a sequence of functions on some subset A of R^n and the f_n's convergence pointwise to a function g defined on A, then g is in K.

(d) K is said to be closed under locally uniform convergence if the following holds. Let f in K represent a sequence of functions on a subset A of R^n. Suppose $G:\omega \to R^n$, which is named in K, and $H:N \to R$, which is in K, code a sequence of open balls of the form $B_m = \{x \text{ in } R^n: d(x, G(m)) < H(m)\}$ which cover A (i.e., every element of A is in one of the balls). (d is just the Euclidean distance function.) Also suppose some $J:N\times N \to R$ is in K, and J(m,r) gives a modulus of uniform convergence of $\{f_k\}$ on the intersection of B_m and A. That is, for x in the intersection of B_m and A, and for p, q > J(m,r), we have $|f_p(x) - f_q(x)| < 1/r$. Then the function $\lim_k f_k$ whose domain is A is in K.

(e) K is said to be closed under inverses if whenever $f:A \to R^n$, where f is one-to-one and A is a subset of R^n, is named in K, then f^{-1} is named in K.

(f) K is said to be closed under extensions if for every f in K with domain A contained in R^n, there is a function g in K such that g equals f on A, but is 0 for x in R^n - A.

(g) K is said to be closed under distance functions if for every A contained in R^n named in K, the function $f:R^n \to R$ is in K, where f(x) = the distance from x to A = inf {d(x,y): y is in A}.

(h) Let the subset A of R^n be closed. Following Cantor, we define the sets A^α, for ordinals α, as follows. A^0 is A. $A^{\alpha+1}$ is the set of all limit points of A^{α}. A^{λ}, for λ a limit, is defined by taking the intersection of the A^{α}'s, for $\alpha < \lambda$. For some countable ordinal α, $A^{\alpha} = A^{\alpha+1}$. The least such ordinal is the Cantor ordinal of A.

(i) An ordinal is said to be named in K if it is the Cantor ordinal of some closed set named in K. The Cantor ordinal of K is the supremum of the set of ordinals named in K.

(j) A class K is said to satisfy axiom α if there is a closed set named in K with Cantor ordinal at least α.

Observe that the intersection of basic classes satisfying any of closure conditions (c) - (g) also satisfies the same conditions. This is not necessarily true for axiom α, however.

2.5 Theorem. Let K_1 be the least basic class closed under locally uniform convergence. The real numbers named in K_1 are precisely the recursive real numbers. The functions from ω^k into ω that are in K_1 are precisely the k-ary (total) recursive functions. The ordinal of K_1 is Church-Kleene ω_1.

2.6 Theorem. Let K_2 be the least basic class closed under pointwise convergence. The real numbers named in K_2 are precisely the arithmetic real numbers. The partial functions from ω^k into ω in K_2 are precisely the partial arithmetic functions. The sets of natural numbers named in K_2 are precisely the arithmetic sets. The ordinal of K_2 is Church-Kleene ω_1.

2.7 Theorem. Let K_3 be the least basic class closed under pointwise convergence and inverses. The real numbers named in K_3 are precisely the hyperarithmetic real numbers. The partial functions from ω^k into ω in K_3 are precisely the partial hyperarithmetic functions. The sets of natural numbers named in K_3 are precisely the hyperarithmetic sets. The ordinal of K_3 is Church-Kleene ω_1.

We conclude by giving an analytic characterization of a notion from set theory.

2.8 Definition. For α a countable ordinal, let $K(\alpha)$ be the intersection of all basic classes satisfying axiom α as well as closure conditions (c) - (g).

2.9 Theorem. A real number is constructible if and only if it is named by some $K(\alpha)$. A function from ω^n into ω is constructible if and only if it is in some $K(\alpha)$. An ordinal is constructibly countable if and only if it is named in some $K(\alpha)$.

"Constructible" refers to Gödel's constructible universe L.

BIBLIOGRAPHY

[1] Robert Byerly, Contributions to axiomatic recursion theory, Ph.D. dissertation, State University of New York at Buffalo, 1979.

[2] Robert Byerly, An invariance notion in recursion theory, Journal of Symbolic Logic, vol. 47(1982), pp. 48 - 66.

[3] R. Byerly, Recursion theory and the lambda-calculus, Journal of Symbolic Logic, vol. 47(1982), pp. 67 - 83.

[4] H. Friedman, An analysis of mathematical texts, and their calibration in terms of intrinsic strength, I - IV, informal notes, SUNY at Buffalo, 1975.

[5] Harvey Friedman, Axiomatic recursive function theory, in: R.O. Gandy and C.M.E. Yates, eds, Logic Colloquium '69, (North-Holland, Amsterdam, 1971), pp. 113 - 137.

[6] Harvey Friedman, Affine representability of r.e. sets, (abstract).

[7] Harvey Friedman, Analytic definitions of notions from modern Logic, (abstract).

[8] A. Grzegorczyk, Axiomatic theory of enumeration, Generalized Recursion Theory, 1972 Oslo Symposium, (North-Holland, Amsterdam, 1974).

[9] J. Hartmanis, A note on natural complete sets and gödel numbers, Theoretical Computer Science, vol. 17(1982), pp. 75-89.

[10] C. Jockusch and R. Soare, Post's problem and his hypersimple set, Journal of Symbolic Logic, vol. 38 (1973), pp. 446 - 452.

[11] S. Kleene, Introduction to Metamathematics, (D. Van Nostrand Company, Princeton, 1952).

[12] S. Kleene, Recursive functionals and quantifiers of finite type I, Transactions of the A.M.S., vol. 91, (1959), pp. 1 - 52.

[13] M. Patterson, Unsolvability in 3 x 3 matrices, Studies in Applied Mathematics, Vol. 49, (1970), pp. 105 - 107.

[14] E. Post, A variant of a recursively unsolvable problem, Bulletin of the American Mathematical Society, vol. 52, (1946), pp. 264 - 268.

[15] Hartley Rogers, Theory of recursive functions and effective computability, (McGraw-Hill, New York, 1967).

[16] Hartley Rogers, Gödel numberings of partial recursive functions, Journal of Symbolic Logic, vol. 23 (1958), pp. 331 - 341.

[17] L. Sasso, A survey of partial degrees, Journal of Symbolic Logic, vol. 30(1979), pp. 120 - 130.

[18] H. Strong, Algebraically generalized recursive function theory, I.B.M. Journal of research and development, vol. 12(1968), pp. 465 - 475.

[19] E. Wagner, Uniformly reflexive structures: on the nature of gödelizations and relative computability, Transactions of the A.M.S., vol. 144(1969), pp. 1 - 41.

HARVEY FRIEDMAN'S RESEARCH ON THE FOUNDATIONS
OF MATHEMATICS, L.A. Harrington et al. (editors)
Elsevier Science Publishers B.V. (North-Holland), 1985

"Big" News From Archimedes to Friedman[1), 2)]

C. Smoryński

"There are some, King Gelon, who think that the number of the sand is infinite in multitude; and I mean by the sand not only that which exists about Syracuse and the rest of Sicily but also that which is found in every region whether inhabited or uninhabited. Again there are some who, without regarding it as infinite, yet think that no number has been named which is great enough to exceed its multitude." So saying (modulo translation by T. L. Heath), Archimedes went on to produce an upper bound on the number of grains of sand in the universe. This might not seem like much of an accomplishment: One has but to pick a lower bound on the size of a grain of sand and an upper bound on the size of the universe... . The real difficulty was notational: In sooth, no number had been named which was great enough to satisfy Archimedes; he had to invent his own method of generating large numbers.

The standard Greek system of enumeration did not go very far. The 27 numbers 1, 2,..., 9; 10, 20,..., 90; 100, 200,..., 900 were denoted by overlined letters of an expanded Greek alphabet, e.g. $\overline{\alpha} = 1, \overline{\theta} = 9$. Every number from 1 to 999 could be alphabetically represented, e.g. 333 is $\overline{\tau}\,\overline{\lambda}\,\overline{\gamma}$. Thousands, from 1000 to 9000 were represented by pre-subscripting a little stroke, e.g. ${}_{\prime}\beta = 2000$. One now has the ability to represent all numbers from 1 to 9999. The next number (10,000) was called a myriad and would be represented by $\overset{\alpha}{M}$; two myriads (20,000) would be denoted $\overset{\beta}{M}$. Alternatively, one could write

1) Reprinted from the Notices of the American Mathematical Society, (1983) Volume 30, No. 3, pp. 251-256, by permission of the American Mathematical Society.

2) Editors' note: Further discussion of this topic may be found in the contributions of S. G. Simpson and R. L. Smith in this volume. The reader may also wish to consult other short articles of C. Smoryński in this volume for background material. Cf. also G. Kolata's article.

$\overline{M\alpha}$ and $\overline{M\beta}$ or even $\ddot{\alpha}$ and $\ddot{\beta}$. However, Greek numeration ended with myriads. In Archimedes' words (translated by Ivor Thomas), "Now we already have names for the numbers up to a myriad and beyond a myriad we can count in myriads up to a myriad myriads." A myriad myriads is just 10^8.

Suppose one has names for all numbers $1,\ldots,X$. One can then call X the "unit" and give names for all numbers up to X^2 by counting units and remainders. In particular, having traditional names for numbers up to a myriad, Archimedes could immediately obtain names for numbers up to a myriad myriads. He called these the numbers of the first order. Using X as a myriad myriads, he concluded the nameability of all numbers up to X^2 (i.e. 10^{16}), and called those between X and X^2 the numbers of the second order. Now, it is obvious that one could use X^2 as a unit and generate names for numbers up to $(X^2)^2 = X^4$. However, Archimedes was a bit more pedestrian and only went from a unit Y (e.g. X or X^2) to $Y \cdot 10^8$. Thus, numbers of the third order were those between X^2 and X^3 (for X the myriad myriads).

Archimedes then noted that, when one has carried out this process a myriad myriads of times (i.e. one has reached X^X for $X = 10^8$), one has obtained a unit P yielding the first period. One can then go on to the first order of the second period (i.e. the numbers between P and $P \cdot X$), the various orders of the third period, and on up to the Xth order of the Pth period, i.e. $P^X = X^{X^2}$, where $X = 10^8$.

This last number cited by Archimedes is probably the largest number ever named in antiquity. It has remained rather large until the present century when it was toppled by the googolplex ($10^{10^{100}}$; Archimedes' number is about $10^{10^{16.903}}$) and also the famous Skewes Number ($10^{10^{10^{34}}}$) and the killion (which I dare not mention). The googolplex was defined for the fun of it and Archimedes' number was generated for the sake of example (the actual bound he produced for the number of grains of sand being only a tiny 10^{63}); the Skewes Number actually appeared in a proof as a bound - it was, until the 1960s I believe, the largest number to ever seriously occur in a mathematical proof.

1. DOMINANCE. There has long been a fascination with large numbers and, consequently, with methods of generating such numbers. These methods are nothing more than rapidly growing functions, with more powerful methods being functions of more rapid growth.

Suppose $(X, <, 0)$ is a linearly ordered set with least element 0 and no greatest element. If $f,g: X \to X$ are two functions, we say f (eventually) dominates g, written $g \ll f$, if $f(x) > g(x)$ from some point on. The dominance relation, $\ll$, is only a partial ordering of X^X. However, it is directed, and can have many linear suborderings, some of which can be used as measures of growth - scales in G. H. Hardy's terminology - for natural classes of functions. For functions of sufficiently slow growth, these measures can be quite meaningful ... but I am getting ahead of myself.

In the late 19th century, Paul du Bois-Reymond initiated a study of the dominance relation restricted to continuous functions on, say, the nonnegative real numbers. This study was furthered and exposited by Godfrey Hardy who, in 1910, published a little booklet thereon entitled Orders of Infinity. Of course, if our goal is to generate large numbers, it is the dominance relation on the set of functions on the nonnegative integers that most interests us. The most interesting results concern the restriction of the dominance relation to "reasonable" collections of functions.

One nice result I know of is due to Hardy. Let LE (for logarithmico-exponential) be the class of functions mapping infinite intervals (r,∞) to similar intervals (r a nonnegative real number) generated as follows:

(i) LE contains log, exp (i.e. $\exp(x) = e^x$), and all constant functions;

(ii) LE is closed under $+, \cdot, \div$, and composition.

Then, LE is linearly ordered by $\ll$.

The arithmetic version of Hardy's Theorem is also worth noting. Define the class EP of exponential polynomials as follows:

(i) the identity functions and the constant functions n, for n a non-negative integer, are exponential polynomials;

(ii) if f,g EP, then $F + g, f \cdot g, f^g \in$ EP.

The functions in EP are obviously the restrictions to the nonnegative integers of functions in LE, whence Hardy's result applies: EP is linearly ordered by the dominance relation. There is more: EP is well-ordered by this relation. (Two things are not known, however: (i) it is yet an open problem if the ordering is decidable; and (ii) the order-type of the ordering, though close to being known, is as yet undetermined.)

2. ITERATION. If the name of the game is BIG NUMBERS the strategy is RAPIDLY GROWING FUNCTIONS, i.e. BIG ELEMENTS of X^X under $\ll$. So, how do we generate these big elements? The simplest, most naive approach is to iterate some method of passing from a function f to a larger one g. In Hardy's already cited exposition of du Bois-Reymond's Infinitärkalkül, he illustrates this as follows: Suppose f is a function satisfying $x < f(x)$ from some point on. Define $f_1 = f, f_{n+1} = f \circ f_n$, i.e. $f_n(x) = f^n(x)$. Then $f = f_1 \ll f_2 \ll \ldots$ and we have successively bigger elements of X^X. Moreover, from a sequence $f_1 \ll f_2 \ll \ldots$, he shows how to find g such that, for all $n, f_n \ll g$; namely, he diagonalises: $g(n) = f_n(n)$. (Actually, since he worked with continuous functions on the reals, he had to do a little more than this. It is probably also worth remarking that du Bois-Reymond's use here of diagonalisation predates Cantor's famous argument.)

Modern logicians prefer to speed this process up a bit. Thus, given a function f satisfying $x < f(x)$, the logician will go directly to the function $g(x) = f^x(x)$, which would only first be obtained after $\omega + 1$ steps in the process mentioned above. Of course, once one realises one can iterate into the transfinite, the logician's impatience really amounts to no more than cutting down the time it takes to hit a certain level of growth by a factor of ω and does not really reach higher levels of growth. Being a logician, however, I prefer the logical approach and shall say a few words about it.

The logician's scale of growth of "slowly rapidly growing" functions is generated by the procedure just described. The logician takes $X = \omega = \{0,1,\ldots\}$ and begins with $F_0(x) = x + 1$. He then defines $F_{n+1}(x) = F_n^x(x)$, the x-fold composition of F_n applied to x. Thus,

$$F_1(x) = 2x, \; F_2(x) = 2^x \cdot x$$

and

$$F_3(x) \text{ is somewhat larger than } 2^{2^{\cdot^{\cdot^{2^x}}}} \Big\} \; x \; 2\text{'s} \; .$$

The functions F_n are cofinal in a natural class of functions, the primitive recursive functions which are just those generated by recursion in one variable. The individual function F_n measures, in a sense, the maximum growth attainable through n applications of recursion.

The difference between the rates of growth of a function F_n and its successor F_{n+1} is so vast that we do not really worry about the ordinary dominance relation. Instead, for a given function f, we generate a class $\mathfrak{F}(f)$ of functions by closure under various forms of explicit definition, e.g. composition and the replacing of variables by constants. Two functions f and g are then said to have roughly the same rate of growth if f is dominated by an element of $\mathfrak{F}(g)$ and vice versa. The emphasis is on the word "rough": F_1 has roughly the same rate of growth as any linear function; F_2 has that of exponentiation, i.e. 2^x or even x^x; and F_3 has that of a linear stack of 2's or even a linear stack of x's. It should be borne in mind, however, that for such rapid growth as exhibited by $F_2, F_3, \ldots$ any finer classification of growth could well be meaningless.

The logician's iteration process is successful relative to this notion of rough equivalence. Each F_{n+1} is demonstrably not of roughly the same rate of growth as F_n. It grows incredibly more rapidly - so much so that it cannot be reached from F_n without the aid of F_{n+1} itself; such was the significance of the closure conditions put on $\mathfrak{F}$.

Of course, if our goal is truly to generate functions of rapid growth, we will want to go beyond this sequence. Following Hardy, we diagonalise $F_\omega(x) = F_x(x)$. Actually, since we are now playing the roles of logicians, we should find it unthinkable to stop here: For countable ordinals α, define F_α by

$$F_\alpha(x) = F_\beta^x(x) \ , \text{ if } \ \alpha = \beta + 1 \ ,$$

$$F_\alpha(x) = F_{\alpha_x}(x) \ , \text{ if } \ \alpha = \lim_{x \in \omega} \alpha_x \ ,$$

where $\alpha_0 < \alpha_1 < \ldots$.

Oops! The definition of F_α , for α a limit ordinal, depends on the path $\alpha_0 < \alpha_1 < \ldots$ by which we choose to reach α : No matter how slowly this sequence grows, F_α will dominate all F_β's for $\beta < \alpha$; but if we pick a rapidly growing sequence, F_α could become a much more rapidly growing function than we intended: It could become, for example, what we intended $F_{\alpha+1}$ to be! In the first problematic case, $\alpha = \omega$, we defined F_ω by choosing $\alpha_n = n$; we could have chosen $\alpha_n = F_n(n)$ $(= F_\omega(n))$ or $\alpha_n = F_{\omega+1}(n)$. With the former choice we would get something much bigger than F_ω relative to dominance, but of roughly the same rate of growth relative to our rough equivalence; with the latter we would violate even the rough equivalence. The idea of this example is this: If we try to get beyond all F_n's by diagonalising and only allow ourselves access to the rates of growth already at hand - i.e. the F_n's - then we will get something like F_ω ; to get beyond F_ω , we must use something of already more rapid growth.

For reasonably small, well-understood ordinals α , like ω and ϵ_0 , we have more-or-less canonical choices for the sequences α_n and, relative to our notion of rough equivalence, it does not matter if we use α_n or $\alpha_{F_\beta(n)}$ for some $\beta < \alpha$. For larger, less studied ordinals, the situation might well be delicate: Two distinct ways of generating an ordinal α could readily yield distinct canonical choices of the sequence α_n , so widely inequivalent definitions of F_α . It suffices to say that $F_\alpha(x)$ can be made well defined (up to rough equivalence) for all "small" α and this suffices for the results we discuss.

But I have begun to digress ... . Our problem is not to show that there is a unique (up to rough equivalence) function we want to call F_α for some particular α , but that we want to generate such a function and admire its rapid growth rate. Thus, our problem is to generate large ordinals. Proof theorists are hard at work on this, but it is a messy undertaking and I do not wish to discuss it here. So we depart from this topic.

3. THEORETICAL GROWTH. One disadvantage to using the transfinite iteration of a given function to generate ever more rapidly growing functions is that it is such a laborious process; it is like generating large numbers by iterating the successor function when exponentiation will get us what we want much more quickly. What plays the role of exponentiation relative to our function iteration? One answer is powerful theories.

Any reasonably powerful theory T will be able to handle numbers and compute all computable functions. Interestingly enough, however, if T is sufficiently sound (i.e. T does not prove too many false arithmetic assertions - the (so-called) ω-consistency of T is sufficient here, though the mere consistency of T is not), T will not be able to prove the computability of all functions it can compute. One can diagonalise on the T-provably computable functions and obtain a function F_T which dominates every T-provably computable function.

For various theories T, F_T is known: For Primitive Recursive Arithmetic, a theory embodying Hilbert's finitism, F_T is roughly F_ω, the provably computable functions of Primitive Recursive Arithmetic being just the primitive recursive ones. For Peano Arithmetic, the usual formulation of number theory (even in a mathematical sense: it is strong enough to prove all results of analytic number theory), F_T is roughly F_{ϵ_0}. Many other examples are known.

Of course, one can now ask about F_T for T being Zermelo-Fraenkel Set Theory (the usual axiomatic set theory) or some extension by new axioms of infinity. And one can now look for progressively more powerful axioms of infinity. There are three problems here. First, with respect to the addition of new axioms, one is required to have consistent axioms; in fact the axioms must yield a theory a bit more than merely consistent. While the study of large cardinals has been an active area of research in logic, and while there is good reason to believe in the consistency of many of the new axioms, there is little evidence for or against their additional soundness.

The second problem is that, once we have F_T , we would like to assign some measure to its size, i.e. we would like to say F_T is roughly equivalent to F_α for some α . As far as I know, the theory allowing this has never been fully worked out. Moreover, the determination of the ordinal α to associate with the theory T has only been made for remarkably weak theories. I can repeat my earlier observation on large ordinals: Proof theorists are hard at work on this, but it is a messy undertaking.

The third problem is an inherent limitation on the theoretical approach. Any decent theory has axioms we can list. All of its provably computable functions are (theoretically) computable, as is their dominating diagonal function. Even the iteration process will get us past the computable functions if we simply go on until we have passed the first noncomputable ordinal. If we decide to allow noncomputable functions, we are again faced with the laborious slowness of the straightforward iteration process and we must look for principles allowing bigger jumps in the dominance ordering. There are such: One diagonalises on functions definable in languages (as opposed to provably computable in theories). However, I am not much interested in noncomputable functions and shall not discuss this further.

4. RAPID GROWTH IN REAL LIFE. The surreal character of the preceding discussion is misleading; rapid growth is not as divorced from mathematical reality as it seems. Large numbers and rapidly growing functions exist in mathematical nature. In calculus, e.g., the difference between convergence and divergence can be so slight that attempts to measure it yield large numbers. Hardy cites the two series

$$A: \sum_{n=3}^{\infty} \frac{1}{n \log n(\log \log n)^2}, \quad B: \sum_{n=3}^{\infty} \frac{1}{n \log n(\log \log n)}.$$

Series A converges to 38.43..., but it converges so slowly that it requires $10^{3.14 \cdot 10^{86}}$ terms to give two-decimal accuracy; series B diverges, but the partial sums exceed 10 only after a googolplex of terms have appeared. It would be interesting to see the full rates of growth of such moduli of convergence and divergence analysed.

Next to logic the most fruitful source of rapid growth is combinatorics, particularly Ramsey Theory (at least: until recently - cf. the next section). The two most celebrated results of Ramsey Theory are Ramsey's Theorem and van der Waerden's Theorem. To state these, we need a few definitions: For a positive integer n, an n-colouring of a set X is just a map $C : X \to n$. If X is a set, we let $[X]^k$ denote the collection of k-element subsets of X.

RAMSEY'S THEOREM. For all m,n,k there is a p so large that: If X is a set of cardinality at least p and $C : [X]^k \to n$ is an n-colouring, there is a subset Y of X of cardinality at least m and greater than k such that $C \quad [Y]^k$ is constant.

More briefly put: We call a subset Y of X homogeneous with respect to the colouring C if $c \quad [Y]^k$ is constant. Ramsey's Theorem asserts that, by making X large enough, there will be a homogeneous subset Y of X of any predetermined size. The sufficient largeness required of X is a function $R(m,n,k)$. The corresponding diagonal function $R(x,x,x)$ is known to be roughly equivalent to F_3, i.e. iterated exponentiation.

VAN DER WAERDEN'S THEOREM. For all m,n, there is a number p so large that for any colouring $C : p \to n$ there is an m-term arithmetic progression $Y \subseteq p$ such that $C \quad Y$ is constant.

Again, van der Waerden's Theorem determines a function $W(m,n)$. The "exact" rate of growth of W, or rather its diagonal $W(x,x)$, is not known. All proofs so far given rely on a double induction, whence the recursion defining $W(m,n)$ is a double recursion - yielding F_ω or a rough equivalent as the immediate upper bound.

Of course, F_ω is only the best known bound for $W(x,x)$. It may well turn out that F_{17} or F_{232} gives a bound. There are, however, other functions of combinatorial interest of definite transfinite growth rates. In fact, around 1977 Jeff Paris and Leo Harrington concocted variants of Ramsey's Theorem, the diagonal functions of which grow roughly as rapidly as F_{ϵ_0}. I shall say a few words about, say, Harrington's variant of Ramsey's Theorem.

HARRINGTON'S VARIANT. For all n,k there is a p so large that : If $q > p$ and $C : [q]^k \to n$, there is a subset Y of q of cardinality of least $\min(Y)$ and greater than k such that $C \upharpoonright [Y]^k$ is constant.

Harrington's Theorem differs from Ramsey's Theorem mainly in that, rather than to require the C-homogeneous set Y to have a prescribed cardinality, it requires Y to be <u>relatively large</u> in that its cardinality is no smaller than its least element. Again, there is an underlying function $H(n,k)$ yielding the value of p. The diagonal function $H(x,x)$ has roughly the same growth rate as F_{ϵ_0}.

Recall that the function F_{ϵ_0} has roughly the rate of growth that F_T has, where T is the theory Peano Arithmetic. Thus $H(x,x)$, as well as the corresponding function for Paris' original (uncited) variant of the Ramsey Theorem, has the rate of growth of the first "natural" function to dominate all functions provably computable in Peano Arithmetic. Originally, in fact, the rate of growth of H and the Paris function was shown to be at least this great by formal considerations about the theory Peano Arithmetic; only more recently have there evolved verifications of this growth rate that proceed by directly relating the functions in question to the F_α-hierarchy. While the verification of rapidity, i.e. the lower bound, is easily (almost effortlessly to the logician with the proper background) established by reference to the formal theory, the ordinal analysis is generally messy.

The work of Paris and Harrington opened the floodgates for many more functions of rapid growth. There is something of a recipe for their concoction: Find a quasi-combinatorial problem that is hard to solve (cf. Clote's paper for hints on this step) and show that the corresponding function dominates all functions provably total in a suitable theory (one generally does this by showing the function to encode incredibly many closure properties). Most often, the theory in question is Peano Arithmetic and the rate of growth is roughly that of F_{ϵ_0}. There are exceptions: Harvey Friedman has pushed this growth rate up quite a bit by working with stronger theories, and among his slower contributions, he, Ken McAloon, and Steve Simpson have even given a variant of Ramsey's Theorem which they find rather simple and which has a corresponding function of rate of growth roughly that of F_{Γ_0}, where Γ_0 is the first countable ordinal which cannot be described without reference (if only oblique) to the uncountable. This variant is not simple enough for me to state here; besides, it is still one of the "first generation" of functions of rapid growth.

5. THE NEW GENERATION OF RAPIDLY GROWING FUNCTIONS. The Paris-Harrington work was a remarkable accident. Paris' Variant of Ramsey's Theorem turned out to be independent of Peano Arithmetic. It followed that its corresponding function grew rapidly. Harrington's Variant was a cosmetically improved version of Paris' Variant and was also independent, etc. The flood of new functions of rapid growth were imitations, disguised versions of the proof-theoretic growth described in §3. Now, however, there is a new approach yielding much more rapidly growing functions.

The new success seems to me as serendipitous as Paris' original success, but it is different in several important respects: (i) the combinatorial problems were found first, with the theories located later; (ii) the independent, combinatorial theorems with corresponding rapidly growing functions are much simpler; and (iii) the connexion of the combinatorial problems with ordinals is more direct.

To build up the suspense, let me first introduce an ordinally poorer result. The underlying theorem is due to R. L. Goodstein and its independence of Peano Arithmetic is due to Lawrence Kirby and Jeff Paris. Let m,n be arbitrary positive integers and write m <u>hereditarily</u> in base n, i.e. write m in base

n , write all exponents occurring in the expression in base n , etc. For example, let $m = 71, n = 2$:

$$71 = 2^{2^{2}+2} + 2^{2} + 2 + 1 .$$

Given this, alternately apply the following two operations:

(a) subtract 1 ;

(b) replace base n by base $n + 1$.

Thus, $71 \to 70 \to 3^{3^{3}+3} + 3^{3} + 3 \approx (2.0589) \cdot 10^{14}$. According to Goodstein, one will eventually reach 0 . It may take a while: Let $G(x)$ denote the supremum over all $m,n \leq x$ of the number of steps it takes to reach 0 . The function G has growth rate roughly at least that of F_{ϵ_0} .

The secret behind Goodstein's Theorem is that the hereditary expression of m in base n mimicks an ordinal notation for an ordinal less than ϵ_0 . The first operation applied reduces the ordinal; the second leaves it fixed. Ordinals are also related to well-quasi-orderings, foremost among which are embeddings of finite trees; hence we see a rich source of rapidly growing functions. The simplest example arises from the following theorem, which is a finite form of Kruskal's Theorem:

FRIEDMAN'S THEOREM. Let $k > 1$ be given. There is a number n such that, for any sequence $T_1, T_2, \ldots, T_n$ of finite trees with T_i having cardinality at most $k + i$, there are $i < j$ such that T_i is homeomorphically embeddable in T_j .

Here, by a finite tree I mean a finite rooted tree, i.e. a finite partially ordered set which (i) branches only upwardly, i.e. the predecessors of any element are linearly ordered (tree property), and (ii) has a minimum element (rootedness). A homeomorphic embedding is a one-to-one inf-preserving map.

The function $F(x)$ which takes one from k to n grows incredibly rapidly. It is, at the time of writing, the most rapidly growing computable function that has been described by a means other than pointing to an ordinal or a theory. One can, of course, point to an ordinal and a theory for F - Friedman has analysed this completely - but I had best not do so: Few readers would recognise either. A modest idea of the growth rate and proof-theoretic strength of Friedman's finitisation of Kruskal's Theorem can be gleaned from my relevant recent contribution to the Mathematical Intelligencer.

In any event, the function F will not long hold its place of honour. Friedman's Theorem is a direct finitisation of a weak form of Kruskal's Theorem; there is the full theorem to contend with - and some strengthenings. Friedman has begun the underlying proof-theoretic analyses of these results and reports that their ordinals and proof-theoretic strengths dwarf those of the above-mentioned finite theorem. When these results are finitised, they will yield functions that dwarf F .

REFERENCES

1. Archimedes, The Sand Reckoner, 214?; reprinted in The Works of Archimedes (T. L. Heath, editor), Dover Publications, Inc., New York.
2. Peter Clote, Anti-basis theorems and their relation to independence results in Peano arithmetic, in Model Theory and Arithmetic (C. Berline, K. McAloon and J. -P. Ressayre, editors), Springer-Verlag, Berlin and New York, 1981.
3. Ian Frazier, The killion, The New Yorker, 6 September 1982.
4. R. L. Goodstein, On the restricted ordinal theorem, Journal of Symbolic Logic, volume 9, 1944, pages 33-41.
5. R. L. Graham, B. L. Rothschild and J. H. Spencer, Ramsey theory, Wiley-Interscience Series in Discrete Mathematics, a Wiley-Interscience Publication, John Wiley & Sons, Inc., New York, 1980.
6. Godfrey H. Hardy, Orders of infinity, 2nd edition, Cambridge University Press, Cambridge, 1924.
7. Lawrence Kirby and Jeff Paris, Accessible independence results for Peano arithmetic, Bulletin of the London Mathematical Society, volume 14, 1982, pages 285-293.
8. Jeff Paris, Some independence results for Peano arithmetic, Journal of Symbolic Logic, volume 43, 1978, pages 725-731.
9. Jeff Paris and Leo Harrington, A mathematical incompleteness in Peano arithmetic, in Handbook of Mathematical Logic (J. Barwise, editor), North-Holland Publishing Company, Amsterdam, 1977.

10. S. Skewes, On the difference $\pi(x) - li(x)$, Journal of the London Mathematical Society, volume 8, 1933, pages 277-283.

11. C. Smoryński, Some rapidly growing functions, Mathematical Intelligencer, volume 2, 1980, pages 149-154.

12. ___________, The varieties of arboreal experience, Mathematical Intelligencer, volume 4, 1982.

HARVEY FRIEDMAN'S RESEARCH ON THE FOUNDATIONS
OF MATHEMATICS, L.A. Harrington et al. (editors)
Elsevier Science Publishers B.V. (North-Holland), 1985

Some Rapidly Growing Functions[1), 2)]

Craig Smorynski

The purpose of this note is pure iconoclasm. I wish to debunk a few mathematical myths about how large "large" is. When the mathematician says "large", the logician is sure to think "small".

The first cliché usually resorted to in discussions of largeness is the Skewes number. For all calculated values, the number theoretic function $\pi(x) - li(x)$ is negative. Thus, when in 1914 J. E. Littlewood nonconstructively proved that this function changed signs infinitely often, curiosities about where it could become positive were aroused. In 1933, on the assumption of the Riemann Hypothesis, S. Skewes gave an upper bound for the first change of sign. This bound was so large (by the standards of the day) that is achieved instant notoriety and even a title - the Skewes Number:

$$S = 10^{10^{10^{34}}} < e^{e^{e^{e^{4.369}}}} .$$

The Skewes number has since been toppled from its position of supremacy. In 1955, Skewes showed how to lower the bound if one still assumed the Riemann Hypothesis; but he saved his reputation by obtaining the even larger upper bound,

$$S' = e^{e^{e^{e^{7.705}}}} ,$$

1) Reprinted from the Mathematical Intelligencer 2 (1980) 149-154, by permission of Springer-Verlag.

2) Editors' note: Further discussion of this topic may be found in the other short articles of C. Smorynski in this volume, as well as in the contributions of S. G. Simpson and R. L. Smith.

when the Riemann Hypothesis was not assumed. While the number S' has not shrunk, its importance has: Smaller bounds for the sign change exist today. For example, in 1966 R. S. Lehman gave the bound

$$L = e^{e^{e^{e^{2.067}}}} .$$

Not even mathematicians would find L large these days: Alan Baker won his Fields medal for, among other things, giving effective bounds in number theory. In his book on Transcendental Numbers, for instance, he cites the bound

$$\max(|x|, |y|) < B = e^{e^{e^{(2H)^{10^{n^{10}}}}}} ,$$

for all integral zeros x,y of an irreducible polynomial f(x,y) (with integral coefficients) of genus 1 , degree n , and height H .

Of course, with so much emphasis on the effectiveness of B , we must assume that mathematicians do not regard B as being very large. What then does the modern mathematician regard as large? Well, in his column in the November 1977 issue of Scientific American, Martin Gardner cites a result of R. L. Graham. According to Gardner, "In an unpublished proof, Graham has recently established an upper bound ... so vast that it holds the record for the largest number ever used in a serious mathematical proof." Intriguing? To see what Graham's number is, we first define a function K by recursion:

$$K(0,y) = y^y$$

$$K(x+1,y) = K(x,K(x,y)) .$$

K(x,y) is "something like" an exponential stack of y's of height x + 2 :

$$K(x,y) \approx y^{y^{y^{\cdot^{\cdot^{\cdot^{y}}}}}} \Big\} x+2 .$$

From K , we define another function G by recursion:

$$G(0) = K(3,3)$$

$$G(x+1) = K(G(x),3) .$$

The growth of G is a bit more difficult to imagine. $G(1)$ is something like a stack of $G(0)$ 3's , i.e.

$$\left. 3^{3^{\cdot^{\cdot^{\cdot^{3}}}}} \right\} 3^{3^{3^{3}}} \left(> 10^{10^{10^{10^{12}}}} \right) .$$

I leave to the reader an estimate of $G(2)$. The bound that Graham gives is

$$G = G(64) .$$

Now this is something the mathematician of today regards as large. I will concede that it dwarfs numbers like S, B, or even say,

$$B^{B^{B^{B^{B}}}} ,$$

for reasonably small values of n and H (say ≤ 100) . I will also concede that it is too large - according to Gardner, the constant that G is an upper bound for is generally believed to be 6 (!) . But I will not concede that G is large. How can any number that is the value of as slowly growing a function as G on so small an argument as 64 be considered large?

To give us some standard for comparison, let me introduce a hierarchy of number theoretic functions. For each natural number $n \in \omega$, define a function $F_n : \omega \to \omega$ as follows:

$$F_0(x) = x + 1$$

$$F_{n+1}(x) = F_n^{x+1}(x) ,$$

i.e. $F_{n+1}(x) = F_n(\ldots (F_n(x))\ldots)$, with $x + 1$ nestings of F_n . Thus, $F_0(x) = x + 1$, $F_1(x) = 2x + 1$, $F_2(x)$ is something like 2^x , and $F_3(x)$ something like

$$2^{2^{\cdot^{\cdot^{\cdot^{2}}}}}\Big\}x\ .$$

F_4, like $G(x)$, is a little harder to describe.

Before making any comparisons, let me quickly explain the rule of comparison that I am going to use: When I say H_1 and H_2 are something like each other, I do not mean to imply any relation nearly as tight as asymptoticity. On the contrary, I have in mind a looser, more liberal, equivalence relation whose looseness grows with the size of the elements of a given equivalence class. Given a function H, one naturally defines a family of functions $\mathfrak{F}(H)$ by taking H along with a few basic functions (e.g. addition) and closing under explicit definability (composition, addition of dummy variables, etc.). I say that H_1 is something like H_2 if the classes $\mathfrak{F}(H_1)$ and $\mathfrak{F}(H_2)$ are cofinal in each other, i.e. if every function of $\mathfrak{F}(H_1)$ is majorised by one of $\mathfrak{F}(H_2)$ and <u>vice versa</u>. This is a very loose measure of equivalence: 2^x, $x!$, and x^x are all something like its own n-fold composition with itself!) This looseness of fit makes the F_n-Hierarchy that much more impressive - for it is a hierarchy: F_{n+1} is nothing like F_n, i.e. F_{n+1} is so large relative to F_n that you need something like F_{n+1} to reach F_{n+1} from F_n. F_n will not do: F_{n+1} eventually majorises every function of $\mathfrak{F}(F_n)$.

The functions K (or, rather, its diagonal) and G fit neatly into the F_n-Hierarchy. $K(x,x)$ is something like F_3; while G is something like F_4. Reflecting on the rapidity of the growth of F_5, F_{236}, or even $F_{G(64)}$, the reader will see that I was not being entirely facetious when I referred earlier to the slow rate of growth exhibited by Graham's function $G(x)$. "Of course", the reader might object, "anyone can produce ever more rapidly growing functions. But the Graham function was used in a "serious mathematical proof." It is true that $F_{G(64)}$ has never been so applied. But F_ω and F_{ϵ_0} have - and we haven't even reached these functions yet!

To discuss the rates of growth of functions used in logic, it is necessary to extend the F_n-Hierarchy into the transfinite. At successor ordinals, $\alpha + 1$, one simply iterates what one did at α:

$$F_{\alpha+1}(x) = F_{\alpha}^{x+1}(x) .$$

At limit ordinals, one diagonalises on what one has done before, e.g.

$$F_{\omega}(x) = F_{x}(x) .$$

The first few values are now lower than those one had before, but the functions defined at limit ordinals do catch up and surpass their predecessors. M. H. Löb and S. Wainer have shown how to carry out this construction to any preassigned countable ordinal. Of course, in logic one doesn't actually need to go that far into the transfinite any more than in analysis one needs to use functions growing as fast as $F_{G(64)}$. The functions we wish to discuss are the F_{α}'s for $\alpha \leq \epsilon_0$. The ordinal ϵ_0 is, the reader might recall, the least fixed point of ordinal exponentiation with base ω, i.e. $\epsilon_0 = \min_{\beta}[\beta = \omega^{\beta}]$. A more intuitive picture of ϵ_0 is given by viewing it as the limit of the sequence,

$$\omega, \omega^{\omega}, \omega^{\omega^{\omega}}, \ldots .$$

The immensity of the step from α to $\alpha + 1$ increases as α increases. Moreover, between ω and ω^{ω} there are not 1, not ω, but ω^{ω} such steps: between ω^{ω} and $\omega^{\omega^{\omega}}$ there are $\omega^{\omega^{\omega}}$ such steps; etc. I don't think that the human mind can comprehend the growth of F_{ϵ_0}. Yet, it is function of definite logical (and, as we shall see: combinatorial) interest. The functions of the set,

$$\bigcup_{\alpha < \epsilon_0} (F_{\alpha}) ,$$

are precisely those provably computable in formal number theory. (Believe it or not, these functions are (theoretically, if not practically) computable.) Thus, F_{ϵ_0} is, in a sense, the first function to eventually majorise all functions provably computable in formal number theory.

Let me interpret this last fact. Suppose we have a sentence of the language of formal number theory of the form

$$\forall x \; \exists y A(x,y) \; , \tag{1}$$

where A is some provably decidable relation. If (1) is provable, then, for some provably computable F,

$$\forall x A(x, F(x)) \tag{2}$$

is also provable; whence, for some $\alpha < \epsilon_0$,

$$\forall x \exists y < F_\alpha(x) \; A(x,y) \tag{3}$$

is provable. [In fact, the exact α can be determined from the proof of (1). Using such considerations, in 1952 G. Kreisel reviewed Littlewood's 1914 paper and noted essentially the same upper bound as Skewes.] If, however, (1) is true but unprovable (a possibility not to be overlooked), the source of its unprovability could be the failure of (2) for all provably computable F, i.e. the failure of (3) for all $\alpha < \epsilon_0$. If this is the case, any function F making (2) true has moments of rapid growth.

Now the situation I have just described is not hypothetical: Jeff Paris and Leo Harrington have recently exhibited such an independent statement - with a somewhat interesting relation $A(x,y)$. Since this is the first not purely logical example of a problem with such nearly astronomical bounds (I comment on a function of truly astronomical growth at the end of this note), I shall discuss it in some detail.

I suppose we should first settle on some notation. If X is a set of natural numbers, we write $[X]^n$ for the collection of all n-element subsets of X. A colouring, C, of $[X]^n$ is simply a map

$$C : [X]^n \to c \; ,$$

where c is some positive natural number and we identify a natural number with its set of predecessors: $c = \{0,\ldots,c-1\}$. In 1929, the economist F. P. Ramsey proved that, if C is a colouring of $[X]^n$ and X is big enough with respect to c and n, then some big subset $Y \subseteq X$ has monochromatic n-element subsets: $C \restriction [Y]^n$ is constant. (To avoid clumsy phrasing, we call such a set Y homogeneous with respect to C.) More specifically, Ramsey proved the following two theorems.

Infinite Ramsey Theorem. Let n, c be positive integers. For any colouring, $C : [\omega]^n \to c$, there is an infinite set $Y \subseteq \omega$ homogeneous with respect to C, i.e. $C \upharpoonright [Y]^n$ is constant.

Finite Ramsey Theorem. Let s (for size), n, c be positive integers, with $s \geq n + 1$. There is a number $R(s, n, c)$ such that, for all $r \geq R(s, n, c)$ and all colourings $C : [r]^n \to c$, there is a homogeneous $Y \subseteq r$ of cardinality s.
(The restriction $s \geq n + 1$ simply rules out trivial cases.)

Neither of these statements is particularly intuitive. The best way to view them is as higher dimensional analogues of Dirichlet's Schubfachprinzip: If $n = 1$, the Infinite Ramsey Theorem just asserts that, if an infinite set is split into a finite disjoint union of subsets, one of the subsets must be infinite - in its more usual formulation: A finite union of finite sets is finite. For $n = 1$ and $s = 2$, the Finite Ramsey Theorem is exactly Dirichlet's Principle: Take $R(2, 1, c) = c + 1$ and test it yourself.

The Finite Ramsey Theorem is a centerpiece of finite combinatorics, with much energy being expended on the calculation of $R(s, n, c)$ and related "Ramsey Numbers". For, though such calculation is conceptually trivial (One merely enumerates all possibilities ...), it is impractically difficult. Still, one can give easy upper bounds: The diagonal $R(x + 1, x, x)$ is bounded by something like F_3 - no longer a very large number by anyone's reckoning. However, by making a subtle (?) change in the statement of the theorem, Paris and Harrington obtain a variant where the function in question exhibits a more respectable rate of growth - the function is something like F_{ϵ_0}.

We need only one more definition to state the Paris-Harrington Theorem: A set $X \subseteq \omega$ is relatively large if its cardinality is not less than its minimum: $card(X) \geq min(X)$.

Paris-Harrington Theorem. Let s, n, c be positive integers with $s \geq n + 1$. There is a number $H(s, n, c)$ such that, for all $h \geq H(s, n, c)$ and all colourings $C : [h]^n \to c$, there is a relatively large homogeneous $Y \subseteq h$ of cardinality at least s.

The Paris-Harrington Theorem is true, but not provable in formal number theory. The function $H(x + 1, x, x)$ (in fact, $H(x + 1, x, 3)$) is something like F_{ϵ_0}. Yet, at least for the novice, the difference between the Finite Ramsey Theorem and the Paris-Harrington Theorem is minimal. This minimality is underlined by the fact that these theorems share a common non-effective proof. It is in their effective proofs, of course, that they differ.

For the curious reader, I outline the non-effective proof. Consider the case of the Finite Ramsey Theorem. Suppose for fixed n, c, and $s \geq n + 1$, the theorem were false. By a monotonicity property it follows that, for every $r \geq s$, there is a colouring $C : [r]^n \to c$ with no size s homogeneous set $Y \subseteq r$. Partially order such colourings by extension: $C_0 \leq C_1$ iff, for some r, $C_0 = C_1 \; [r]^n$. This partially ordered set turns out to be an infinite finitely branching tree. Choose an infinite path $C_0 < C_1 < \ldots$ (by König's Lemma), take $C = \cup_i C_i$, and invoke the Infinite Ramsey Theorem to obtain an infinite set X homogeneous with respect to $C : |\omega|^n \to c$. Letting Y consist of the first s elements of $X, r = \max(Y) + 1$, and $C' = C \; [r]^n$ readily gives one a contradiction. The non-effective proof of the Paris-Harrington Theorem is entirely analogous.

The original proof of the independence of the Paris-Harrington Theorem over formal number theory is very appealing to the logician. Combinatorial constructions of nonstandard models of arithmetic inside given such models can be used to prove i. the independence of the Theorem, ii. the equivalence of the Theorem with a strong expression of faith in the system (i.e. a stronger-than-usual assertion of consistency), and iii. the eventual majorisation of all provably computable functions by $H(x + 1, x, x)$. Combined with a familiar proof theoretic analysis of formal number theory, this gives information about the growth of $H(x + 1, x, x)$ in terms of the F_α-Hierarchy.

As I say, this proof appeals to the logician. It might not appeal to others. It also might not do much for the understanding by either of what exactly makes $H(x + 1, x, x)$ grow so rapidly. Enter Robert M. Solovay. Having heard about the Paris-Harrington Theorem and its independence, but not having seen the proof and, consequently, unaware that one could read off information on this growth from their proof, he set out to establish this growth directly. He succeeded with the lower bounds; but had to resort to the proof theory to obtain the upper bounds. Later, Jussi Ketonen gave direct proofs of the upper bounds and the two of them went on to give rather sharp estimates of the growth of the function $H(x + 1, x, 7)$ and some of its variants (e.g. $H(x + 1, x, 7)$) . For example, they showed that, for $x \geq 20$,

$$F_{\epsilon_0}(x - 3) < H(x + 1, x, x) < F_{\epsilon_0}(x - 1) .$$

The Ketonen-Solovay elementary proof, like elementary proofs of theorems of analytic number theory, is somewhat longer than the Paris-Harrington proof and I certainly cannot present it in the space allotted here. However, I can give a bit of the flavour of their proof by showing that $H(x + 1, x, x)$ eventually majorises all functions F_n for finite n .

As is always the case in such matter, I must first pause to give a definition and comment on notation. First the definition: Let us say that a colouring $C : |\omega|^2 \to c$ captures a function F if every non-trivial relatively large set Y homogeneous with respect to C satisfies i . $\min(Y) \geq 3$ and ii. if $x, y \in Y$ and $x < y$, then $Fx \leq y$. Condition i is technical; condition ii just asserts that Y grows at least as rapidly as F .

As for notation, I shall write $Y = y_0 < y_1 <\ldots< y_{k-1}$. If Y is relatively large, then $k \geq y_0$. Further, if $x < y$, it is customary to write $C(x,y)$ instead of $C(\{x,y\})$.

<u>Lemma</u>. For each n , there is a colouring $C_n : [\omega]^2 \to 4 \cdot 3^n$ that captures F_n .

Proof: By induction on n .

<u>Basis</u>. A moment's thought will reveal that condition ii of the definition of capturing is always satisfied. To capture condition i , simply define

$$C_0(x,y) = \begin{cases} 0, & x = 0 \\ 1, & x = 1 \\ 2, & x = 2 \\ 3, & x \geq 3 \,. \end{cases}$$

If $Y = y_0 < y_1 < \ldots < y_{k-1}$ is relatively large and homogeneous with respect to C_0, then $C_0(y_0, y_1) = C_0(y_1, y_2)$ entails $C_0(y_0, y_1) = 3$, i.e. $y_0 \geq 3$.

<u>Induction step</u>. Suppose $C_n : |\omega|^2 \to 4 \cdot 3^n$ captures F_n. Define C_{n+1} by

$$C_{n+1}(x,y) = \begin{cases} \langle C_n(x,y),\, 0\rangle, & y < F_n^{x-1}(x) \\ \langle C_n(x,y),\, 1\rangle, & F_n^{x-1}(x) \leq y < F_n^{x+1}(x) \\ \langle C_n(x,y),\, 2\rangle, & F_n^{x+1}(x) \leq y \,. \end{cases}$$

Let $Y = y_0 < y_1 < \ldots < y_{k-1}$ be relatively large and homogeneous with respect to C_{n+1}. Projecting on the first coordinate of $C_{n+1}(x,y)$, we note that Y is homogeneous for C_n - whence i. C_{n+1} captures F_n, and ii. we can notationally ignore the first coordinate and just write $C_{n+1}(x,y) = 0,1$, or 2.

Since C_{n+1} captures F_n, we have

$$y_{k-1} \geq F_n(y_{k-2}) \geq \ldots \geq F_n^{k-1}(y_0) \,.$$

But $k = \mathrm{card}(Y) \geq \min(Y) = y_0$ and F_n is monotone, so

$$y_{k-1} \geq F_n^{k-1}(y_0) \geq F_n^{y_0-1}(y_0) \,,$$

and we see that $C_{n+1}(y_0, y_{k-1}) \neq 0$. [Perhaps I should point out that this is the part of the proof in which relative largeness is used.]

Since $C_{n+1}(y_i, y_j) \neq 0$.

$$y_2 \geq F_n^{y_1-1}(y_1) > F_n^{y_0-1}(y_0) \,.$$

But

$$y_1 \geq F_n^{y_0-1}(y_0)$$

and F_n is monotone, whence

$$y_2 \geq F_n^{y_0-1}(F_n^{y_0-1}(y_0)) = F_n^{2y_0-2}(y_0)$$

$$\geq F_n^{y_0+1}(y_0) = F_{n+1}(y_0) .$$

[Here we use condition i : $y_0 \geq 3$ implies $2y_0 - 2 \geq y_0 + 1$.] From this inequality, we conclude $C(y_i, y_j) = C(y_0, y_2) = 2$, and thus

$$F_{n+1}^{y_i+1}(y_i) \leq y_i+1 ,$$

i.e. C_{n+1} captures F_{n+1} . Q.E.D.

Theorem. $H(x + 1, 2, x)$ eventually majorises each F_n .

Proof: For a given n , let $C_n : [\omega]^2 \to 4 \cdot 3^n$ capture F_n . Assume also that $x \geq 4 \cdot 3^n$ (so that we can view C_n as mapping into x) . Now suppose $Y = y_0 < y_1 <\ldots< y_{k-1} \leq H(x + 1, 2, x)$ is relatively large, homogeneous with respect to C_n , and has cardinality at least $x + 1$.

Since $y_0 \geq 3$, we have $y_i \geq 3 + i$. In particular, $y_{x-1} > x$. But C_n captures F_n , whence $y_x \geq F_n(y_{x-1}) > F_n(x)$. Collecting our inequalities, we have

$$H(x + 1, 2, x) \geq y_{k-1} \geq y_x > F_n(x) ,$$

for all $x \geq 4 \cdot 3^n$.

We remark that this proof shows $H(x + 1, 2, 4 \cdot 3^x)$ to majorise F_ω . Since $H(x + 1, 2, 4 \cdot 3^x)$ is something like $H(x + 1, 2, x)$, we can conclude

$H(x + 1, 2, x)$ to lie at least at level ω of the F_α-Hierarchy. In fact, this is exactly its level. $H(x + 1, 3, x)$ is something like F_{ω^ω} : $H(x + 1, 4, x)$ is something like $F_{\omega^{\omega^\omega}}$; etc.

Proponents of rapid growth do not wish to stop here. They seek ever more rapidly growing functions and ever more powerful principles to produce such functions. Defining, for example, a finite set X to be relatively F-large if $card(X) \geq F(min(X))$, one can iterate the Paris-Harrington construction and generate more rapidly growing functions. This is, however, a pedestrian approach. A seemingly more exciting approach is to look for a new statement of the form,

$$\forall x \exists y A(x,y) ,$$

which is independent over a considerably stronger theory than formal number theory and hope that the corresponding choice function F ,

$$\forall x \ A(x, F(x)) ,$$

eventually majorises all provably computable functions of the new theory. This has been done, but not with very interesting relations $A(x,y)$. In all, the problem of getting a function which is at once more rapidly growing than F_{ϵ_0} and of as great an interest as that of $H(x + 1, x, x)$ seems to be open.

Actually, what I should have said was that the problem of getting interesting computable functions of more rapid growth than that of F_{ϵ_0} is open. If we drop the requirement of computability - well, there is Tibor Rado's Busy Beaver Function. This irrepressible little fellow grows more rapidly than any theoretically computable function - a feat of not particularly great magnitude: It is an easy matter to eventually majorise all computable functions by diagonalisation; and the Busy Beaver is the result of such a diagonalisation. But it is a delightful diagonalisation and I must comment on it here.

A Turing machine (named after Alan Turing, an enigmatic cryptanalyst) is an imaginary device for computing functions. Each Turing machine M has a finite number x of states and a two-way infinite tape divided into consecutive congruent squares, each of which can either bear the symbol 1 or be blank. M has the great flexibility of, on reading what is on the particular square it is scanning and on considering the state it is in, being able to erase the 1 if it is there, print a 1 if there is none, or let it stand; to move the tape one square to the left or to the right, or to continue staring at the same one; and to change states or remain in the same state. Being a machine M is, of course, not allowed any free will - its actions are completely determined by its basic program.

Turing machines do not sound particularly useful; but they are powerful enough to compute all theoretically computable functions. (Of course, with only 1's and blanks, this requires a coding. One usually represents a natural number x by a string of x + 1 1's.) Moreover, they have all the inherent difficulties associated with more sophisticated machines - most notably the difficulty in determining from the machine's program whether its computations will ever finish. This problem has no effective solution.

The Busy Beaver poses himself the following problem: Suppose M is a machine with x states. If one feeds M a blank tape, what is the maximum number of 1's that M can print on the tape before halting? This function,

$$BB(x) = \max(\{\#(1\text{'s } M \text{ can print}) : M \text{ has exactly } x \text{ states}\}) ,$$

is not computable and, in fact, it eventually majorises every computable function. [The obvious computation procedure is not effective: One would like to simply enumerate those machines with x states, feed them all blank tapes, wait until they finish their computations, and then see how many 1's each one prints. The problem is that some of these machines might never finish - and there is no way of knowing which ones these are. But rather than explain the result, let me prove it.] The proof is quite simple: Let F be any monotone computable function and let

$$F'(x) = F(2x + 3) .$$

Suppose M computes F' and has x_0 states. For each x , we can find a machine M_x with x + 1 states that will print x + 1 1's and halt when fed a blank tape. The hook-up of M_x with M results in a machine with $x + x_0 + 2$ states which, when fed a blank tape will output F'(x) + 1 1's . If $x \geq x_0$,

$$F(2x+2) \leq F(2x+3) = F'(x) < BB(2x+2) \leq BB(2x+3) \ ,$$

which was to be proven.

A bit more on Turing machines and the Busy Beaver can be found in Enderton's contribution to the Handbook of Mathematical Logic. We note merely that the few known values of the function,

$$BB(0) = 0, \ BB(1) = 1, \ BB(2) = 4 \ ,$$
$$BB(3) = 6, \ BB(4) = 13 \ ,$$

do not exactly reflect later developments - even Graham's function starts out better than this.

REFERENCES

[1] Herb Enderton, Elements of recursion theory, in: J. Barwise, ed. Handbook of Mathematical Logic (North-Holland, Amsterdam 1977).

[2] Paul Erdös and George Mills, Some bounds for the Ramsey-Paris-Harrington numbers, to appear.

[3] Jussi Ketonen and Robert M. Solovay, Rapidly growing Ramsey functions, to appear.

[4] Martin H. Löb and Stan Wainer, Hierarchies of number-theoretic functions, I, II, Archiv f. math. Logik 13 (1970), pp. 39-51, 97-113.

[5] Jeff Paris and Leo Harrington, A mathematical incompleteness in Peano arithmetic, in: J. Barwise, ed., Handbook of Mathematical Logic (North-Holland, Amsterdam 1977).

[6] Frank P. Ramsey, On a problem of formal logic, Proc. London Math. Soc. (2) 30 (1929), pp. 264-286.

[7] Tibor Rado, On non-computable functions, Bell System Tech. J. 41 (1962), pp. 877-884.

HARVEY FRIEDMAN'S RESEARCH ON THE FOUNDATIONS
OF MATHEMATICS, L.A. Harrington et al. (editors)
Elsevier Science Publishers B.V. (North-Holland), 1985

The Varieties of Arboreal Experience[1), 2)]

C. Smoryński

Not long ago I graced these columns with a slightly tongue-in-cheek account of a recent independence result about which logicians were quite excited. This result, the Paris-Harrington Theorem, is a variant of the Finite Ramsey Theorem which, though provable in ordinary set theory (and so a theorem of ordinary mathematical practice), is not provable in formal number theory. This independence being of a character different from that of Gödel's undecidable sentence or that associated with independence results in set theory, the result has been much touted by logicians seeking social acceptance by the rest of the mathematical world as something closely akin to the coming of the Millennium. This is unfortunate as the result really is remarkable - philosophically because of this new character of independence, and mathematically for reasons requiring a new paragraph.

The, at first sight subtle, difference between the usual Finite Ramsey Theorem and the Paris-Harrington Theorem is surprisingly consequential: Both concern computationally hairy combinatorial problems with unfeasible bounds; but the former's unfeasible bounds are readily intelligible, while the latter's are not. Indeed, as I noted in my article, the function witnessing the validity of the Paris-Harrington Theorem was the most rapidly growing function of obvious mathematical interest ever to appear. This is no longer true.

Harvey Friedman, who has the most original mind in logic today, has shown a simple finite form of Kruskal's Theorem to be independent of a theory much

---

1) Reprinted from the Mathematical Intelligencer 4 (1982) 182-189, by permission of Springer-Verlag.

2) Editors' note: Further discussion of this topic may be found in the other short articles of C. Smorynski in this volume, as well as in the contributions of S. G. Simpson and R. L. Smith.

stronger than formal number theory and to have an associated witnessing function of a rate of growth simply dwarfing that of the Paris-Harrington function. The astonishing thing about this is that Friedman's Finite Form (FFF) of Kruskal's Theorem is genuinely natural: It is a direct finitisation of Kruskal's Theorem and does not have any hidden subtlety to it. This naturalness, coupled with the phenomenal growth-rate involved, makes the result doubly pleasing to logicians. And, if the acceptance of the Paris-Harrington Theorem by combinatorists [Cf. in this respect the recent book by Graham, Rothschild and Spencer] is any indication, this result will be pleasing to them as well. Other mathematicians might at least share some vicarious pleasure in the result - particularly those who identify with the logician Darth Vader who is ruthlessly mining Luke Skywalker's combinatorial field with little black holes. Thus, in the name of mathematical hedonism, I offer the following account of Friedman's work.

KRUSKAL'S THEOREM AND FRIEDMAN'S FINITISATION

By a finite _tree_ we shall understand a finite _rooted_ tree, i.e. a partially ordered set $(T, <)$ in which i. (tree property) every nonminimal element has a unique immediate predecessor, and ii. (rootedness) there is a unique minimal element. We usually suppress mention of the ordering and write T for $(T, <)$.

A _homeomorphic embedding_ of a tree T_1 into another tree T_2 is a one-to-one inf-preserving map $h : T_1 \to T_2$. [N. B. : From the preservation of infs follows immediately the preservation of order.]

In 1960, J. B. Kruskal proved something a bit more general than the following:

KRUSKAL'S THEOREM. If $T_1, T_2, \ldots$ is an infinite sequence of finite trees, then there exist i and j such that $i < j$ and T_i is homeomorphically embeddable in T_j.

Kruskal's Theorem (or, perhaps more accurately: its full version about _labelled_ trees) is a powerful tool. To cite just one application: A. Ehrenfeucht applied it to G. H. Hardy's result on the linear ordering of his "logarithmico-exponential scales" to conclude the well-ordering of the set of integral exponential polynomials (with hereditarily nonnegative coefficients) under the ordering given by eventual domination.

Because of the power of Kruskal's Theorem, when he first learned of it, Friedman immediately recognised its potential for siring a powerful finite version. Surprisingly, this is the first (nontrivial) finitisation that would come to mind:

FRIEDMAN'S FINITE FORM (FFF) OF KRUSKAL'S THEOREM. For each $k > 1$, there is an n such that, if $T_1,\ldots,T_n$ is a sequence of trees with the cardinality of T_i at most $k + i$, then there are $i < j \leq n$ such that T_i is homeomorphically embeddable in T_j.

How does this differ from Kruskal's Theorem? Well, one simply puts an initial finiteness condition on it (T_1 has cardinality at most $k + 1$.) and adds the smallest quickly describable growth rate consistent with not trivialising the outcome. The conclusion is correspondingly strengthened to give a bound n depending only on k to the place where some T_i is homeomorphically embeddable in a later T_j. In short, there is no subtlety: FFF is a direct finitisation of Kruskal's Theorem.

As a direct finitisation of Kruskal's Theorem, FFF should be a direct corollary thereto. It is: If for some k no uniform bound n existed, then a standard appeal to a compactness phenomenon would produce a counterexample to Kruskal's Theorem.

As with the Paris-Harrington Theorem, the proof cited is highly nonconstructive. A direct constructive proof is a bit more difficult. ... Quite a bit more difficult.

There are several ways of measuring (or, if this is too strong a word: indicating) the difficulty of giving a direct constructive proof of a result. The simplest thing to point to is the strength of the assertion when it is used as an axiom: If it has unexpected consequences, it must be difficult to prove (and if it has new consequences it must be unprovable - from the axioms at hand). The Paris-Harrington Theorem (PH) described in my earlier column is a good example: Over formal number theory, say FNT, PH is equivalent to a strengthened assertion of the consistency of FNT. In particular, over FNT, PH implies the consistency of FNT and, by Gödel's Second Incompleteness Theorem, PH is not provable in FNT. Now, FFF also implies the consistency of FNT and is unprovable in FNT. But it also implies the consistency of FNT + PH and so is stronger than PH (both proof-theoretically insofar as it proves the consistency of FNT + PH and mathematically insofar as it proves PH).

[How powerful is FFF? Well, pretty powerful. Suppose we iterate the consistency statement, i.e. we define a transfinite sequence of theories $T_0, T_1, \ldots,$ with $T_0 = \text{FNT}$, $T_{\alpha+1} = T_\alpha + \text{Consistency of } T_\alpha$, and $T_\lambda = \bigcup_{\alpha<\lambda} T_\alpha$ for λ a limit. (This can be done - up to a point.) Then FNT + PH has the same proof-theoretic strength as $T_{{}_0}$, where ${}_0$ is the ordinal familiar from analysing FNT . Now FNT + FFF goes far beyond this. It goes beyong $T_1, T_2, \ldots, T_{{}_0}, \ldots$ and far beyond the theories $T_{\Gamma_0}, T_{\Gamma_1}, \ldots$ with ordinals to be described below. In fact, FNT + FFF has the proof-theoretic strength of a theory T_α, where α is a countable ordinal whose size is not readily described without reference to the first uncountable ordinal!]

In logical form, FFF is a statement $\forall k \; \exists n \; A(k,n)$. While it is independent of FNT, each instance $\exists n \; A(0,n)$, $\exists n \; A(1,n), \ldots$ is provable in FNT. One way of measuring the difficulty of proving FFF directly is to see how difficult it is to derive such provable instances. Friedman has shown that, giving FNT the obvious axiomatisation (via the usual logical axioms and rules of inference, unicity of the successor function, recursion equations for addition and multiplication, and the full schema of induction), any formal derivation of the provable assertion $\exists n \; A(10,n)$ in FNT must consist of at least

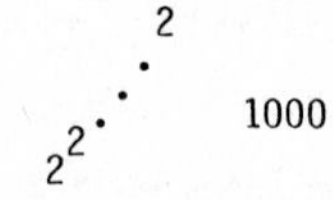

symbols. [Compare the Skewes number to this!]

And, of course, for an assertion of the form $\forall k \; \exists n \; A(k,n)$, a good measure of the difficulty of providing a direct proof of the assertion is the rate of growth of the function assigning n to k . As we saw when I discussed the Paris-Harrington Theorem, for functions of sufficiently rapid growth, we can assign ordinal numbers as rates of growth: The bigger the ordinal assigned, the more rapidly the function grows; the more phenomenonally large the ordinal, the more phenomenonally rapid the growth.

Of the three measures just described - proof-theoretic strength, length of proof, and rate of growth - the most readily intelligible one is rate of growth. [It is also the least controversial: Respect for proof-theoretic strength requires some familiarity with formal systems, and length of proof is highly intensional - it depends (albeit usually only linearly) on one's exact choice of language and axioms.]

Now, as I said in my column on the Paris-Harrington Theorem, the rate of growth involved is so large that it cannot be intuitively grasped. But it can be measured. We can assign ordinals to certain rates of growth and, if we do so, we will ultimately assign to the function witnessing FFF some ordinal α. The exact ordinal α is not known, but we will see nonetheless that it is quite remarkably large.

First, however, I wish to digress to discuss the connexion between Kruskal's Theorem and ordinals.

WELL-QUASI-ORDERINGS

The relation between Kruskal's Theorem and ordinals is fairly close. In fact, Kruskal's Theorem is most memorably stated, if not in terms of ordinals and well-ordering, at least in terms of well-quasi-ordering.

A binary relation $\leq$ is a _quasi-ordering_ if it is reflexive and transitive. A quasi-ordering misses being a partial ordering by failing to identify equivalent things, where objects a,b in the field of the relation are equivalent if $a \leq b$ and $b \leq a$. One can, of course, pass to the induced partial ordering of the equivalence classes; but sometimes it is less inconvenient not to have anti-symmetry than to deal with equivalence classes, whence the notion of a quasi-ordering. But I digress.

The notion of a well-quasi-ordering is a suitable generalisation of that of a well-ordered set to the context of a quasi-ordering. A more familiar such generalisation is given by the notion of well-foundedness: A quasi-ordering $\leq$ is well-founded if there are no infinite descending sequences in $\leq$, i.e., if there are no sequences $a_0 > a_1 > \ldots$ (where $>$ is the strict form of $\geq$). This is a useful generalisation, but a liberal one. It allows bad infinite sequences to avoid running out of things to descend to by making many lateral moves. In the linearly ordered case, such lateral moves are impossible and the well-foundedness condition can be restated as follows:

(W) If $a_0, a_1, \ldots$ is an infinite sequence of elements of the field of $\leq$, then there are $i < j$ such that $a_i \leq a_j$.

A quasi-ordering is called a <u>well-quasi-ordering</u> if it satisfies condition (W).

Kruskal's Theorem can now be simply expressed.

KRUSKAL'S THEOREM. The quasi-ordering of trees under homeomorphic embeddability is a well-quasi-ordering.

With this statement of Kruskal's Theorem, the fact that there is a connexion with ordinals should surprise noöne. The direct connexion has been studied by logicians. D. H. J. de Jongh and R. Parikh have, for example, calculated the ordinals of linear extensions of certain well-partial-orderings connected with a theorem of G. Higman closely related to Kruskal's Theorem. And Diana Schmidt, in her Heidelberg Habilitationsschrift, calculated several such ordinals connected with Kruskal's full theorem (i.e. the generalisation, not discussed here, to labelled finite trees). The ordinals she obtained were quite large.

ORDINALS

The exact ordinal associated with FFF is not now known. Both Friedman's and Schmidt's work give vast upper bounds: and Friedman has given moderately huge lower bounds. Now, very large ordinals are difficult to describe. [In fact, Schmidt proves a theorem to this effect: If one chooses to generate large ordinals by a class of functions, then from a certain size on the functions can no longer be simultaneously monotonic and increasing (where $\phi(\alpha_1,\ldots,\alpha_n)$ is said to be monotonic if, for all ordinals $\alpha_1,\ldots,\alpha_n,\beta_1,\ldots,\beta_n,\alpha_1 \leq \beta_1,\ldots,\alpha_n \leq \beta_n$ imply $\phi(\alpha_1,\ldots,\alpha_n) \leq \phi(\beta_1,\ldots,\beta_n)$; and ϕ is increasing iff, for all $\alpha_1,\ldots\alpha_n$, $\alpha_i \leq \phi(\alpha_1,\ldots,\alpha_n))$.] Thus, I shall describe here only the lower bounds. I think they are quite large enough to impress the reader with the rate of growth they represent.

First, I shall describe a hierarchy of functions mapping countable ordinals to countable ordinals. Certain small values of these functions are famous ordinals. These allow us to reach larger ordinals, which allow us to reach even larger ordinals, which allow us to reach ever larger ordinals. The first ordinal we cannot reach from below is called Γ_0 . This is still too small.

The most familiar infinite ordinal is ω , the order type of the set of natural numbers. It is followed successively by

$$\omega + 1,\ \omega + 2,\ \omega + 3,\ldots,\omega + n,\ldots\quad n \in \omega .$$

The limit of this sequence is denoted $\omega + \omega$ or $\omega \cdot 2$. There are also

$$\omega \cdot 3,\ \omega \cdot 4 \cdot 5,\ldots,\omega \cdot n,\ldots\quad n \in \omega .$$

The limit of this sequence is $\omega \cdot \omega$ or ω^2 . Again we can define the sequence

$$\omega,\ \omega^2,\ \omega^3,\ldots,\omega^n,\ldots\quad n \in \omega .$$

The limit of this sequence is ω^ω . But it is not the limit of this sequence we wish to consider - rather, its extension into the transfinite: Define, for all ordinals $\alpha,\phi(0,\alpha) = \omega^\alpha$. Thus $\phi(0,0)$, $\phi(0,1),\ldots$ is the transfinite sequence:

$$1,\omega\ \omega^2,\ldots,\omega^\omega,\ \omega^{\omega+1},\ldots;\ \omega^{\omega^2};\ldots;\ \omega^{\omega^\omega};\ldots .$$

Usually α is a lot smaller than ω^α ; but occasionally - in fact uncountably often - the ordinals α do catch up: For example, if ϵ_0 is the limit of sequence,

$$\omega^{\omega^{\omega^{\cdot^{\cdot^{\omega}}}}} n,\ n \in \omega ,$$

then $\epsilon_0 = \omega^{\epsilon_0}$, i.e. $\phi(0,\epsilon_0) = \epsilon_0$.

From $\phi(0,-)$ we proceed to the function $\phi(1, -)$ which enumerates all fixed points of $\phi(0, 0)$:

$$\phi(1,0) = \epsilon_0, \phi(1,1) = \epsilon_1, \ldots, \phi(1, \epsilon_0) = \epsilon_{\epsilon_0}, \ldots$$

This new function $\phi(1, -)$ itself has uncountably many fixed points. We let $\phi(2, -)$ enumerate these. [The number $\phi(2, 0)$ is called the <u>first critical epsilon number</u> and is larger than all of

$$\epsilon_0, \epsilon_1, \ldots, \epsilon_\omega, \ldots, \epsilon_{\epsilon_0}, \ldots, \epsilon_{\epsilon_1}, \ldots, \epsilon_{\epsilon_{\epsilon_0}}, \text{ etc.}]$$

In general, given a function $\phi(\alpha, -)$, we obtain a new function $\phi(\alpha + 1, -)$ by specifying that $\phi(\alpha + 1, -)$ enumerates the fixed points of $\phi(\alpha, -)$. If α is a limit number, we let $\phi(\alpha, -)$ enumerates all <u>common</u> fixed points of $\phi(\beta, -)$ for $\beta < \alpha$. The functions $\phi(\alpha, -)$ are total, i.e. $\phi(\alpha,\beta)$ is defined for all countable α,β . Moreover, each $\phi(\alpha, -)$ is strictly increasing [in the usual sense. In the sense of our earlier parenthetic remark, the function $\phi(-, -)$ is both monotonic and increasing.]

Now, $\phi(0, -)$ is a very rapidly growing function. The next function, $\phi(1, -)$, is even more rapidly growing and starts from a point further down the ordinal line. Ditto for $\phi(2, -)$, $\phi(3, -)$, $\phi(\omega, -)$, $\phi(\epsilon_0, -)$, $\phi(\phi(\epsilon_0, 0), -)$, etc. Thus, it seems hardly likely that an ordinal α can ever be in the range of $\phi(\alpha, -)$. But, such ordinals exist in abundance. For example, if we define

$$\gamma_0 = \phi(0,0), \gamma_{n+1} = \phi(\gamma_n, 0)$$

and set $\Gamma_0 = \sup\{\gamma_n : n \in \omega\}$, we obtain just such an ordinal: $\Gamma_0 = \phi(\Gamma_0, 0)$.

ABOUT Γ_0 ; A PHILOSOPHICAL DIGRESSION

Let me say a few words about the size of Γ_0 . First, let me cite a simple lemma:

Lemma. Let $\Gamma = \phi(\Gamma, 0)$. If $\alpha,\beta < \Gamma$, then $\phi(\alpha,\beta) < \Gamma$.

By this lemma, Γ_0 cannot be obtained from below by means of the powerful principle used in generating the ϕ-hierarchy of ordinal functions. In this respect, it is an analogue to the first uncountable ordinal Ω . Indeed, the existence of Γ_0 presupposes that of Ω . This is most clearly illustrated by another existence proof: One can extend the hierarchy ϕ into the uncountable, notice that $\phi(\Omega, 0) = \Omega$, and conclude on general logical grounds the existence of countable solutions to this equation.

The use of Ω , or, at least, some uncountable object, in the relatively simple existence proof for Γ_0 I first gave is a bit more subtle. If we ask ourselves how we know $\phi(\alpha + 1, -)$ is always defined, we can answer ourselves either nonconstructively or constructively. Nonconstructively, we can note that $\phi(\alpha, \Omega) = \Omega$ and appeal to general logical grounds to conclude $\phi(\alpha,\beta) = \beta$ is satisfied uncountably often among the countable ordinals. Constructively, we can simply iterate $\phi(\alpha, -)$: For any ordinal β_0 , the supremum of the sequence $\{\beta_n\}_{n \in \omega}$, where β_n is defined from β_0 by $\beta_{n+1} = \phi(\alpha,\beta_n)$, is a fixed point $\beta > \beta_0$ of $\phi(\alpha, -)$. Thus, the sequence of fixed points to $\phi(\alpha, -)$ is given by $\phi(\alpha + 1, -)$ in the same manner in which that of fixed points to $\phi(-, 0)$ is given by $\Gamma_0, \Gamma_1, \ldots$. Certainly the latter grows more rapidly, but it is after all constructed by the same principle as $\phi(\alpha + 1, -)$ is from $\phi(\alpha, -)$, is it not, and why, therefore, am I making such a fuss about Ω and its role in believing in Γ_0 ? Well, they are _not_ given by the same principle applied, as it seems, to a different argument of the binary function $\phi(-, -)$! The step from $\phi(\alpha, -)$ to $\phi(\alpha + 1, -)$ is a process of iteration: $\phi(\alpha + 1, -)$ is, basically, an iteration of the _function_ $\phi(\alpha, -)$. The construction of $\phi(-, 0)$ is also a process of iteration - but an iteration of the process of iteration itself(!); or, in more mathematically acceptable terms: an iteration of the _functional_ taking us from $\phi(\alpha, -)$ to $\phi(\alpha + 1, -)$. Thus, this proof of the existence of Γ_0 presupposes as domain of this functional a class of functions on ordinals and thus (on close examination) the uncountable.

The point I am aiming for is philosophical, but hardly subtle or unmathematical: At the turn of the century there were many conservative reactions to the threatening anarchy of set theory and its paradoxes. One defense against the Cantorian incursion was the semi-intuitionism of the French - Borel, Lebesgue, Poincaré, etc. Their response was predicativity - the belief that only those objects that could be directly defined (as opposed to the full intuitionists' demand that such be constructed) could be considered to exist. In Poincaré's words (taken from Eric Temple Bell's Men of Mathematics), "the important thing is never to introduce entities not completely definable in a finite number of words."

Of course, there is a certain vagueness to Poincaré's words. One must specify a language, rules of definition, etc., to obtain any real precision. But, even without precision, one can draw a few conclusions. First, directness of definition should prescribe circularities and require concepts to be built up from below. Second, even if we allow the natural numbers as initially given, a finite number of words will never completely define an uncountable object.

Lest I digress even further and repeat the usual platitudes about the foundational crisis unleashed on mathematics by Cantor, let me quickly recall Γ_0. This ordinal is impredicative. Referring to it as the least ordinal Γ satisfying $\Gamma = \phi(\Gamma, 0)$ is indirect and somewhat circular - the word "least" involves reference to all ordinals, including Γ_0. The "constructive" definition as the result of an iteration of a functional again violates, albeit more subtly, a similar aspect of predicativity. In the 1950's S. Feferman and K. Schütte gave convincing analyses of the notion of predicativity and demonstrated Γ_0 to be the first impredicative ordinal, i.e. the first ordinal not predicatively definable. Moreover, in connexion with his work on finitising Kruskal's Theorem, Friedman has convincingly shown a commitment to Γ_0 to entail a commitment to the uncountable.

Now, why this lengthy digression on Γ_0, predicativity, and the French semi-intuitionists and their attitude toward set theory? Expositions of this subject are sufficiently abundant, and besides, people seem finally to have grown tired of "the foundational crisis of the turn of the century." [Witness, for example, the occasional recent attempts to "revive" the philosophy of mathematics. My explanation of this revivalist fervour is that the difficulties in set theory were so interesting and important that they (and the ensuing battle between Brouwer and Hilbert) monopolised the Philosophy of Mathematics for decades and it is only recently that everyone has bored of that crisis and, looking around, has discovered this monopoly to have left the Philosophy of Mathematics without any other leadership or direction.] [Lakatos fans will no doubt dispute this theory.]

Where was I? I was about to explain that I digressed to discuss predicativity and the impredicativity of Γ_0 for the purpose of making two points about FFF and its (as yet unknown) associated ordinal.

The first point is mathematical. The status of Γ_0 as the first impredicative ordinal makes its size rather qualitatively impressive. Any function assigned Γ_0 as a growth rate is indeed rapidly growing. Now the ordinal of FFF, though not yet known, is known to be larger than Γ_0. In fact, it is larger than (say) $\Gamma_1, \Gamma_\omega, \Gamma_{\epsilon_0}$, and even Γ_{Γ_0}, where Γ_α is the α-th solution Γ to $\phi(\Gamma,0) = \Gamma$. Thus, the rate of growth of the function witnessing the truth of FFF is strongly impredicative and thus impressively and qualitatively immense.

The second point is philosophical: There are many incompleteness and independence results in modern logic. Even if we restrict our attention to those results where the proofs of undecidability determine the truth of the undecided assertions, we find there are still several types of such results. Gödel's undecidable sentence, for example, is purely universal, i.e. of the form $\forall k\, A(k)$, with $A(k)$ relatively trivial. The undecidability of $\forall k\, A(k)$ is occasioned by the _non-uniformity_ of the proofs of $A(0), A(1),\ldots$. [Indeed, G. Kreisel has conjectured that an assertion $\forall k\, A(k)$ is provable in FNT iff there is a uniform bound on the number of lines used in the individual proofs of $A(0), A(1),\ldots$. There are some positive partial results in this direction.] The exciting thing about the Paris-Harrington Theorem (PH) [is not, as many logicians have asserted, that it is the first genuinely mathematical statement shown independent of FNT - I regard Hilbert as a mathematician and, thus, his beloved consistency statement as mathematical - but] is that it is of the form $\forall k\, \exists n\, A(k,n)$ and is undecidable for a different reason. The construction needed to establish it is too complex; in fact, the function given by the con-struction grows so fast that the theory FNT cannot handle it. The independence of FFF shares this feature with that of PH - with, of course, the difference that FFF is independent of far stronger theories, has a far more complex construction underlying its validity, and is witnessed by a far more rapidly growing function.

Now to the really philosophical part of the point: Gödel's Theorems have philosophical, as well as mathematical, significance: Each Theorem (i.e. both the First Incompleteness Theorem on the existence of true unprovable assertions of the form $\forall k\ A(k)$ and the Second Incompleteness Theorem on the unprovability of consistency) convincingly refutes Hilbert's philosophy of mathematics. The unprovabilities of PH and FFF in FNT do not similarly offer definitive refutations of Hilbert's philosophy and programme. [Briefly: Hilbert distinguished between real and ideal statements and his philosophy concerned the former. As PH and FFF are ideal, their independence refutes nothing. However, these statements are equivalent to much strengthened consistency - like assertions and their unprovabilities could have dealt quite a blow to confidence in Hilbert's philosophy and programme. But I digress.] Nonetheless, through its unprovability in theories of strength greater than Γ_0, i.e. the impredicative nature of any proof of it, FFF illustrates beautifully the fallacy of predicativity: FFF is a concrete assertion about finite objects instantly understandable to any predicativist (predicatician?); but any proof of it must appeal to impredicative principles. In short, FFF would have been meaningful to Poincaré, but he would not have been able to prove it, disprove it, or accept any proof of it given to him.

[One last parenthetical comment before exiting this overlong digression: Notice that I said FFF <u>illustrates</u>, not <u>demonstrates</u>, the failure of predicativity. The definitive refutation of a philosophy cannot be given externally. It must be given internally so that the holder of philosophy can accept the argument and admit defeat. I am tempted to illustrate this point by saying that Poincaré would not even have understood (or admitted to have understood) the attempted refutation of predicativity via FFF. Theoretically this is true and I can safely replace "Poincaré" in this slogan by "the idealised predicativist" or "the idealised French semi-intuitionist." Talking about Poincaré himself requires a deeper psychological understanding of him than I am prepared to acquire; and talking about the actual French semi-intuitionists requires greater historical understanding - and here I note that, if not they themselves, their successors only too gladly embraced the uncountable. Indeed, modern descriptive set theorists are the most notorious producers and consumers of impredicative mathematics today!]

KRUSKAL'S THEOREM AND Γ_0

To those who accepted my word that I was writing this article for hedonistic purposes and unwittingly ventured into the preceding digression in which I attempted a short course in the Philosophy of Mathematics, let me offer by way of recompense a delightful little piece of mathematics. [The Philistine who has skipped my philosophical excursion gets a free treat. This is unfair; but we must live with Life's little injustices.] This offering is the exhibition of a simple direct connexion between Kruskal's Theorem and Γ_0, a connexion revealed by a fairly standard method of the area: One assigns to each finite tree T an ordinal $h(T) < \Gamma_0$ and shows the map to i. be onto and ii. preserve order, i.e. if T_1 is homeomorphically embeddable in T_2, then $h(T_1) \leq h(T_2)$. Exactly what this does I shall explain later.

The key to the assignment of ordinals is the following

<u>Lemma</u>. [Generalisation of the Cantor Normal Form). Let $\alpha < \Gamma_0$. There are unique $n \in \omega$ and $\alpha_1, \ldots, \alpha_n, \beta_1, \ldots, \beta_n < \alpha$ such that

i. $\phi(\alpha_1, \beta_1) \geq \phi(\alpha_2, \beta_2) \geq \ldots \geq \phi(\alpha_n, \beta_n)$ and

ii. $\alpha = \phi(\alpha_1, \beta_1) + \ldots + \phi(\alpha_n, \beta_n)$.

The definition of $h(T)$ is a simple recursion. As the basis of this recursion, we define $h(T) = 0$ for the trivial tree consisting solely of a root. When T has more nodes, the root has successors and the definition of $h(T)$ depends on the number of successors the root has:

Case 1. The root has only one successor. Then T is constructed from another tree T_1 by attaching a new root below that of T_1. Simply define $h(T) = h(T_1)$.

Case 2. The root has exactly two successors. Then T is constructed from smaller trees T_1 and T_2 by putting them over a common new root. Without loss of generality we can assume $h(T_1) \geq h(T_2)$ and define $h(T) = h(T_1) + h(T_2)$.

Case 3. The root has three successors. Then T is constructed from smaller trees T_1, T_2, T_3 by putting them over a common new root. Again, we can assume $h(T_1) \geq h(T_2 \geq h(T_3)$ and define $h(T) = \phi(h(T_2), h(T_1))$.

Case 4. The root has at least four successors. Then T is constructed from smaller trees $T_1,\ldots,T_m$, with $m \geq 4$, by putting these over a common new root. Again, we can assume $h(T_1) \geq \ldots \geq h(T_m)$ and define $h(T) = \phi(h(T_1), h(T_2))$.

The first thing to note is that h maps onto Γ_0:

<u>Lemma</u>. Let $\alpha < \Gamma_0$. Then $\alpha = h(T)$ for some finite tree T.

Proof: By transfinite induction. If $\alpha = 0$, it is the ordinal of the trivial tree.

If $\alpha > 0$, by the Normal Form Theorem there are $\alpha_1, \beta_1 < \alpha$ such that either $\alpha = \alpha_1 + \beta_1$ and $\alpha_1 \geq \beta_1$ or $\alpha = \phi(\alpha_1, \beta_1)$. In either case, we can assume the existence of trees T_1, T_2 such that $h(T_1) = \alpha_1$ and $h(T_2) = \beta_1$ and construct T from them and the trivial tree accordingly. Q. E. D.

The crucial fact that h preserves order requires several trivialities which are collected together in the following lemmas:

<u>Lemma</u>. (Assume all ordinals exhibited to be $< \Gamma_0$).

i. If $\alpha_1 \leq \alpha_2$ and $\beta_1 \leq \beta_2$, then $\phi(\alpha_1, \beta_1) \leq \phi(\alpha_2, \beta_2)$

ii. If $\alpha \leq \beta$, then $\alpha \leq \beta \leq \alpha + \beta \leq \beta + \alpha \leq \phi(\alpha,\beta) \leq \phi(\beta,\alpha)$.

Lemma. i. Let T be a finite tree, $a \in T$, and T_a the truncation of T to those nodes $b \geq a$. Then $h(T_a) \leq h(T)$.

ii. Let T_1 be homeomorphically embedded in T_2 via a map F and let $a \in T_1$. If a has at least m immediate successors in T_1, then $F(a)$ has at least m immediate successors in T_2.

Each of these lemmas is relatively simple - the former requiring perhaps some effort for those not thoroughly familiar with ordinals. Regardless of whether one proves them or accepts them, when one puts them together one finds the following conclusion an easy exercise:

Theorem. Let T_1, T_2 be finite trees. If T_1 is homeomorphically embeddable in T_2, then $h(T_1) \leq h(T_2)$.

CONCLUDING REMARKS

What does this Theorem do for us? Well, at the very least, it shows any linear extension of the well-quasi-ordering of finite trees under homeomorphic embeddability to have order type at least Γ_0. But, if we look at the construction of the map just given through the eyes of a logician, we see that it really doesn't use much. First, one doesn't really need to deal with ordinal numbers and transfinite induction; one can use notations for objects one believes to be ordinals and their finite generation. If one does this, one then sees that one obtains a proof in a proof-theoretically weak theory that Kruskal's Theorem implies there is no descending sequence of ordinals below Γ_0, i.e. Γ_0 is an ordinal, i.e. transfinite induction holds for Γ_0.

The direct provability of transfinite induction on large ordinals of proof-theoretically significant: As G. Gentzen originally showed, one can often obtain proof-theoretic analyses of formal theories, and conclude consistency thereby, by appeal to transfinite induction on appropriate ordinals. FNT can, for example, be analysed by ϵ_0, and so-called predicative analysis by Γ_0. Thus, Kruskal's Theorem readily implies the consistency of predicative analysis and so transcends predicative principles.

And what about FFF? Well, Friedman observed that the proof-theoretic analyses of formal systems do not generally require the full power of the non-existence of descending sequences of ordinals and, upon careful analysis, verified that FFF is already strong enough to rule out the offending sequences. Thus, e.g., FFF implies over a relatively weak system the consistency of predicative analysis and so transcends predicative principles.

A TABLE OF SOME GROWTH RATES

Ordinal	Comment
0	Successor function: $S(x) = x + 1$
1	Doubling: $D(x) = 2x$
2	Exponentiation: $E(x) = 2^x$
3	Super-exponentiation: $F(x) = 2^{2^{\cdot^{\cdot^{\cdot^{2}}}}}\ x$ The ordinary Ramsey functions occur at this level. N. Kalton and L. Rubel have encountered this growth rate in the study of gap-interpolation theorems for entire functions.
4	Graham's little function $G(x)$ described in my column on PH occurs here.
ω	The Ackermann function. A big function of Graham appears here. In unpublished work, Stroehl has discovered this growth rate in dealing with finite trees. Moreover, the currently best-known bounds for van der Waerden's Theorem occur at this level.
$_0$	The Paris-Harrington function. There are now several function with this growth rate. A most amusing one arising directly from ordinal notations will be published by L. Kirby and J. Paris.
Γ_0	The first impredicative ordinal. H. Friedman, K. McAloon and S. Simpson have produced a Ramsey-type theorem whose witnessing function occurs at this level.
?	The ordinal of Friedman's finitisation of Kruskal's Theorem. [Note added in proof: Friedman has since determined this ordinal; but it

is not one of the better known ordinals and has no catchy name to put into the left column.]

REFERENCES

First, let me simply update the references to my column on the Paris-Harrington Theorem: The papers of Erdös and Mills and of Ketonen and Solovay have since appeared, the former in the Journal of Combinatorial Theory, series A, vol. 30 (1981), pp. 53-70, and the latter in the Annals of Mathematics, vol. 113 (1981), pp. 267-314. In addition, I note the book Ramsey Theory (J. Wiley, 1980) by R. Graham, B. Rothschild and J. Spencer has a nice exposition of a variant of the Ketonen-Solovay work.

Friedman's work reported above is not, at the time of writing, in manuscript form and I can only cite some background material. With respect to Kruskal's Theorem and ordinals, I suggest:

1. J. B. Kruskal, Well-quasi-ordering, the tree theorem, and Vázsonyi's conjecture, Trans. AMS 95 (1960), pp. 210-225.
2. C. St. J. A. Nash-Williams, On well-quasi-ordering finite trees, Proc. Cambridge Phil. Soc. 59 (1963), pp. 833-835.
3. D. Schmidt, Well-partial-orderings and their maximal order types, Habilitationsschrift, Heidelberg, 1978.

Finally, a good reference on Γ_0, ordinals, and proof theory is K. Schütte, Proof Theory (Springer-Verlag, 1977).

HARVEY FRIEDMAN'S RESEARCH ON THE FOUNDATIONS
OF MATHEMATICS, L.A. Harrington et al. (editors)
Elsevier Science Publishers B.V. (North-Holland), 1985

Does Gödel's Theorem Matter to Mathematics?[1), 2)]

Gina Kolata

The recent discovery of two natural but undecidable statements indicates that Gödel's theorem is more than just a logician's trick.

In the 1930's, Kurt Gödel shook the world of mathematics by showing that there are statements in every logical system whose truth or falsehood simply cannot be determined by staying within the system. And if you try to fix up a logical system by calling the undecidable statements axioms and thereby declaring them to be true, new undecidable statements will crop up.

This result made mathematicians wonder about many of the famous unsolved problems that plague them. Could it be that some of these problems are undecidable? "What people would really like is to take a big unsolved problem like Fermat's Last Theorem and show it is undecidable. That would be spectacular," says Joel Spencer of the State University of New York in Stony Brook.

But, so far, that has not happened and mathematicians have engaged in philosophical debates over whether it ever will, whether Gödel's theorem applies to statements that matter. Many think it does not. Craig Smoryński, a logician at Ohio State University, remarks, "It is fashionable to deride Gödel's theorem as artificial, as dependent on a linguistic trick." Logician Robert Solovay of the University of California at Berkeley adds, "The feeling was that Gödel's theorem was of interest only to logicians."

A few years ago, however, two logicians found an example of a "natural"

[1)] Reprinted from Science vol. 218, pp. 779-780, 19 November 1982, by permission of the American Association for the Advancement of Science.

[2)] Editors' note: The reader may wish to consult short articles by C. Smoryński in this volume for more discussions of this topic.

statement, involving only finite quantities, that cannot be proved true within the normal axiomatic structure of finite mathematics. In a sense, this statement just missed being provable. Now, another logician has found an even more "natural" statement that cannot be proved true in an even stronger system of axioms. The proof of this second statement requires structure far beyond the mathematical system used for finite quantities, raising it to a much higher level of undecidability. These two results are leading a number of mathematicians to believe that Gödel's theorem does in fact apply to problems that matter.

The first of these undecidable statements was discovered by Jeff Paris of Manchester University and Leo Harrington of the University of California at Berkeley. The statement involves combinatorics and is "natural" because it is not the sort of concocted statement that only logicians would devise. "The Paris-Harrington theorem looks like a natural mathematical question with no trace of logic about it. That's what's spectacular - it is natural and combinatorial in character," says Solovay.

The theorem is a statement in Ramsey theory, which is a branch of mathematics dedicated to the proposition that "complete disorder is impossible," according to Ronald Graham of Bell Laboratories. If you choose a big enough set, you are bound to find structure in it. The question is, however, how big must the set be? In the case of the Paris-Harrington theorem, the size of the set grows so large so quickly that the function describing its growth simply cannot be shown to be well defined in Peano arithmetic, which is the ordinary axiomatic system used in mathematics to talk of finite things. Peano arithmetic, says Spencer, is "the accepted bedrock of mathematics."

A special case of what is known as Ramsey's theorem is the party problem: How many guests must you have at your party to be assured that a certain number of them either all know each other or all are strangers to each other? If you want to be sure that at least three guests are mutual acquaintances or mutual strangers, you must have at least six people at your party. If you want four guests all to know each other or all to be strangers you need at least 18 people at the party. But no one knows the minimum number of guests you need at the party to guarantee a similar group of five. The number is somewhere between 42 and 55. Says Graham, "It is hopeless to try and compute the exact number. It is way beyond our present computing power."

The Paris-Harrington theorem is a slight variation of Ramsey's theorem. According to Ramsey's theorem, if you have an infinite set and you assign a color, say red or blue, arbitrarily to each pair of members of the set, then you can find an infinite subset, all of whose pairs are red or all of whose pairs are blue. More generally, if you pick numbers r and k and if you have an infinite set and you assign one of r colors arbitrarily to each k-element subset of the set, then there is an infinite subset, all of whose k-tuples have the same color.

Paris and Harrington devised a finite version of Ramsey's theorem. They started out by defining a "large" set of integers to be one that has at least as many elements as its smallest integer. For example, the set 3, 15, 25, 26 , is "large" because it has at least three elements. The set 100, 102, 104, 106, 108 is not "large" because it has fewer than 100 elements. Then Paris and Harrington showed if you take a big enough set of integers and assign colors, such as red or blue, to each pair of integers you can find a "large" set all of whose pairs are red or all of whose pairs are blue. Of, more generally, if you choose r and k and assign r colors to the k-element subsets of your initial set, you can always find a "large" subset, all of whose k-tuples are the same color.

How big must your original set be? It depends on how many colors and how you partition the subsets, but Solovay found that the lower bound on the size of the set grows so fast that it isn't even well defined in Peano arithmetic. Says Graham, "The lower bound for how large the initial set must be grows fast. It is hard to grasp how fast it grows. It grows so quickly that the numbers somehow begin to lose all meaning."

The way the lower bound grows is analogous to the way a function, called the Ackermann function, grows. This is a function of two variables that is recursively defined: $f(a,b) = f[(a-1), f(a,b-1)]$ where $f(1,b) = 2b$ and $f(a,1) = a$ for a greater than 1 . With this function

$$f(3,2) = 2^{2^{2}} = 16 \ ,$$

$$f(3,4) = 2^{2^{2^{2}}} = 2^{65536} ,$$

a number with more than 19,000 digits. (When evaluating towers of exponents, mathematicians work from the top of the tower down.) The term f(6,6) is so large that if you wanted to evaluate it you couldn't write it on a piece of paper. And these are just the initial values of the function - the values that are very close to the origin.

Paris and Harrington used model theory, a standard method of mathematical logic, to show that their theorem is undecidable in Peano arithmetic. Harrington explains that they produced two models for Peano arithmetic - two equivalent sets of axioms. In one of these models, the theorem was true and in the other it was not true, indicating that the theorem is undecidable. The analogy is with the axioms for geometry. In Euclidean geometry, one model, parallel lines never meet. In non-Euclidean geometry, a different model, they can meet.

After proving that the Paris-Harrington theorem is undecidable in Peano arithmetic, Harrington wrote a letter to Solovay noting that they had obtained this result but not saying how they got it. Solovay and his colleague Jussi Ketonen then devised their own proof that the theorem is undecidable, a proof that is more combinatorial in nature.

Solovay and Ketonen showed that because the lower bound on the size of the initial set grows so fast, you need a structure just beyond Peano arithmetic to prove it is well defined and thus to prove the Paris-Harrington theorem.

Peano arithmetic contains all the integers up to "infinity," which is denoted ω . Then, after the first copy of the integers, it continues with the terms $\omega + 1$, $\omega + 2$, and so on up to 2ω . (Spencer likens the system to the children's book On Beyond Zebra which goes on after the alphabet ends at z for zebra.) But Peano arithmetic does not end at 2ω . It continues to

$$\omega^{\omega^{\omega^{\cdot^{\cdot}}}}$$

Yet even the exponential tower of ω's is not large enough to deal with the Paris-Harrington theorem. What is needed is ϵ_0 , defined as the limit to which

$\omega^{\omega^{\omega^{\cdot^{\cdot^{\cdot}}}}}$ converges. Asked how you know that the tower of ω's converges, Spencer replies, "It takes a leap of faith."

Very recently, Harvey Friedman of Ohio State University found a second undecidable theorem, but his theorem involves a function that grows so fast that it dwarfs the function of the Paris-Harrington theorem. Even $_0$ is not enough to prove Friedman's theorem. "The Paris-Harrington theorem lies just barely beyond Peano arithmetic," says Spencer. "Friedman's theorem is much farther out."

Friedman's theorem is a finite version of a well-known result discovered by Joseph B. Kruskal of Bell Laboratories. Kruskal's theorem involves "trees," which are sets of points connected by lines and containing no cycles. Evolutionary biologists draw trees when they describe the ancestors of species and geneologists draw family trees.

Collections of trees can be infinite as well as finite and, if they are infinite, complete disorder is impossible, according to Kruskal's theorem. The theorem states that if you have an infinite collection of finite trees ordered in any arbitrary way, then at least one of those trees must fit into a later one so that the branches of the first fit inside those of the second. Kruskal's theorem forms the basis for a branch of combinatorics called "well-quasi-ordering."

To make a finite version of Kruskal's theorem, Friedman said that you don't need an infinite collection of trees. All you need is a sufficiently large finite collection of trees. How large is large? That is where the enormous function comes in. "It is _gigantic_. I mean it's _really_ gigantic," says Graham. According to Somoryński, it is "the most rapidly growing computable function that has ever been described."

The observation that Friedman's theorem is far beyond the reach of Peano arithmetic demonstrates, to Spencer at least, that Kruskal's theorem is indeed a deep one. "Friedman's result bears this out since he shows that if you turn Kruskal's theorem into a finite theorem, the proof is beyond the normal methods of finite mathematics." Harrington is impressed by the extreme naturalness of Friedman's theorem. It is the sort of theorem, he says, that could have arisen in combinatorics with no reference to mathematical logic and undecidability. "I found it easy to convince myself that combinatorialists could have thought of this, "Harrington says.

Friedman also demonstrated that if you take a large enough collection of finite trees and ask whether a finite tree of a particular size (as opposed to any arbitrarily chosen size) must fit into another tree, you can use Peano arithmetic to show that it must. But the proof requires an enormous number of steps. Friedman showed that if, for example, you want to prove that a tree containing ten nodes must fit into another tree, the proof would require more than

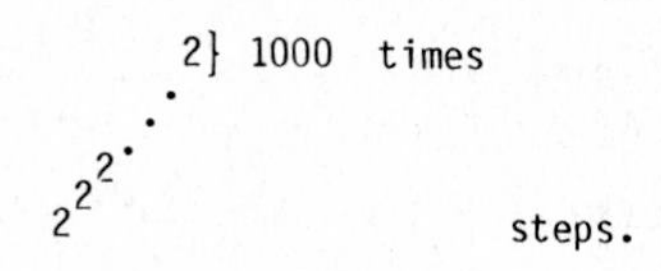

steps.

Smorynski predicts that the enormous function Friedman has described is just the beginning. Friedman has only dealt with a weak form of Kruskal's theorem. Hs is now working on a finite version of the full form of Kruskal's theorem and, according to Smoryński, "when these results are finitized, they will yield functions that dwarf F [the function Friedman has so far described]."

The more natural but undecidable theorems that are found, of course, the more willing mathematicians are to believe that Gödel's theorem might apply to important results. The recent discoveries of the Paris-Harrington and Friedman theorems might also lead mathematicians to a greater appreciation of mathematical logic, including infinite objects such as ω^{ω} and ϵ_0. "It shows the mathematical reasonableness of these weird objects," says Harrington.

HARVEY FRIEDMAN'S PUBLICATIONS

Model Theory:

1. Beth's Theorem in Cardinality Logics, Israel J. Math., Vol. 14, No. 2, (1973), pp. 205-212.

2. Countable Models of Set Theories, Lecture Notes in Mathematics, Vol. 337, Springer-Verlag, (1973), pp. 539-573.

3. On Existence Proofs of Hanf Numbers, J. of Symbolic Logic, Vol. 39, No. 2, (1974), pp. 318-324.

4. Adding Propositional Connectives to Countable Infinitary Logic, Mathematical Proceedings of the Cambridge Philosophical Society, Vol. 77, No. 1, (1975), pp. 1-6.

5. On Decidability of Equational Theories, J. of Pure and Applied Algebra, Vol. 7, (1976), pp. 1-3.

6. The Complexity of Explicit Definitions, Advances in Mathematics, Vol. 20, No. 1, (1976), pp. 18-29.

7. On the Naturalness of Definable Operations, Houston J. Math., Vol. 5, No. 3, (1979), pp. 325-330.

Proof Theory and Intuitionism:

8. Bar Induction and π^1_1 - CA, J. of Symbolic Logic, Vol. 34, No. 3, (1969), pp. 353-362.

9. Iterated Inductive Definitions and Σ^1_2 - AC, Intuitionism and Proof Theory, North-Holland, (1970), pp. 435-442.

10. The Consistency of Classical Set Theory Relative to a Set Theory with Intuitionistic Logic, J. of Symbolic Logic, Vol. 38, No. 2, (1973), pp. 315-319.

11. Some Applications of Kleene's Methods for Intuitionistic Systems, Lecture Notes in Mathematics, Vol. 337, Springer-Verlag, (1973), pp. 113-170.

12. The Disjunction Property Implies the Numerical Existence Property, Proc. Natl. Acad. Sci., Communicated by K. Godel, Vol. 72, No. 8, (August 1975), pp. 2877-2878.

13. Some Systems of Second Order Arithmetic and Their Use, Proceedings of the 1974 International Congress of Mathematicians, Vol. 1, (1975), pp. 235-242.

14. Subsystems of Second Order Arithmetic with Restricted Induction I, II abstracts, J. of Symbolic Logic, Vol. 41, No. 2, (1976), pp. 557-559.

15. Set Theoretic Foundations for Constructive Analysis, Annals of Mathematics, Vol. 105, (1977), pp. 1-28.

16. On the Derivability of Instantiation Properties, J. of Symbolic Logic, Vol. 42, No. 4, (1977), pp. 506-514.

17. Classically and Intuitionistically Provably Recursive Functions, Higher set theory, Springer Lecture Notes, Vol. 669, (1978), pp. 21-27.

18. A Strong Conservative Extension of Peano Arithmetic, Proceedings of the 1978 Kleene Symposium, North Holland, (1980), pp. 113-122.

19. (with K. McAloon and S. Simpson), A Finite Combinatorial Principle Equivalent to the 1-consistency of Predicative Analysis, Patras Logic Symposion, ed. G. Metakides, North-Holland, (1982), pp. 197-230.

20. (with S. Simpson and R.L. Smith), Countable Algebra and Set Existence Axioms, Annals of Pure and Applied Logic, 25 (1983), pp. 141-181.

21. (with A. Scedrov), Set Existence Property for Intuitionistic Theories with Dependent Choice, Annals of Pure and Applied Logic, 25, (1983), pp. 129-140, and corrigendum, 26, (1984), p. 101.

22. (with A. Scedrov), Large Sets in Intuitionistic Set Theory, Annals of Pure and Applied Logic 27 (1984), pp. 1-24.

23. (with A. Scedrov), Arithmetic Transfinite Induction and Recursive Well Orderings, Advances in Math., to appear.

24. (with A. Scedrov), Intuitionistically Provable Recursive Well Orderings, Annals of Pure and Applied Logic, to appear.

25. (With A. Scedrov), The Lack of Definable Witnesses and Provably Recursive Functions in Intuitionistic Set Theories, Advances in Math., to appear.

26. (with A. Scedrov), On the Quantificational Logic of Intuitionistic Set Theory, Math. Proc. of the Cambridge Philosophical Society, to appear.

27. (with R. Flagg) Epistemic and Intuitionistic Formal Systems, Annals of Pure and Applied Logic, to appear.

28. (with M. Sheard) An Axiomatic Approach to Self-referential Truth, Annals of Pure and Applied Logic, to appear.

29. (with P. Freyd and A. Scedrov), Lindenbaum Algebras of Intuitionistic Theories and Free Categories, Math. Proc. of the Cambridge Philosophical Society, to appear.

30. (with J. Pearce), Fragments of Admissible Set Theory and Bar Induction, to appear.

31. (with R. Flagg), Maximality in Modal Logic, to appear.

Recursion Theory:

32. (with R. Jensen), Note on Admissible Ordinals, The Syntax and Semantics of Infinitary Languages, Springer-Verlag Lecture Notes in Mathematics, Vol. 72, (1968), pp. 77-79.

33. Axiomatic Recursive Function Theory, Logic Colloquium '69, North-Holland, (1971), pp. 113-137.

34. Algorithmic Procedures, Generalized Turing Algorithms, and Elementary Recursion Theory, Logic Colloquium '69, North-Holland, (1971), pp. 361-389.

35. (with H. Enderton), Approximating the Standard Model of Analysis, Fundamenta Mathematicae, LXXII, (1971), pp. 175-188.

36. Borel Sets and Hyperdegrees, J. of Symbolic Logic, Vol. 38, No. 3, (1973), pp. 405-409.

37. Minimality in the Δ^1_2- degrees, Fundamenta Mathematicae, LXXXI, (1974), pp. 183-192.

38. Equality Between Functionals, Logic Colloquium, Springer Lecture Notes, Vol. 453, (1975), pp. 22-37.

39. Recursiveness in π^1_1 Paths Through $\mathcal{O}$, Proceedings of the AMS, Vol. 54, (January 1976), pp. 311-315.

40. Provable Equality in Primitive Recursive Arithmetic With and Without Induction, Pacific J. of Math., Vol. 57, No. 2, (1975), pp. 379-392.

41. Uniformly Defined Descending Sequences of Degrees, J. of Symbolic Logic, Vol. 41, No. 2, (1976), pp. 363-367.

Set Theory:

42. A More Explicit Set Theory, Axiomatic Set Theory AMS Symposium Pure Mathematics. Vol. XIII, Part I, (1971), pp. 49-65.

43. Higher Set Theory and Mathematical Practice, Annals of Math. Logic, Vol. 2 No. 3, (1971), pp. 325-357.

44. Determinateness in the Low Projective Hierarchy, Fundamenta Mathematicae, LXXII, (1971), pp. 79-95.

45. On Closed Sets of Ordinals, Proceedings of the AMS, (1974), pp. 190-192.

46. PCA Well-orderings of the Line, J. of Symbolic Logic, Vol. 39, No. 1, (1974), pp. 79-80.

47. Large Models of Countable Height, Transactions of the AMS, Vol. 201, (1975), pp. 227-239.

48. A Definable Non-separable Invariant Extension of Lebesgue Measure, Illinois J. Math., Vol. 21, No. 1, (1977), pp. 140-147.

49. Categoricity with Respect to Ordinals, Higher Set Theory, Springer Lecture Notes, Vol. 669, (1978), pp. 17-20.

50. A Proof of Foundation from the Axioms of Cumulation, Higher Set Theory, Springer Lecture Notes, Vol. 669, (1978), pp. 15-16.

51. A Consistent Fubini-Tonelli Theorem for Nonmeasureable Functions, Illinois J. Math., Vol. 24, No. 3, (1980), pp. 390-395.

52. On Definability of Nonmeasurable Sets, Canadian J. Math., Vol. XXXII, No. 3, (1980), pp. 653-656.

53. On the Necessary Use of Abstract Set Theory, Advances in Math., Vol. 41, No. 3, (September 1981), pp. 209-280.

54. Unary Borel Functions and Second Order Arithmetic, Advances in Math., Vol. 50, No. 2, November 1983, pp. 155-159.

55. Necessary Uses of Abstract Theory in Finite Mathematics, to appear.

Miscellaneous:

56. One Hundred and Two Problems in Mathematical Logic, J. of Symbolic Logic, Vol. 40, No. 2, (1975), p. 113-129.

57. A Cumulative Hierarchy of Predicates, Zeitschrift fur Mathematische Logik und Grundlagen der Mathematik, Bd. 21, (1975), pp. 309-314.

58. (with M. Talagrand), Un Ensemble Singulier, Bull. Sci. Math. 104, (1980), pp. 337-340.

59. (with Ker-I Ko), Computational Complexity of Real Functions, J. of Theoretical Comp. Science, 20, (1982), pp. 323-352.

60. The Computational Complexity of Maximization and Integration, Advances in Math., Vol 53, No. 1, 1984, pp. 80-98.

61. On the Spectra of Universal Relational Sentences, Information and Control, to appear.

RANDALL LIBRARY-UNCW
3 0490 0028301 U